Computer-Assisted Structural Analysis and Modeling

Computer-Assisted Structural Analysis and Modeling

Marc Hoit
University of Florida

PRENTICE HALL, Englewood Cliffs, New Jersey 07632

Library of Congress Cataloging-in-Publication Data

Hoit, Marc Ira.

    Computer-assisted structural analysis and modeling / Marc Hoit.

       p. cm.

    Includes index.

    ISBN 0-02-355972-1

    1. Structural analysis (Engineering)—Data processing.

  2. Computer-aided engineering.   I. Title.

  TA647.H63   1995

  624.1'71—dc20   94-25071

                                   CIP

Editor: William Stenquist

Production Supervisor: Elaine W. Wetterau

Production Manager: Francesca Drago

Text and Cover Designer: Brian Deep

Illustrations: Accurate Art, Inc.

© 1995 by Prentice-Hall, Inc.

A Simon & Schuster Company

Englewood Cliffs, New Jersey 07632

The author and publisher of this book have used their best efforts in
preparing this book. These efforts include the development, research,
and testing of the theories and programs to determine their effectiveness.
the author and publisher shall not be liable in any event for incidental
or consequential damages in connection with, or arising out of, the
furnishing, performance, or use of these programs.

Printed in the United States of America

10 9 8 7 6 5 4 3 2 1

ISBN 0-02-355972-1

PRENTICE-HALL INTERNATIONAL (UK) LIMITED, *London*
PRENTICE-HALL OF AUSTRALIA PTY LIMITED, *Sydney*
PRENTICE-HALL CANADA, INC., *Toronto*
PRENTICE-HALL HISPANOAMERICANA, SA, *Mexico*
PRENTICE-HALL OF INDIA PRIVATE LIMITED, *New Delhi*
PRENTICE-HALL OF JAPAN, INC., *Tokyo*
SIMON & SCHUSTER ASIA PTE. LTD., *Singapore*
EDITORIA PRENTICE-HALL DO BRASIL, LTDA., *Rio de Janeiro*

To Fay, Sarah, Mom, Dad, and Berkeley

Contents

PREFACE   xiii

1 BASIC CONCEPTS   2
   1.1 Structural Modeling   4
   1.2 Forces and Corresponding Displacements   5
   1.3 Independent and Dependent Displacements   6
   1.4 Independent and Dependent Forces   7
   1.5 Conditions Governing Structural Behavior   7
        1.5.1 Static Equilibrium   8
        1.5.2 Constitutive or Stress–Strain Law   8
        1.5.3 Boundary Conditions   8
        1.5.4 Compatibility   9
   1.6 Linear Elastic Structural Behavior   9
   1.7 Matrix Methods of Structural Analysis   11
        1.7.1 Stiffness Method Spring Example   11
        1.7.2 Flexibility Method Spring Example   12
   1.8 Flexibility Coefficients   13
   1.9 Stiffness Coefficients   14
   1.10 Relation Between Stiffness and Flexibility   15
   1.11 Constrained Displacements   16
   1.12 Constrained Forces   16
   1.13 Symmetry of Flexibility and Stiffness   17
   1.14 Static and Kinematic Indeterminacy   18

2 MATRIX METHODS   20
   2.1 Basic Linear Algebra   21
   2.2 Solution of Equations   25
   2.3 Banded Solution Method   28
   2.4 Equation Renumbering   31

2.5 **Profile Solution Method**   34
2.6 **CAL-90: Matrix Interpretive Language**   36
2.7 **Problems**   38

3 VIRTUAL WORK AND CONSISTENT DEFORMATIONS   42
3.1 **Use for Derivations**   43
3.2 **Internal Work for Trusses**   45
3.3 **Internal Work for Beam Elements**   46
3.4 **Example of Virtual Forces (Dummy Unit Load)**   48
3.5 **Consistent Deformations**   51
   3.5.1 Consistent Deformations Example   54
3.6 **Consistent Deformations for Trusses**   57
   3.6.1 Consistent Deformation Truss Example   58
3.7 **Multiple-Redundant Structures**   62
   3.7.1 Multiple-Redundant Structure Example   63
Condensed Example: Two-Redundant Truss Structure   67
3.8 **Temperature Deformations Using Virtual Work**   70
   3.8.1 Temperature Deformations in Trusses   70
   3.8.2 One-Redundant Truss with Temperature Example 71
   3.8.3 Temperature Deformations in Beams   72
   3.8.4 One-Redundant Beam with Temperature Example 74
3.9 **Support Displacements Using Consistent Deformations**   75
3.10 **Shear Deformations in Beams**   76
3.11 **Problems**   79

4 STIFFNESS METHODS   88
4.1 **Slope Deflection**   89
4.2 **Application of Slope Deflection**   94
   4.2.1 Simple Slope Deflection Example   96
4.3 **Modified Slope Deflection**   99
   4.3.1 Modified Slope Deflection Example   100
4.4 **Drawing Displaced Shapes**   102
4.5 **Frames by Slope Deflection**   107
   4.5.1 Frame Example by Slope Deflection   107
4.6 **Frame with Slanted Member**   110
4.7 **Stiffness Matrices by Definition**   116
   4.7.1 Stiffness by Definition Example   119
Condensed Example: Stiffness by Definition—Frame   130
Condensed Example: Stiffness by Definition—Truss   133
4.8 **Structural Loading**   136
4.9 **Distributed Loads**   137
   4.9.1 Distributed Load Example   138
4.10 **Problems**   140

5 DIRECT STIFFNESS   150
5.1 **Stiffness Transformation Matrix**   151
   5.1.1 Definition of the **a** Matrix   153

5.2   Derivation of Stiffness by Direct Stiffness   156
5.3   Two-Dimensional Beam Element Stiffness Derivation   157
5.4   Direct Stiffness Solution Procedure   159
5.5   Beam Element Loads   159
5.6   Beam Element Force Recovery   160
5.7   Final Direct Stiffness Solution Procedure   162
5.8   Direct Stiffness Example   162
Condensed Example: Frame Using the Full a Matrix   168
*Condensed Example: Truss by Direct Stiffness Using
the Full a Matrix*   173
5.9   The a Matrix Revisited   176
    5.9.1 Inclusion of Axial Deformations   179
    5.9.2 Derivation of Rotational Transformation   180
    5.9.3 Derivation of the Location Transformation   182
    5.9.4 Stiffness Assembly   185
    5.9.5 Force Recovery Using Transformations   188
    5.9.6 Split Transformation Analysis Procedure   188
5.10 Stiffness Analysis Using CAL-90   189
    5.10.1 Direct Stiffness Operations   190
    5.10.2 Example of CAL-90 Direct Stiffness Commands   191
Condensed Example: Frame Analysis Using CAL-90   195
*Condensed Example: Truss by Direct Stiffness Using
the Split a Matrix*   199
5.11 Three-Dimensional Beam Elements   201
    5.11.1 Rotational Transformation in Three Dimensions   203
5.12 Problems   207

6 GENERAL STRUCTURAL ANALYSIS PROGRAMS   214
6.1   General Program Structure   215
    6.1.1 Set Up the Problem Geometry   215
    6.1.2 Form All Element Matrices   217
    6.1.3 Assemble the Global Stiffness and Loads   217
    6.1.4 Solve for the Unknown Displacements   217
    6.1.5 Recover the Element Forces   218
6.2   Load Cases   218
6.3   Element Types   219
6.4   SSTAN Capabilities   219
6.5   Use of STANPLOT   221
    6.5.1 Running STANPLOT   222
6.6   SSTAN Examples   227
    6.6.1 Two-Dimensional Truss Example   228
    6.6.2 Continuous Beam Example   231
    6.6.3 Two-Story Braced Frame   237
    6.6.4 Three-Dimensional Example   245
6.7   Solution Errors and Model Correctness   249
    6.7.1 Solution Errors   249
    6.7.2 Model Correctness   250

	6.8	Result Interpretation	251
		Condensed Example: Zero-Compression Members	252
	6.9	Problems	255

7 STRUCTURAL MODELING — 266
7.1	Load Projection	267
7.2	Boundary Conditions	270
	7.2.1 Spring Supports	270
	7.2.2 Inclined Supports	270
	7.2.3 Hinges	273
	Condensed Example: Hinged Member	275
7.3	Imposed Displacements	277
7.4	Symmetry and Antisymmetry	279
7.5	Rigid-End Offsets	284
	Condensed Example: Rigid-End Offsets	287
7.6	Geometric Stiffness: $P-\Delta$	289
	7.6.1 Consistent Geometric Stiffness	292
	7.6.2 Geometric Stiffness Example	295
	Condensed Example: $P-\Delta$ Analysis	299
7.7	Problems	301

8 FINITE ELEMENTS — 310
8.1	Finite-Element Theory	311
	8.1.1 Simple FEM Theory	311
	8.1.2 Available Elements	314
8.2	Shape Functions	316
	8.2.1 Tapered Extensional Example	317
	8.2.2 Shape Function Accuracy	319
8.3	Numerical Integration	319
	8.3.1 Gauss Quadrature	320
	8.3.2 Mapping Errors	321
	8.3.3 Two-Dimensional Extension	323
8.4	Problems	323

9 MEMBRANE ELEMENT — 324
9.1	Introduction	325
9.2	Membrane Theory	325
9.3	Two-Dimensional Shape Functions	327
9.4	Cantilever Beam Example Using Membranes	329
	9.4.1 Four-Node-Element Model	330
	9.4.2 Nine-Node-Element Model	333
9.5	Shear Locking	337
9.6	Mesh Correctness and Convergence	338
	9.6.1 Stress Difference	338
	9.6.2 Element Meshing	339

9.7   **Distributed Loads in Membranes**   340
    9.7.1 Four- and Nine-Node Load Factors   340
Condensed Example: Notched Beam—Membrane Elements   341
9.8   **STANPLOT: Stress and Displacement Contours**   344
9.9   **Problems**   345

10 AXISYMMETRIC SOLIDS   352
  10.1   **Introduction**   353
  10.2   **Axisymmetric Theory**   353
  10.3   **SSTAN Axisymmetric Element**   356
  10.4   **Axisymmetric Circular Plate Example**   356
  10.5   **Loads on Axisymmetric Models**   357
    10.5.1 Nodal Loads for Trapezoidal Line Loads   358
  10.6   **SSTAN Axisymmetric Plate Example**   358
  10.7   **Axisymmetric Results Evaluation**   362
Condensed Example: Steel Collar—Axisymmetric Elements   362
  10.8   **Problems**   366

11 FLAT PLATE AND SHELL ELEMENTS   370
  11.1   **Introduction**   371
  11.2   **Plate Theory**   371
    11.2.1 Kirchhoff Theory   371
    11.2.2 Mindlin Theory   372
    11.2.3 Generalized Stress   374
  11.3   **Surface Load Distribution Factors**   374
  11.4   **Cantilever Plate Example**   375
  11.5   **Flat Slab Example**   378
    11.5.1 STANPLOT Stress and Displacement Contours   383
  11.6   **Flat Shell Elements**   383
    11.6.1 Bridge Example   384
  11.7   **Problems**   393

12 SOLID ELEMENTS   396
  12.1   **Introduction**   397
  12.2   **Solid Element Behavior and DOFs**   397
  12.3   **Modeling Capabilities and Results**   397
  12.4   **Solid Element Loads**   398
  12.5   **Solid Element Cantilever Example**   399
  12.6   **Problems**   403

Appendix A: CAL-90 Users' Guide   406

Appendix B: SSTAN Users' Guide   422

Index   437

Preface

The primary goal of this book is to teach the student structural behavior and modeling. A secondary goal is to teach how to use computer structural analysis programs competently. To do this, it is felt that the student must have a complete background in the techniques used within most standard analysis programs. The focus of this book is to prepare the student by teaching the concepts that underlie all structural analysis programs. The book avoids lengthy theoretical derivations and excludes many traditional hand solution techniques. All derivations are done using virtual work stressing matrix forms. Emphasis is placed on the physical interpretation of matrices, solution procedures, and interpretation of results. Coverage includes finite elements from an element usage standpoint. Most of the homework involves using the included computer programs CAL-90 and SSTAN.

CAL-90 is a matrix interpretive language that is designed to perform the matrix operations required for matrix structural analysis. SSTAN is a general finite-element analysis program designed to teach the techniques required to use modern analysis programs. SSTAN also has a graphics postprocessor for displaying the structure and its results. Both programs are included with the book for use on PC/MSDOS-based systems.

The book is intended to be used as an undergraduate textbook in the second structural analysis course, commonly referred to as indeterminate or matrix structural analysis. It could be used as a graduate textbook for a first graduate course in structural analysis if the instructor provides supplementary material.

The text is organized in what I have found a natural progression for teaching structural analysis. Chapter 1 gives a general introduction to the concepts of analysis, including degrees of freedom, flexibility, stiffness, and determinacy. Chapter 2 defines the basic linear algebra required for the book. It also includes the concepts of equation solving and sparse matrix techniques, since these are used in all general computer programs. Sections of Chapter 2 may be skipped and referred back to when the content is required. Chapter 3 reviews virtual work for trusses and beams and consistent deformations for one-degree indeterminate structures. These topics should have been covered in the previous course, but I find that most students need the review. The end of Chapter 3 covers consistent deformations for multiply indeterminate structures where the use of matrix notation is first introduced. This material may be new to some students.

Chapter 4 starts the main body of the book by developing the groundwork for stiffness analysis methods. The slope deflection method is reviewed and demonstrated as the first stiffness technique. It is derived, as well as all stiffness terms, using the method of consistent deformations and virtual work. Slope deflection is used to teach an understanding of degrees of freedom as well as displaced shapes and constraints. If slope deflection is taught in the previous class, this part may be used as a review for the student. The second part of Chapter 4 introduces general stiffness methods by developing stiffness matrices using the definition of a stiffness matrix. Using the definition gives the student a good feel for where a stiffness term comes from and what members contribute to its value. All required force–displacement equations come directly from the slope deflection equations.

Chapter 5 introduces the direct stiffness technique. This is shown as an automated method of creating the required stiffness and load matrices. Through the use of transformation matrices, the process of creating a stiffness matrix can be simplified to a series of matrix manipulations. The process is simplified still further by breaking the transformation matrix up into its rotation and assembly portions. All of the direct stiffness manipulations are performed using the matrix language provided by CAL-90. This allows the student to analyze more complex structures without the tedium of large numbers of hand calculations. Chapter 5 completes the fundamentals required for a good background in matrix structural analysis.

The first five chapters are expected to require about four to six weeks to complete. Although this is a very fast pace, a large portion of the material is assumed to be a review. Some instructors prefer to skip consistent deformations and slope deflection entirely, starting with the second half of Chapter 4, stiffness by definition. Doing this requires that the stiffness terms be taken on faith by the students (which for many may be prudent). Another way to reduce some of the material is to skip portions of Chapter 4. A large portion of Chapter 4 covers slope deflection and stiffness by definition using a frame with a slanted member. The slanted member introduces many complications. These complications lead to increased understanding of complex structures and their behavior. However, for many students this type of depth may not be necessary.

The remainder of the course, 10 to 12 weeks, is intended to cover structural analysis and finite elements using general computer programs. This is intended to be the main body of the course. Here is where structural behavior and modeling using advanced computer programs is covered. The introduction of the basic finite elements (membrane, plate, axisymmetric solid, and solid) is intended to give a working knowledge of finite elements.

Once the students understand assembly and transformations, they are ready for general analysis programs. Chapter 6 covers the use of general analysis programs. SSTAN is introduced as the final simplification in the analysis automation progression. Analysis examples are given and comparisons are drawn about the general process and the direct stiffness method.

Chapter 7 covers basic structural modeling techniques using frame and truss structures. Time is spent discussing how to model a structure within the confines of an analysis program. Methods for specifying nonstandard boundary conditions, rigid-end offsets, hinges, symmetry, and settlements are covered. $P-\Delta$ analysis and simple nonlinear analysis are covered briefly, since many building codes require them.

Chapters 8 through 12 cover finite elements. These chapters are considered crucial to the course. Chapter 8 provides a very simple introduction to finite-element theory. The theory is presented using an axial member (truss). A tapered member in one dimension is used to show what approximations are used in finite elements and the errors they cause. Emphasis is placed on a physical understanding of finite elements. Chapters 9, 10, 11, and 12 cover membrane, plate, axisymmetric, and solid elements, respectively. Element behavior is stressed and very little theory is covered in these chapters.

It is hoped that this book can provide the required understanding to develop competent structural engineers. The low cost of computers and finite-element computer programs have made the use of these tools common in small and large engineering firms. However, many engineers have not been trained in the use of matrix analysis and finite elements. Yet the curriculum has no more room for additional courses. This book presents a solution by removing many of the traditional hand methods and concentrating on behavior and modeling. A natural extension of this book is a rigorous finite-element course. Graduate students should also consider more rigorous courses in continuum mechanics, finite elements, and elasticity for a more complete understanding of structural behavior and modeling.

I would like to thank the manuscript reviewers: Dr. Robert R. Archer of the University of Massachusetts, Dr. Panos D. Kiousis of the University of Arizona, and Dr. Kuo-Kuang Hu of the Kansas State University for their helpful comments and corrections. Thanks also go to Dr. Cliff Hays of the University of Florida for his help, ideas, and many constructive comments. I give special thanks to Gary Consolazio of the University of Florida who reviewed the manuscript as well as taught our matrix analysis class using preliminary copies of the book. Gary's excellent teaching, programming assistance, and care in correcting the manuscript have been invaluable.

M. H.

Computer-Assisted Structural Analysis and Modeling

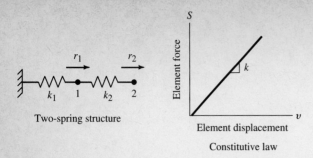

Two-spring structure

Element force

Element displacement

Constitutive law

BASIC CONCEPTS

This book is intended to teach the concepts used in modern structural analysis. Most current structural analysis is done on the computer using the direct stiffness method, a matrix method that traditionally includes the use of beam and truss members. The stiffness matrix and resulting methods for these elements can easily be derived on a physical behavior basis. In addition, these elements model the most common components used in designing structures.

Stiffness methods became extremely popular with the advent of the computer. Prior to that, these methods were not used, due to the large number of simultaneous equations that need to be solved. In the late 1950s and early 1960s many researchers published papers describing how matrix methods could be used to analyze complex structures. This period coincides with the development of modern digital computers. Matrix methods were a natural development for use on the computer because of the computer's excellent ability to manage systematic and repetitive calculations.

Quickly, the use of matrix methods became a dominant area of research. Development of the finite-element method helped to increase the use of matrix methods by increasing the scope of the type of behavior that could be modeled using matrices. Membrane, plate, and solid elements were developed soon afterward that allowed analysts to model more complex and detailed structures. Soon afterward, nonlinear and dynamic analysis techniques became available as standard tools.

As a result of the rapid development of inexpensive computers and powerful matrix techniques, matrix-based computer programs for the analysis of structures are considered the standard of practice in the structural engineering field. This has caused an evolution in the teaching and understanding of structural analysis and behavior. Many of the traditional hand methods developed to simplify a structure are no longer required. Computers can be used to analyze a structure more accurately and rapidly than ever before.

While computers can do a more detailed job than traditional hand methods, they can also create pseudoexperts who trust the results implicitly. There is much concern in industry and academia that many new engineers do not understand structural behavior and the ranges for stresses and displacements that in the past were developed using traditional hand methods. Although these concerns are valid, the use of computers does not exclude the understanding of structural behavior.

The same physical understanding can be derived from stiffness and flexibility matrices as from a moment distribution method. As a result, in this book the only classical method used is virtual work. Virtual work is used only as a method to derive matrix methods. Once the matrices are derived, all emphasis will be placed on understanding structural

3

behavior through the use of matrices. The goal of this book is to teach the student structural behavior and analysis techniques using matrix methods.

1.1 Structural Modeling

Structural modeling is the process of replacing a complex real structural system with a simpler model that retains the pertinent characteristics of the real system for the purpose of analysis. The model may be either analytical or experimental. The purpose of constructing the model is to allow some sort of rational analysis, as the real system is usually too complex for direct analysis. The analyst must keep in mind that the goal is to approximate the behavior and must not get lost in the mathematical procedures.

Figure 1.1 shows a relatively simple industrial type of framed structure. The structure is a rigid frame building constructed of many tapered rigid frames connected by girts and

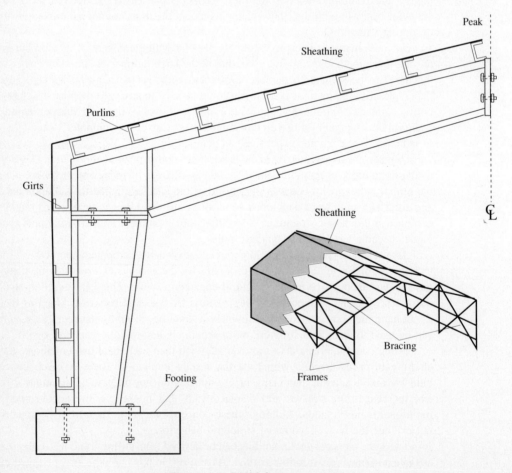

Figure 1.1 Rigid frame structure.

purlins. Tapered rigid frames are geneally custom made, consisting of changing flange and web thickness throughout the tapered section. On top of the girts and purlins is attached a sheet metal skin to act as the outside walls and roof of the structure. In addition to the girts and purlins, bracing members are used between the rigid frames in the roof and sidewalls. These structures are generally attached to the ground by bolting the rigid frames to a foundation.

When attempting to perform an analysis of the system described above, the designer is faced with a number of decisions:

1. Can the rafters and columns be represented by single straight-line elements with variable moments of inertia, or should the model include the change of taper at every change in thickness?

2. Do the purlins and girts offer bending stiffness to the system?

3. Does the sheet metal skin affect the strength of the structure?

4. Is the attachment to the footing, even though bolted, to be modeled as pinned, fixed, or semirigid?

5. Should the foundation stiffness be estimated and included in the analysis?

6. Do three-dimensional effects come into play, or is a two-dimensional analysis sufficient?

7. Should a more refined finite-element model be used?

These are only a few of the decisions that need to be made before an analysis can be done. Clearly, development of computer methods of analysis has increased the number of decisions facing the analyst. Fortunately, these methods make it possible to investigate the effects of the analyst's modeling in an economical manner and to quantify these effects for improved analysis.

1.2   Forces and Corresponding Displacements

Matrix methods rely heavily on modeling a structure as a combination of elements that can be joined at a finite number of points called *nodes*. It is at the nodes that the behavior of an element as well as the entire structure is described. It is also at the nodes where the relationships that describe how elements interact are enforced. To describe behavior and interaction, we use a concept of discrete displacements and forces that exist in one-to-one correspondence. It is through this discretization process that we develop a set of simultaneous equations that represent the behavior of the system being modeled.

A force, R, and a displacement, r, are said to correspond if when the structure moves, work is done by the force moving through the displacement. Thus the concentrated force R will have as its corresponding displacement a displacement parallel to R and at the same point of application as R. We generalize the term *force* to include a moment, and the displacement that corresponds to a moment is a *rotation*. Figure 1.2 shows a simple structure with a set of forces, R_i, and the corresponding displacements, r_i.

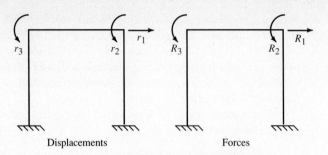

Figure 1.2 Corresponding displacements and forces.

A system of forces that describe the entire structure is represented by a vector of the form

$$\mathbf{R} = \begin{bmatrix} R_1 \\ R_2 \\ R_3 \end{bmatrix} \qquad (1.1)$$

The displacements that correspond to the system of forces are also described by a vector of the form

$$\mathbf{r} = \begin{bmatrix} r_1 \\ r_2 \\ r_3 \end{bmatrix} \qquad (1.2)$$

To describe the relationship between elements that connect at a node, nodes are usually assigned a set of degrees of freedom (DOFs). The DOFs of a system are defined as the total set of independent forces or displacements used in analyzing a structure.

1.3 Independent and Dependent Displacements

A number of displacements can be used in modeling the behavior of a structure. Displacements that are not related or constrained to each other are called *independent*. Displacements that are related or constrained to another displacement are considered to be *dependent*. For example, looking at the frame shown in Figure 1.3, if axial deformations of an element are not allowed, displacements r_1 and r_4 are constrained to be equal, due to the beam. Only one of them is independent; the other is considered dependent. Displacements r_2 and r_5 are constrained to be zero, due to no axial deformations in the columns. Displacements r_3 and r_6 are independent.

A physical check on independence is the following: If two displacements are independent, one can remain zero while the other moves, without violating any assumptions. Clearly, you cannot hold r_1 to be zero and move r_4 without stretching the beam element. Hence r_1 and r_4 are constrained to each other. If axial deformations are allowed, all six displacements would be independent.

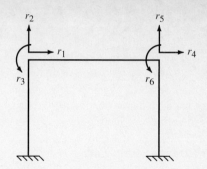

Figure 1.3 Displacement degrees of freedom.

1.4   Independent and Dependent Forces

Forces that are not related or constrained to each other are called *independent*. Forces that are related or constrained to other forces are considered to be *dependent*. Obviously, all loads on a stable structure can be applied arbitrarily with respect to each other and hence are independent. The support reactions on a structure can be a combination of dependent and independent forces. There can be at most three dependent forces on a two-dimensional structure. This corresponds to the three equations of equilibrium that can be used to solve for support reactions. Looking at Figure 1.4, we can see that the force R_5 depends on forces R_1, R_2, R_3, and R_4 and the support reactions. The dependent forces for a structure are the set of forces required to make the structure stable.

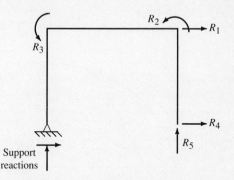

Figure 1.4 Force degrees of freedom.

A different set of dependent and independent forces can be chosen. For example, R_1 and the support reactions can be chosen as the dependent forces and R_2, R_3, R_4, and R_5 as the independent forces. This will generate a different set of equations, but the final results will be identical.

1.5   Conditions Governing Structural Behavior

To analyze a structure, four fundamental principles are used to relate the behavior of the elements to give the overall structural behavior.

1.5.1   Static Equilibrium

The summation of all forces must be zero. This is usually divided into the summation of vertical, horizontal, and rotational forces. It is statics that generate the equations of overall structural behavior for a stiffness analysis. Static equilibrium of forces generates the stiffness matrix, which contains the equations relating the displacements to forces. When analyzing under dynamic conditions, the structure must be in dynamic equilibrium, which includes the effects of inertia.

1.5.2   Constitutive or Stress–Strain Law

A valid relationship must exist between stress and strain. There are many types of relationships, some of which are shown in Figure 1.5.

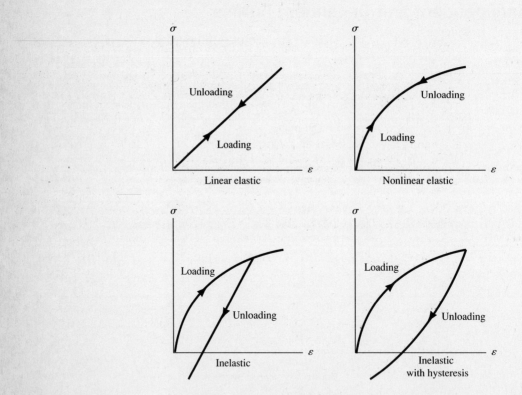

Figure 1.5   Some possible stress–strain relationships.

1.5.3   Boundary Conditions

There must exist a specified set of forces and/or displacements for the structure at every node. In a two-dimensional system, each node has two translations and one rotation. In a three-dimensional system, each node has three translations and three rotations, one about each of the coordinate axes (Figure 1.6). Every DOF in a structural system must have either a force or a displacement boundary condition (BC) specified. In a stiffness analysis, specified displacements act as supports. Specified forces are either the applied loads or

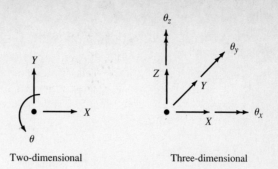

Figure 1.6 Degrees of freedom for a node.

zero. Stiffness methods solve for all the displacements that are left unspecified (coincident with the DOFs at which forces are specified).

Boundary conditions are generally shown as having three possibilities in two dimensions. Figure 1.7 shows the support conditions used in this book. These support conditions are a combination of displacement and force specifications on a node. The fixed condition specifies that the node has a specified zero displacement for X and Y translations and zero rotation. This has the result of saying that the corresponding forces are unspecified and are a result of the analysis. The pinned condition specifies a zero displacement for the X and Y translations but releases the rotation. The moment (rotational force) is also specified to have a value of zero. The final support condition is the roller. The roller specifies that the Y displacement is zero and that the X force and moment are zero. The roller is assumed to mean that there is *no* vertical displacement allowed (either up or down). These boundary conditions have their equivalents in three dimensions. The fixed condition is completely analogous. The pinned condition equates to a ball and socket. The roller equates to movement allowed in a plane and all three rotations allowed.

DOFs released

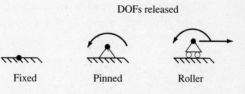

Fixed   Pinned   Roller

Figure 1.7 Structural support conditions.

1.5.4 Compatibility

The structure must deform smoothly except at hinges. No cracks can form and elements cannot tear apart from the nodes. Although nonlinear analysis methods exist to handle this type of behavior, they are not covered in this book.

1.6 Linear Elastic Structural Behavior

When the stress–strain law for the material is linear and the displacements are small enough that the equilibrium of the structure can be formulated on the undeformed structure,

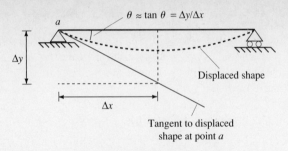

Figure 1.8 Small-angle theory.

the structural response is considered to be *linear*. The assumption of a linear stress–strain law is used throughout this book. In addition, small-deflection theory is also used. The use of small-deflection theory also implies that small-angle theory can be used. Throughout the book, deflections and angles are drawn greatly exaggerated. This is done to clarify the displacements, which are drawn many orders of magnitude larger than in a real structure. The use of small deflections and small angles gives the relationships shown in Figure 1.8 and given by the equation

$$\theta \approx \tan \theta = \frac{\Delta Y}{\Delta X} \tag{1.3}$$

From Figure 1.8 we can see that the rotation of the member at point *a* is assumed to be equal to the tangent at point *a*.

The use of linear stress–strain and small angles also indicates that superposition applies. *Superposition* means that the results from two different loadings on a structure can be added to get the same results as if the loadings were applied simultaneously. It also implies that if the loads are doubled, the stresses will also double. This is shown in Figure 1.9. For a statically determinate structure, the reactions and internal force distribution can be found without using the stress–strain relationships.

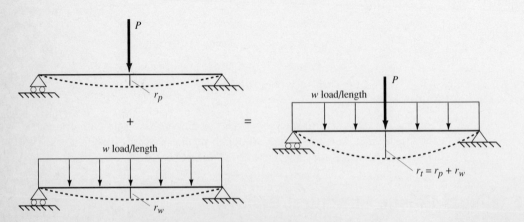

Figure 1.9 Superposition of loads and displacements.

1.7 Matrix Methods of Structural Analysis

Most matrix methods of analysis can be described as methods of combining elements with known behavior to describe the behavior of a structure that is a collection of such elements. A simple example that describes the use of the stiffness method is a two-spring problem.

1.7.1 Stiffness Method Spring Example

The problem consists of two attached springs fixed at one end (Figure 1.10). There are two DOFs for the system, labeled r_1 and r_2. The two springs behave identically, as represented by the linear constitutive law. The equation forms of the constitutive laws for the two spring elements are

$$S_1 = k_1 v_1 \qquad (1.4)$$

$$S_2 = k_2 v_2 \qquad (1.5)$$

Note that v_1 and v_2 are the axial deformations of each spring and S_1 and S_2 are the forces in each spring. The compatibility equations for the problem relate the element displacements to the global nodal displacements. The equations of compatibility are

$$v_1 = r_1 \qquad (1.6)$$

$$v_2 = r_2 - r_1 \qquad (1.7)$$

If we use a free body of each node and equilibrium, a summation of forces in the X direction, we can generate an equation at each node relating the internal forces of the spring to the external applied force at the node. Note that the directions assumed for the spring element forces are tension, defined as positive.

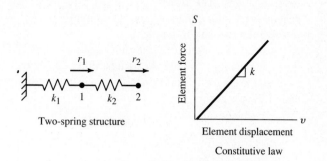

Figure 1.10 Basic matrix stiffness example.

Using Figure 1.11, we get the equilibrium equations,

$$-S_1 + S_2 + R_1 = 0 \qquad (1.8)$$

$$-S_2 + R_2 = 0 \qquad (1.9)$$

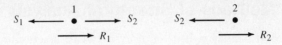

Figure 1.11 Nodal force equilibrium for spring example.

By combining the constitutive laws [equations (1.4) and (1.5)] with the compatibility equations [(1.6) and (1.7)] we can express the element forces as functions of the unknown displacements. This gives us

$$S_1 = k_1 v_1 = k_1 r_1 \tag{1.10}$$

$$S_2 = k_2 v_2 = k_2(r_2 - r_1) \tag{1.11}$$

If we now substitute these element forces [equations (1.10) and (1.11)] into the equilibrium equations [(1.8) and (1.9)], we get a set of two equations written in terms of the unknown displacements:

$$R_1 = (k_1 + k_2)r_1 - k_2 r_2 \tag{1.12}$$

$$R_2 = -k_2 r_1 + k_2 r_2 \tag{1.13}$$

By solving equations (1.12) and (1.13) for any given loads (R_1 and R_2), we can find the displacements for the entire structure. If the individual spring forces are required, these can be recovered from equations (1.10) and (1.11). Notice that equations (1.12) and (1.13) form a set of simultaneous equations in two unknowns. These can be rewritten in matrix form, creating a matrix method for stiffness analysis which has displacements as unknowns.

1.7.2 Flexibility Method Spring Example

The stiffness method uses displacements as unknowns and equilibrium to generate the final equations. The flexibility method uses forces as unknowns and compatibility to generate the final equations. As an example of use of the flexibility method, we will re-solve the two-spring problem.

Instead of using the equilibrium equations (1.8) and (1.9) as the basis for generating the final equations, we use the compatibility equations (1.6) and (1.7). Solving equations (1.4) and (1.5) for v_1 and v_2 and then substituting into the compatibility equations (1.6) and (1.7), we get

$$\frac{S_1}{k_1} = r_1 \tag{1.14}$$

$$\frac{S_2}{k_2} = r_2 - r_1 \tag{1.15}$$

Substituting r_1 from equation (1.14) into (1.15) and rearranging, we get

$$\frac{S_2}{k_2} + \frac{S_1}{k_1} = r_2 \tag{1.16}$$

Also, by substituting equation (1.9) into (1.8), we can rewrite equilibrium equation (1.9) to contain only S_1:

$$S_1 = R_1 + R_2 \qquad (1.17)$$

Substituting the values for S_1 and S_2 from equilibrium equations (1.9) and (1.17) into the rewritten compatibility equations (1.14) and (1.16), we get

$$\frac{R_1 + R_2}{k_1} = r_1 \qquad (1.18)$$

$$\frac{R_2}{k_2} + \frac{R_1 + R_2}{k_1} = r_2 \qquad (1.19)$$

If we rearrange to factor out the forces, we get the final equations:

$$\frac{1}{k_1}R_1 + \frac{1}{k_1}R_2 = r_1 \qquad (1.20)$$

$$\frac{1}{k_1}R_1 + \left(\frac{1}{k_1} + \frac{1}{k_2}\right)R_2 = r_2 \qquad (1.21)$$

Equations (1.20) and (1.21) now have the loads (R_1 and R_2) as unknowns and the displacements as the right-hand side. This method, using the compatibility equations to form a matrix method, forms a flexibility analysis that has forces as unknowns.

1.8 Flexibility Coefficients

Let's consider a structure that has N independent loads, $R_1, R_2 \ldots, R_N$, and N corresponding displacements. Then we can define the flexibility coefficients as:

f_{ij} is the displacement at DOF i, corresponding to the force at DOF i, due to a unit force applied at DOF j.

The flexibility coefficients for the cantilever beam shown in Figure 1.12 are determined from the displaced shapes shown. The displaced shapes are caused by a unit load applied at each of the N independent load locations. These displacements can be calculated by any method desired. The final displacements are given in Figure 1.12.

Clearly, using superposition, we can say that the final displacements of the cantilever due to an arbitrary applied load is the sum of the displacements caused by the unit loads scaled by the amount of the applied load. In equation form for a general case with three unknown forces, we have

$$\begin{aligned}
r_1 &= f_{11}R_1 + f_{12}R_2 + f_{13}R_3 \\
r_2 &= f_{21}R_1 + f_{22}R_2 + f_{23}R_3 \\
r_3 &= f_{31}R_1 + f_{32}R_2 + f_{33}R_3
\end{aligned} \qquad (1.22)$$

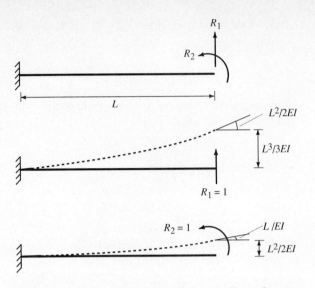

Figure 1.12 Flexibility terms for a cantilever beam.

In matrix form, we have

$$\mathbf{r} = \mathbf{FR} \tag{1.23}$$

where $\mathbf{F}$ is the matrix of the f_{ij} coefficients given above. For the cantilever example, the flexibility matrix is

$$\mathbf{F} = \begin{bmatrix} \dfrac{L^3}{3EI} & \dfrac{L^2}{2EI} \\[2ex] \dfrac{L^2}{2EI} & \dfrac{L}{EI} \end{bmatrix} \tag{1.24}$$

1.9 Stiffness Coefficients

Now, let's consider a linear elastic structure with N independent displacements, $r_1, r_2, \ldots, r_N$, and the corresponding forces, $R_1, R_2, \ldots, R_N$. We define the stiffness coefficients as:

K_{ij} is the force required at DOF i, corresponding to the displacement at DOF i, needed to maintain the structure in equilibrium due to a unit displacement at DOF j and zero displacements at all other DOFs.

For the same cantilever as that used before, we can show the stiffness coefficients for the two displaced shapes corresponding to a unit displacement at each DOF. Here the displaced shapes are caused by applying a unit displacement at each of the N independent displacements and keeping the other DOF displacements equal to zero. These forces can be calculated using any method desired. The method of virtual work, one such method, is discussed in Chapter 3.

The superposition of forces is still valid, so we can say that the final force at a DOF is equal to the sum of the forces caused by a unit deformation scaled by the amount of

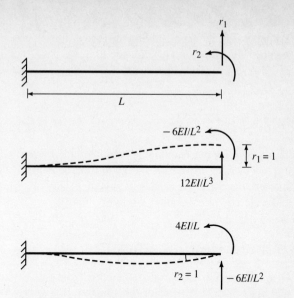

Figure 1.13 Stiffness terms for a cantilever beam.

applied displacements. In equation form, for a structure with three unknown displacements, we have

$$
\begin{aligned}
R_1 &= k_{11}r_1 + k_{12}r_2 + k_{13}r_3 \\
R_2 &= k_{21}r_1 + k_{22}r_2 + k_{23}r_3 \\
R_3 &= k_{31}r_1 + k_{32}r_2 + k_{33}r_3
\end{aligned}
\tag{1.25}
$$

In matrix form we have

$$
\mathbf{R} = \mathbf{Kr} \tag{1.26}
$$

where $\mathbf{K}$ is the matrix of coefficients given above. For the cantilever structure given in Figure 1.13, the stiffness matrix is

$$
\mathbf{K} =
\begin{bmatrix}
\dfrac{12EI}{L^3} & -\dfrac{6EI}{L^2} \\[2ex]
-\dfrac{6EI}{L^2} & \dfrac{4EI}{L}
\end{bmatrix}
\tag{1.27}
$$

1.10 Relation Between Stiffness and Flexibility

Using the cantilever example, we can multiply equation (1.26) by the inverse of the stiffness ($\mathbf{K}^{-1}$), to get

$$
\mathbf{K}^{-1}\mathbf{R} = \mathbf{r} \tag{1.28}
$$

This is identical to equation (1.23) except that $\mathbf{F}$ is replaced by $\mathbf{K}^{-1}$. As a result, we see that the flexibility is equal to the inverse of the stiffness. We can also do the same process starting with equation (1.23). Therefore, we have shown that the flexibility and stiffness are inverses of each other:

$$\mathbf{K}^{-1} = \mathbf{F} \qquad (1.29)$$

$$\mathbf{F}^{-1} = \mathbf{K} \qquad (1.30)$$

This is true for our example because an independent set of forces and the corresponding displacements were used to derive the coefficients for the stiffness. It is possible to derive stiffness matrices that cannot be inverted. These matrices represent an unstable structure and have no corresponding flexibility matrix.

1.11 Constrained Displacements

Displacements that are constrained are defined as having an additional condition applied that links them together. For the structure in Figure 1.14, assuming no axial deformation links the displacements at the ends of the beam member, r_1 and r_3 are identical. Displacements that are constrained are not independent, hence no stiffness matrix exists. No stiffness matrix can ever exist that includes dependent displacements. Remember, to find k_{ij} we need to apply a unit displacement to one displacement, keeping all others equal to zero. Obviously, both constrained displacements must move at the same time and are dependent. Although there is no stiffness matrix that includes both these constrained displacements, there will be a flexibility matrix. The flexibility matrix for the forces corresponding to all four of these displacements could be formed, but it will not be invertible.

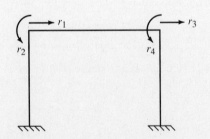

Figure 1.14 Constrained displacements.

1.12 Constrained Forces

Forces that are constrained are defined as having an additional condition applied that links them. As an example, Figure 1.15 shows a simply supported beam. The reactions R_1 and R_3 are linked to the applied loads through statics. Forces that are constrained are not independent and no flexibility matrix exists. Remember, to find f_{ij} we need to apply a unit force and find the displacements at all other DOFs. At the reactions R_1 and R_3

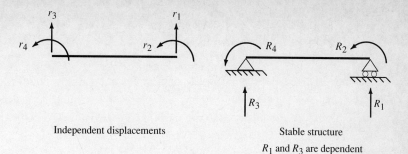

Independent displacements

Stable structure

R_1 and R_3 are dependent

Figure 1.15 Constrained force example.

there will be no displacement. There will be a stiffness matrix for the corresponding displacements, but it will not be invertible.

Consider the single member shown in Figure 1.15 with the four independent displacements. Clearly, we could apply a unit displacement one at a time to the DOFs and develop the stiffness matrix. However, this stiffness matrix does not have an inverse. This is because the structure is unstable and requires at least two support forces to be stable (shown in the force side).

1.13 Symmetry of Flexibility and Stiffness

Maxwell's reciprocal law states:

The displacement at a point a in a structure due to a unit load at another point, b, is equal to the displacement at point b due to a unit load at point a.

Note that the displacements are measured in the location and direction of the applied unit loads. Graphically, this is shown in Figure 1.16.

The flexibility coefficients are displacements due to unit forces. The terms are calculated by applying a unit load at one location and calculating displacements at other locations. Maxwell's law clearly applies and the flexibility coefficients measured at j due to a unit load at i must be equal to the coefficient measured at i due to a unit load at j.

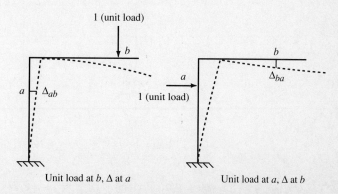

Unit load at b, Δ at a

Unit load at a, Δ at b

Figure 1.16 Maxwell's reciprocal law.

This can be stated in equation form as

$$f_{ij} = f_{ji} \tag{1.31}$$

The stiffness matrix can be thought of as the inverse of the flexibility matrix. A theorem of linear algebra proves that the inverse of a symmetric matrix is always symmetric. Therefore, the stiffness matrix is also symmetric. This can also be stated in equation form as

$$k_{ij} = k_{ji} \tag{1.32}$$

Only roughly one-half of the coefficients of a matrix needs to be calculated and stored due to the symmetry of stiffness and flexibility. However, calculating the symmetric terms gives an excellent method for checking the correctness of a matrix. Another property that is very useful for checking a structure is that the diagonal elements (k_{ii} and f_{ii}) must be positive. A negative value would imply that if you pushed on a structure, it would move toward you.

1.14 Static and Kinematic Indeterminacy

The degree of static indeterminacy of a structure is considered to be the number of unknown forces that must be found in order to calculate the remainder of the forces in the structure by statics. The degree of kinematic indeterminacy is the number of unknown displacements that must be found in order to calculate all the member end forces. For example, if the displacements r_1, r_2, and r_3 are known for the frame shown in Figure 1.17, we could calculate the end member forces for all the members. We could use slope deflection to get the member end moments and then use statics to get the shear and axial forces. Thus the frame is said to be *kinematically indeterminate to the third degree*. If we look at the support reactions for this same structure, we see that there are six unknowns. From statics we know that there are three independent equations of equilibrium in two dimensions. Therefore, the structure is statically indeterminate to the third degree.

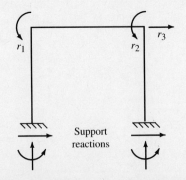

Figure 1.17 Kinematic indeterminacy.

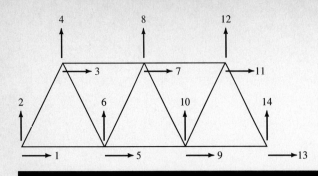

MATRIX
METHODS

2.1 Basic Linear Algebra

As we increase our ability to handle larger analysis problems, so does the complexity increase in solving the equations generated. With only a few unknowns, simple substitution can be used to solve simultaneous equations. However, as the number of unknowns gets large, we need more systematic methods for representing, generating, and solving the equations. This is where matrix methods become useful.

Matrix methods are a subset of linear algebra. Linear algebra covers a wide range of topics well outside the scope of this book. It is strongly recommended that anyone interested in going beyond the level of this book take a linear algebra course. We explain all the matrix procedures required to understand the methods used within this book, but we do not discuss the accuracy or stability of the methods or other such topics.

The techniques developed in this book are matrix based. Therefore, matrix notation will be used in many of the equations. These equations will be manipulated using standard linear algebra techniques. We first need to discuss the definition of a matrix. A matrix is a collection of numbers organized into rows and columns. A matrix has a size that is defined by its number of rows and columns. For example, in the matrix

$$\mathbf{A} = \begin{bmatrix} 1.2 & -2.3 \\ 5.4 & 1.0 \\ 0.0 & -0.2 \end{bmatrix} \tag{2.1}$$

the size of matrix $\mathbf{A}$ is three rows by two columns. In common language this is stated as "matrix $\mathbf{A}$ is 3 by 2." The number of rows is given first, then the number of columns. Any term in the matrix is referred to by its row and column position. The row position is always given first [i.e., $\mathbf{A}(2, 2) = 1.0$]. The diagonal of a matrix is defined as all terms where the row number and the column number are the same [i.e., $\mathbf{A}(1, 1)$ and $\mathbf{A}(2, 2)$].

Matrices can be added. Addition has the requirement that the matrices being added *must* be the same size. The result of adding two matrices is a matrix of the same size. The values consist of the sum of the corresponding terms of the two matrices to be added. As an example, given

$$\mathbf{B} = \begin{bmatrix} -1.2 & 9.2 \\ 3.4 & -2.1 \\ 3.1 & 6.5 \end{bmatrix} \tag{2.2}$$

the equation

$$\mathbf{A} + \mathbf{B} = \mathbf{C} \tag{2.3}$$

gives the result

$$\mathbf{C} = \begin{bmatrix} 0.0 & 6.9 \\ 8.8 & -1.1 \\ 3.1 & 6.3 \end{bmatrix} \tag{2.4}$$

Subtraction is handled in the same way except that the terms are subtracted. As an example, the equation

$$\mathbf{D} = \mathbf{A} - \mathbf{B} \tag{2.5}$$

gives the result

$$\mathbf{D} = \begin{bmatrix} 2.4 & -11.5 \\ 2.0 & 3.1 \\ -3.1 & -6.7 \end{bmatrix} \tag{2.6}$$

Matrices can be multiplied but not divided. Matrix multiplication is *not* commutative. This means that it is order dependent. The terms of the result consist of multiplying a row of the first matrix times a column of the second matrix. As an example, given the equation

$$\mathbf{G} = \mathbf{A} \cdot \mathbf{F} \tag{2.7}$$

and the matrix

$$\mathbf{F} = \begin{bmatrix} 3.4 & -2.1 \\ 7.3 & 1.0 \end{bmatrix} \tag{2.8}$$

the 1, 1 term (row 1, column 1) of $\mathbf{G}$ is

$$G(1, 1) = 1.2(3.4) + (-2.3)(7.3) = -12.71 \tag{2.9}$$

The other terms in the result come from multiplying the other rows times other columns in the same fashion. The location where the result goes is dictated by the row and column number being multiplied. The result $\mathbf{G}(3, 2)$ comes from multiplying row 3 of $\mathbf{A}$ times column 2 of $\mathbf{F}$. The final matrix $\mathbf{G}$ is

$$\mathbf{G} = \begin{bmatrix} -12.71 & -4.82 \\ 25.66 & -10.34 \\ -1.46 & -0.20 \end{bmatrix} \tag{2.10}$$

The size requirement of the matrices being multiplied can be seen by carefully examining the process. The number of terms in a row of **A** must be equal to the number of terms in a column of **F**. The number of rows in **A** and the number of columns in **F** are arbitrary. The size of the result, **G**, will be the number of rows in **A** by the number of columns in **F**. Stated another way, if **A** is N by M, and **F** is M by L, the result, **G**, will be N by L.

$$\begin{array}{ccccc} \mathbf{G} & = & \mathbf{A} & \cdot & \mathbf{F} \\ (N \times L) & & (N \times M) & (M \times L) & \end{array}$$
$$[\text{must be same}]$$
$$(2.11)$$

Multiplication is associative and distributive. Therefore, the following equations are both true:

$$\mathbf{A} \cdot \mathbf{B} \cdot \mathbf{C} = (\mathbf{A} \cdot \mathbf{B}) \cdot \mathbf{C} = \mathbf{A} \cdot (\mathbf{B} \cdot \mathbf{C}) \qquad (2.12)$$

$$(\mathbf{A} + \mathbf{B}) \cdot \mathbf{C} = \mathbf{A} \cdot \mathbf{C} + \mathbf{B} \cdot \mathbf{C} \qquad (2.13)$$

Matrices have a multiplicative identity just like regular multiplication. That is, there exists a matrix that can be multiplied times any matrix and not change the matrix. This is called the *identity matrix*, **I**. It consists of a matrix that has ones on the diagonal and zeros everywhere else (off-diagonal terms). The identity matrix is always square (it has the same number of rows as columns). For example, a 3×3 identity matrix looks like

$$\mathbf{I} = \begin{bmatrix} 1.0 & 0.0 & 0.0 \\ 0.0 & 1.0 & 0.0 \\ 0.0 & 0.0 & 1.0 \end{bmatrix} \qquad (2.14)$$

The identity matrix has the property

$$\mathbf{A} \cdot \mathbf{I} = \mathbf{I} \cdot \mathbf{A} = \mathbf{A} \qquad (2.15)$$

That is, you can postmultiply or premultiply a matrix by the identity matrix and not change the original matrix. Note that **A** does not have to be square. In that case, the identity matrix used for premultiplication will be of a different size than that used for postmultiplication.

The inverse of a matrix is one of the more important properties of a matrix. The *inverse of a matrix* is the matrix that when multiplied times the original matrix will give the identity matrix. The inverse is designated by a superscript -1. For example, the inverse of matrix **A** is designated as $\mathbf{A}^{-1}$. For example,

$$\mathbf{A} \cdot \mathbf{A}^{-1} = \mathbf{A}^{-1} \cdot \mathbf{A} = \mathbf{I} \qquad (2.16)$$

A matrix *must* be square to be inverted. Not all matrices can be inverted. If a matrix can be inverted, it is called *nonsingular* or *invertible*. If it cannot be inverted, it is called *singular*. Matrix inversion is very useful in explaining the solution of equations. A set of equations such as

$$
\begin{aligned}
5x_1 - 4x_2 + x_3 \quad\quad\quad &= \quad 0 \\
-4x_1 + 6x_2 - 4x_3 + x_4 &= \quad 1 \\
x_1 - 4x_2 + 6x_3 - 4x_4 &= \quad 0 \\
x_2 - 4x_3 + 5x_4 &= -2
\end{aligned}
\tag{2.17}
$$

can easily be represented as the multiplication of two matrices equated to a third. If we define a matrix $\mathbf{A}$ as the matrix of the coefficients of the x terms, the matrix (vector) $\mathbf{x}$ as the matrix of the unknown x terms, and the matrix $\mathbf{b}$ as the matrix of the right-hand side,

$$
\mathbf{A} =
\begin{bmatrix}
5 & -4 & 1 & 0 \\
-4 & 6 & -4 & 1 \\
1 & -4 & 6 & -4 \\
0 & 1 & -4 & 5
\end{bmatrix}
\quad
\mathbf{x} =
\begin{bmatrix}
x_1 \\ x_2 \\ x_3 \\ x_4
\end{bmatrix}
\quad
\mathbf{b} =
\begin{bmatrix}
0 \\ 1 \\ 0 \\ -2
\end{bmatrix}
\tag{2.18}
$$

The set of equations in (2.17) is represented by

$$
\mathbf{Ax} = \mathbf{b}
\tag{2.19}
$$

This set of equations can be solved schematically by the same process as in any algebraic equation. Multiply both sides by the inverse of $\mathbf{A}$: the inverse times a matrix is the identity; the identity times a matrix is the matrix.

$$
\mathbf{A}^{-1}\mathbf{Ax} = \mathbf{A}^{-1}\mathbf{b}
\tag{2.20}
$$

$$
\mathbf{Ix} = \mathbf{A}^{-1}\mathbf{b}
\tag{2.21}
$$

$$
\mathbf{x} = \mathbf{A}^{-1}\mathbf{b}
\tag{2.22}
$$

In practice, the inverse is never calculated; however, it is quite frequently used during the manipulation of equations. There are two types of solution methods used in practice: (1) direct and (2) iterative. Direct solutions can be thought of as variations of the Gauss elimination method. The iterative schemes can be thought of as variations of the Gauss–Seidel solution technique.

Another important manipulation needed is called a *transpose*. The transpose is defined as switching the rows with the columns. Since the rows and columns are switched, the size of the transpose is also just the switch of the original size. The transpose of a matrix is designated by a superscript T. For example, the transpose of matrix $\mathbf{H}$ is $\mathbf{H}^{\mathrm{T}}$:

$$
\mathbf{H} =
\begin{bmatrix}
-1.2 & 5.4 & 3.3 & 0.0 \\
4.3 & 1.1 & -3.1 & -7.8 \\
1.0 & 0.0 & 8.2 & 1.9
\end{bmatrix}
\quad
\mathbf{H}^{\mathrm{T}} =
\begin{bmatrix}
-1.2 & 4.3 & 1.0 \\
5.4 & 1.1 & 0.0 \\
3.3 & -3.1 & 8.2 \\
0.0 & -7.8 & 1.9
\end{bmatrix}
\tag{2.23}
$$

$$
(3 \times 4) \quad\quad\quad\quad\quad\quad (4 \times 3)
$$

The transpose can also be taken of a set of multiplied matrices. This is equivalent to reversing the order of the multiplications and individually transposing the matrices:

$$(\mathbf{H} \cdot \mathbf{A} \cdot \mathbf{D})^{\mathrm{T}} = \mathbf{D}^{\mathrm{T}}\mathbf{A}^{\mathrm{T}}\mathbf{H}^{\mathrm{T}} \tag{2.24}$$

One final definition needs to be covered and that is for a symmetric matrix. A matrix is *symmetric* when the rows are equal to the corresponding column. In other words, the matrix is equal to its transpose. As an example, the following matrix is symmetric:

$$\mathbf{M} = \begin{bmatrix} 6.0 & 2.0 & 1.0 & 0.0 & -1.0 & 0.0 \\ 2.0 & 5.0 & -1.0 & 2.0 & 0.0 & 0.0 \\ 1.0 & -1.0 & 6.0 & 1.0 & 2.0 & -1.0 \\ 0.0 & 2.0 & 1.0 & 4.0 & 1.0 & 2.0 \\ -1.0 & 0.0 & 2.0 & 1.0 & 5.0 & 1.0 \\ 0.0 & 0.0 & -1.0 & 2.0 & 1.0 & 5.0 \end{bmatrix} \tag{2.25}$$

2.2 Solution of Equations

Structural analysis using matrix methods generates a set of simultaneous linear equations. The solution of this set of equations is required to find the response of the structure. When using stiffness methods, the response is the nodal displacements. These displacements are then transformed into member end forces and the structure is designed. Efficient methods for solving these equations are required since this process requires about 70 percent of the numerical effort in analyzing a linear static structure. Thus the amount of computer power required to analyze a structure depends on the effort required for the solution of the equations.

The solution of the generated equations is represented by the inverse. That is, the stiffness equation, given as

$$\mathbf{Kr} = \mathbf{R} \tag{2.26}$$

can be solved to yield the displacements by multiplying both sides of the equation by **K** inverse.

$$\mathbf{r} = \mathbf{K}^{-1}\mathbf{R} \tag{2.27}$$

As stated before, in practice we never form the actual inverse when solving a set of equations. There are two types of methods for solving a set of equations: direct and indirect. Indirect methods are also referred to as *iterative*. These techniques work by making repetitive corrections to an approximate solution for **r**. These include methods such as Gauss–Seidel, successive overrelaxation (SOR), and conjugate gradient. Indirect methods are very good for three-dimensional structures and structures where good approximations for the displacements exist. They are also very good for implementing on vector or parallel processing computers.

Direct solution techniques are the most common ones used in structural analysis. Direct techniques involve a set number of steps to convert the applied load to structural displacements. The inversion and multiplication is a direct method but not very efficient. Other direct methods include Gauss elimination and the various forms of triangular factorization.

Gauss elimination can be thought of as the fundamental method for solving a set of equations. It is a process by which the coefficient matrix and the right-hand side are converted into a set of equations that are easier to solve. Gauss elimination can be thought of as converting equation (2.26) into the form

$$\mathbf{Ur} = \mathbf{Y} \tag{2.28}$$

where $\mathbf{U}$ is called the upper triangular form of the matrix $\mathbf{K}$ and $\mathbf{Y}$ is the converted form of the load matrix $\mathbf{R}$. $\mathbf{U}$ has the special form of having only zero values below the diagonal:

$$\begin{bmatrix} u_{11} & u_{12} & u_{13} & u_{14} & u_{15} & u_{16} & u_{17} \\ 0 & u_{22} & u_{23} & u_{24} & u_{25} & u_{26} & u_{27} \\ 0 & 0 & u_{33} & u_{34} & u_{35} & u_{36} & u_{37} \\ 0 & 0 & 0 & u_{44} & u_{45} & u_{46} & u_{47} \\ 0 & 0 & 0 & 0 & u_{55} & u_{56} & u_{57} \\ 0 & 0 & 0 & 0 & 0 & u_{66} & u_{67} \\ 0 & 0 & 0 & 0 & 0 & 0 & u_{77} \end{bmatrix} \begin{bmatrix} r_1 \\ r_2 \\ r_3 \\ r_4 \\ r_5 \\ r_6 \\ r_7 \end{bmatrix} = \begin{bmatrix} Y_1 \\ Y_2 \\ Y_3 \\ Y_4 \\ Y_5 \\ Y_6 \\ Y_7 \end{bmatrix} \tag{2.29}$$

As can be seen, the solution for the value of r_7 is trivial. Once r_7 is found, r_6 can be calculated, and so on. This process is called *backward substitution*. Overall, the process takes one-third the amount of computer time as is taken by a full matrix inversion. As an example of the process, we will solve the following set of equations:

$$\begin{bmatrix} 4 & -1 & 2 \\ -1 & 3 & 1 \\ 2 & 1 & 4 \end{bmatrix} \begin{bmatrix} r_1 \\ r_2 \\ r_3 \end{bmatrix} = \begin{bmatrix} 4 \\ -2 \\ 8 \end{bmatrix} \tag{2.30}$$

The process of converting the stiffness matrix to its triangular form involves two principles of linear algebra. The first is that the solution of a set of equations is not changed if both sides of any one equation are multiplied by a constant. The second principle is that the solution to a set of equations is not changed by performing linear combinations of the equations. As a result, we can multiply equation (row) 1 by 0.25 and add it to equation (row) 2. The result is that the first term in row 2 becomes zero. Through multiple applications of this process we can convert the stiffness to its triangular form. The right-hand side must also be converted during this process.

First, a couple of definitions must be given. The process proceeds on a column basis. The diagonal term of the current column being reduced is called the *pivot*. The row in which the pivot exists is called the *pivot row*.

$$\text{Pivot for column } 1 = 4 \text{ (the 1, 1 term)}$$

$$\begin{bmatrix} 4 & -1 & 2 \\ -1 & 3 & 1 \\ 2 & 1 & 4 \end{bmatrix} \begin{bmatrix} r_1 \\ r_2 \\ r_3 \end{bmatrix} = \begin{bmatrix} 4 \\ -2 \\ 8 \end{bmatrix} \qquad (2.31)$$

A formalized procedure to do Gauss elimination is as follows:

1. For each column (I) of the coefficients except the last (column N) and for the columns of the right-hand side:

For each row (J) of the current column below the diagonal:

For each column (K) from the current column to the end, including the right-hand-side columns:

(a) Form the factor to use on the current row J:

$$\text{factor} = \frac{\text{term to make zero (column } I, \text{ row } J)}{\text{pivot}}$$

(b) Multiply the pivot row by the factor.
(c) Subtract the factored pivot row from the current row J.

2. Back substitute when factorization is complete.

An example of the process is shown using the following matrix and right-hand side. We first need to convert the (2, 1) term to zero. The factor is $-\frac{1}{4}$. Multiplying the factor times the first row and subtracting, we get

$$\begin{bmatrix} 4 & -1 & 2 \\ 0 & \frac{11}{4} & \frac{3}{2} \\ 2 & 1 & 4 \end{bmatrix} \begin{bmatrix} 4 \\ -1 \\ 8 \end{bmatrix} \qquad (2.32)$$

Next, we need to convert the (3, 1) term. The factor is $\frac{2}{4}$. Multiplying and subtracting, we get

$$\begin{bmatrix} 4 & -1 & 2 \\ 0 & \frac{11}{4} & \frac{3}{2} \\ 0 & \frac{3}{2} & 3 \end{bmatrix} \begin{bmatrix} 4 \\ -1 \\ 6 \end{bmatrix} \qquad (2.33)$$

We now need to change to the second column. Here we only need to convert the (3, 2) term. The factor is $\frac{\frac{3}{2}}{\frac{11}{4}}$. Multiplying and subtracting, we get

$$\begin{bmatrix} 4 & -1 & 2 \\ 0 & \dfrac{11}{4} & \dfrac{3}{2} \\ 0 & 0 & \dfrac{24}{11} \end{bmatrix} \begin{bmatrix} 4 \\ -1 \\ \dfrac{72}{11} \end{bmatrix} \tag{2.34}$$

We now have the desired upper triangular form and are ready to perform the backward substitution. First we can solve for r_3.

$$r_3 = \frac{\frac{72}{11}}{\frac{24}{11}} = \frac{72}{24} = 3.0 \tag{2.35}$$

Next, we can solve for r_2.

$$r_2 = \frac{-1 - \frac{3}{2}r_3}{\frac{11}{4}} = \frac{-1 - \frac{3}{2}(3.0)}{\frac{11}{4}} = -2.0 \tag{2.36}$$

Finally, we can solve for r_1.

$$r_1 = \frac{4 - 2r_3 - (-1)r_2}{4} = \frac{4 - 2(3.0) - (-1)(-2.0)}{4.0} = -1.0 \tag{2.37}$$

This method is very easy to implement in the computer but is not the most efficient. Gauss elimination was the first solution method used during the late 1950s and early 1960s in the first matrix methods/structural analysis computer programs. Other direct methods take better advantage of computer memory, machine architectures, and the shape of the stiffness matrix. Some of these methods are discussed in the following sections.

2.3 Banded Solution Method

Although the basic Gauss elimination method is very simple to codify, it requires that the full matrix be stored in memory. As structures become larger, so do the stiffness matrices (or equations that need to be solved). During the early to middle 1960s, computers became more prevalent and engineers wanted to solve larger problems. During this time, people began to notice special properties about the matrices used to analyze structures. The first property was that the matrices generated for structural analysis were always symmetric. A second property was that the matrices had a structure that is called *sparse*. This means that many of the terms in the matrix have a value of zero. In addition, the sparsity has a pattern to it called *banded*.

A banded matrix is one that is sparse and has the nonzero terms clustered about the diagonal. An example of a banded matrix is

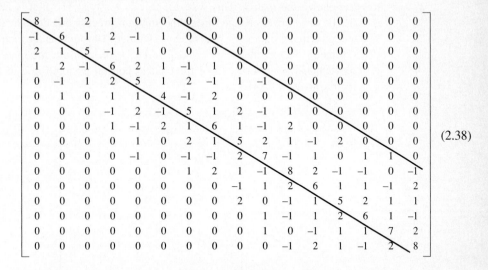

$$(2.38)$$

A symmetric banded matrix can be characterized by a property called the half-bandwidth of the matrix. The *half-bandwidth* can be defined as

$$\text{half-bandwidth} = \underset{\text{all columns } J}{\text{maximum}} (I - J + 1) \qquad (2.39)$$

where I is the row number of the first nonzero term encountered in a column starting at the top going down. J is the column number. What this physically means is that the half-bandwidth is equal to the maximum height of all the columns measuring from the diagonal up to the highest nonzero term. It can also be characterized as finding a line parallel to the diagonal that encloses all nonzero terms. This can be shown as

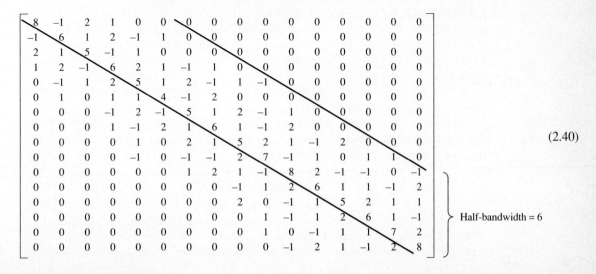

$$(2.40)$$

Half-bandwidth = 6

All the numbers outside the bandwidth are zero by definition. It was found that due to symmetry and this banded property, only half of the band needs to be saved in order to triangularize the matrix. This was due to two reasons. The first is because the zeros outside the band never change during the solution process and do not need to be stored. The second is that a symmetric matrix stays symmetric throughout the factorization process if a symmetric factorization is used (e.g., LDL^T). As a result, the matrix is stored in a different manner. Only one-half of the banded portion is stored in a rectangular matrix. The matrix above is stored as

$$\begin{bmatrix}
8 & -1 & 2 & 1 & 0 & 0 \\
6 & 1 & 2 & -1 & 1 & 0 \\
5 & -1 & 1 & 0 & 0 & 0 \\
6 & 2 & 1 & -1 & 1 & 0 \\
5 & 1 & 2 & -1 & 1 & -1 \\
4 & -1 & 2 & 0 & 0 & 0 \\
5 & 1 & 2 & -1 & 1 & 0 \\
6 & 1 & -1 & 2 & 0 & 0 \\
5 & 2 & 1 & -1 & 2 & 0 \\
7 & -1 & 1 & 0 & 1 & 1 \\
8 & 2 & -1 & -1 & 0 & -1 \\
6 & 1 & 1 & -1 & 2 & \times \\
5 & 2 & 1 & 1 & \times & \times \\
6 & 1 & -1 & \times & \times & \times \\
7 & 2 & \times & \times & \times & \times \\
8 & \times & \times & \times & \times & \times
\end{bmatrix} \tag{2.41}$$

The new matrix is the number of equations, NEQ, by the half-bandwidth, B. The $\times$'s represent space that is wasted because it is not needed for storing the matrix but must exist when defining a rectangular matrix. It is important to realize that this change is only a rearrangement of the way the matrix is stored. A change also has to be made for equation solving since the lower portion of the matrix is not stored. This change in solution process is not very difficult since all terms required from the lower portion are found in the upper portion of a symmetric matrix, even during factorization.

Some other beneficial results occur as a result of the new storage. The first is the fact that the amount of storage required by the coefficient matrix is greatly reduced. The required storage for the full and banded matrices is

$$\text{full storage} = \text{NEQ}^2 \tag{2.42}$$

$$\text{half-banded storage} = \text{NEQ} \times B \tag{2.43}$$

A second benefit is the amount of effort, or time, required to solve the set of equations. Since the terms outside the bandwidth are zero and remain so throughout the solution,

Table 2.1

NEQ	B (Half-Band)	Full Matrix Solution (Number of Operations)	Banded Matrix Solution (Number of Operations	Factor Improved
59	16	68,460	7,552	9.1
72	15	124,416	8,100	15.4
198	17	2,587,464	28,611	90.4
221	104	3,597,954	1,195,168	3.0
310	29	9,930,334	130,355	76.2
492	45	40,429,125	498,150	81.2
607	328	74,549,514	360,406	206.8
918	108	257,873,544	5,353,776	48.2
1007	49	340,382,448	1,208,904	281.6

they do not need to be factored. The banded storage requires that these terms be skipped since they do not exist. Numerical effort, which can be converted directly into time for solution, is usually measured as the number of multiplies and divides required to perform an algorithm. The required number of operations for both methods (full and banded) are

$$\text{number of operations} = \tfrac{1}{3}\text{NEQ}^3 \qquad \text{(full elimination)} \qquad (2.44)$$
$$\text{number of operations} = \tfrac{1}{2}\text{NEQ} \times B^2 \qquad \text{(banded solution)} \qquad (2.45)$$

Some typical sizes for the values of NEQ and B as well as the number of operations are given in Table 2.1. It is clear that the banded solution scheme can offer two orders of magnitude of savings in storage and four orders of magnitude in savings of numerical effort for solution. This savings is important because the solution of the equilibrium equations is usually 70 percent of the effort required in statically analyzing a linear structure.

2.4 Equation Renumbering

The advantages of a banded solution prompted a lot of research into methods for reducing the bandwidth of the equations generated when analyzing a structure. The bandwidth of a structure is directly related to the way in which the degrees of freedom (DOFs) of the

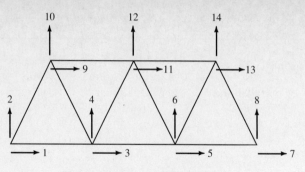

Figure 2.1 Poor DOF numbering.

structure are numbered. As an example, the truss structure shown in Figure 2.1 and the resulting stiffness matrix shape and bandwidth are given. The DOFs for the structure are numbered in two patterns and the results shown (Figures 2.1 and 2.2). The ×'s represent nonzero terms in the stiffness matrix. The resulting matrix is

$$
\begin{bmatrix}
X & X & X & X & 0 & 0 & 0 & 0 & X & X & 0 & 0 & 0 & 0 \\
X & X & X & X & 0 & 0 & 0 & 0 & X & X & 0 & 0 & 0 & 0 \\
X & X & X & X & X & X & 0 & 0 & X & X & X & X & 0 & 0 \\
X & X & X & X & X & X & 0 & 0 & X & X & X & X & 0 & 0 \\
0 & 0 & X & X & X & X & X & X & 0 & 0 & X & X & X & X \\
0 & 0 & X & X & X & X & X & X & 0 & 0 & X & X & X & X \\
0 & 0 & 0 & 0 & X & X & X & X & 0 & 0 & 0 & 0 & X & X \\
0 & 0 & 0 & 0 & X & X & X & X & 0 & 0 & 0 & 0 & X & X \\
X & X & X & X & 0 & 0 & 0 & 0 & X & X & X & X & 0 & 0 \\
X & X & X & X & 0 & 0 & 0 & 0 & X & X & X & X & 0 & 0 \\
0 & 0 & X & X & X & X & 0 & 0 & X & X & X & X & X & X \\
0 & 0 & X & X & X & X & 0 & 0 & X & X & X & X & X & X \\
0 & 0 & 0 & 0 & X & X & X & X & 0 & 0 & X & X & X & X \\
0 & 0 & 0 & 0 & X & X & X & X & 0 & 0 & X & X & X & X
\end{bmatrix}
\qquad (2.46)
$$

Half-bandwidth = 10

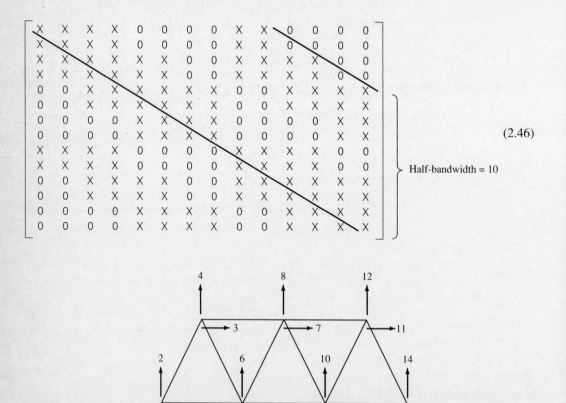

Figure 2.2 Optimal DOF numbering.

The resulting matrix for optimal numbering is

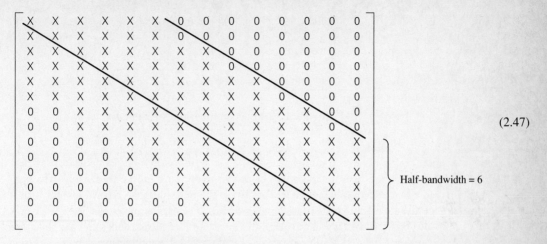

$$(2.47)$$

Half-bandwidth = 6

Even for the small structures shown, the bandwidth is reduced considerably by a good choice of numbering of the DOFs. This recognition of "best" ordering prompted research into the automation of equation numbering. Many computer programs have automatic nodal renumbering algorithms that incorporate these techniques. These renumbering schemes allow the user to specify any nodal numbering when defining the structure and the program will rearrange them to a near-optimal order. Other computer programs assign the DOF numbers sequentially without renumbering. The problem of ordering the equations to produce a minimum bandwidth is left to the user. This reduces the problem of achieving a small bandwidth to numbering the nodes in optimal order. Equation numbers are then usually given to the nodes' active DOFs in sequence.

There are some common rules of thumb for numbering the nodes of a structure. The

Figure 2.3 Proper numbering for skinny frame.

4	8	12	16	20	24	28	32
3	7	11	15	19	23	27	31
2	6	10	14	18	22	26	30
1	5	9	13	17	21	25	29

Figure 2.4 Proper numbering for short frame.

first rule is always to number along the short (in the number of nodes sense) side of the structure first. See, for example, the proper numbering for the frame shown in Figure 2.3. Note that we could equally have started at the bottom. If the frame is wider than it is tall (number of bays versus number of stories), the numbering would go up (or down) the side (Figure 2.4). This same rule of thumb is valid for bridge decks and flat plate structures. The next rule of thumb is for pipe structures. The numbering should go around the circumference and then through the depth (Figure 2.5). This is, of course, assuming that there are more sections in the depth direction than around the circumference.

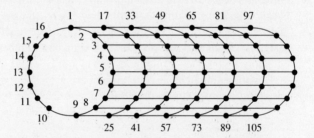

Figure 2.5 Optimal numbering for a pipe mesh.

While these rules of thumb are very good for simple structures, there are no simple rules for complex structures. As a result, there has been a large effort into the automatic renumbering of equations for optimal bandwidth. There are a number of good methods, some of which include reverse Cuthill–McKee, Gibbs–Pool, and Stockmeyer and PFM. All major analysis programs include one of these techniques so that the user need not worry about how to number the nodes.

2.5 Profile Solution Method

The next evolution in equation solving came when it was realized that there were a number of zero terms included in the half-bandwidth that also never changed during the solution process. It was found that only the terms below and including the first nonzero term in a column was required for solution. The first nonzero terms defined a boundary called the *profile* or *skyline* of a matrix. For the given matrix, the profile is

$$
\begin{bmatrix}
8 & -1 & 2 & 1 & 0 & 0 & 0 & 0 & 0 & 0 & 0 & 0 & 0 & 0 & 0 & 0 \\
-1 & 6 & 1 & 2 & -1 & 1 & 0 & 0 & 0 & 0 & 0 & 0 & 0 & 0 & 0 & 0 \\
2 & 1 & 5 & -1 & 1 & 0 & 0 & 0 & 0 & 0 & 0 & 0 & 0 & 0 & 0 & 0 \\
1 & 2 & -1 & 6 & 2 & 1 & -1 & 1 & 0 & 0 & 0 & 0 & 0 & 0 & 0 & 0 \\
0 & -1 & 1 & 2 & 5 & 1 & 2 & -1 & 1 & -1 & 0 & 0 & 0 & 0 & 0 & 0 \\
0 & 1 & 0 & 1 & 1 & 4 & -1 & 2 & 0 & 0 & 0 & 0 & 0 & 0 & 0 & 0 \\
0 & 0 & 0 & -1 & 2 & -1 & 5 & 1 & 2 & -1 & 1 & 0 & 0 & 0 & 0 & 0 \\
0 & 0 & 0 & 1 & -1 & 2 & 1 & 6 & 1 & -1 & 2 & 0 & 0 & 0 & 0 & 0 \\
0 & 0 & 0 & 0 & 1 & 0 & 2 & 1 & 5 & 2 & 1 & -1 & 2 & 0 & 0 & 0 \\
0 & 0 & 0 & 0 & -1 & 0 & -1 & -1 & 2 & 7 & -1 & 1 & 0 & 1 & 1 & 0 \\
0 & 0 & 0 & 0 & 0 & 0 & 1 & 2 & 1 & -1 & 8 & 2 & -1 & -1 & 0 & -1 \\
0 & 0 & 0 & 0 & 0 & 0 & 0 & 0 & -1 & 1 & 2 & 6 & 1 & 1 & -1 & 2 \\
0 & 0 & 0 & 0 & 0 & 0 & 0 & 0 & 2 & 0 & -1 & 1 & 5 & 2 & 1 & 1 \\
0 & 0 & 0 & 0 & 0 & 0 & 0 & 0 & 0 & 1 & -1 & 1 & 2 & 6 & 1 & -1 \\
0 & 0 & 0 & 0 & 0 & 0 & 0 & 0 & 0 & 1 & 0 & -1 & 1 & 1 & 7 & 2 \\
0 & 0 & 0 & 0 & 0 & 0 & 0 & 0 & 0 & 0 & -1 & 2 & 1 & -1 & 2 & 8
\end{bmatrix} \quad \text{Profile} \tag{2.48}
$$

Now, only the terms within the profile need to be stored. The profile method of storage places the columns within the profile sequentially in a one-dimensional array. A companion integer array is also required as a pointer to the diagonal terms of the matrix. The storage would look as follows:

one-dimensional array
$$
\begin{aligned}
= \;& 8 \;-1\; 6\; 2\; 1\; 5\; 1\; 2\; -1\; 6\; -1\; 1\; 2\; 5\; 1\; 1\; 1\; 4\; -1\; 2\; -1\; 5\; 1\; -1\; 2\; 1\; 6\; 1\; 0\; 2\; 1 \\
& 5 \;-1\; 0\; -1\; -1\; 2\; 7\; 1\; 2\; 1\; -1\; 8\; -1\; 1\; 2\; 6\; 2\; 0\; -1\; 1\; 5\; 1\; -1\; 1\; 2\; 6\; 1\; 0\; -1 \\
& 1\; 1\; 7\; -1\; 2\; 1\; -1\; 2\; 8
\end{aligned}
$$

integer pointer $= 1 \quad 3 \quad 6 \quad 10 \quad 14 \quad 19 \quad 24 \quad 28 \quad 33 \quad 39 \quad 44 \quad 48 \quad 53 \quad 58 \quad 64 \quad 70$

This further reduces the amount of storage required for a stiffness matrix. For the example above, the storage required for the banded scheme is 96 terms, while the profile method requires only 70 terms. It also reduces the number of operations to solve a set of equations, but no simple formula can be written.

The profile scheme is the most common and generally considered to be fastest for solving a set of equations in a sequential direct method. A variation of this method called the *frontal method* is similar in that it operates only on the nonzero terms within the profile; however, it also combines the benefits of substructuring. The frontal method assembles and solves simultaneously so that the full stiffness (coefficient) matrix is never completely assembled. This process requires the same number of operations as the profile scheme. If the profile scheme is blocked, the matrix broken into smaller parts, it is equivalent to the same amount of storage and effort as the frontal scheme. Renumbering schemes have been developed that take specific advantage of the profile storage without concern for the bandwidth.

2.6 CAL-90: Matrix Interpretive Language

CAL-90 is an interactive matrix manipulation program developed specifically for matrix structural analysis. It is based on the original CAL-78 matrix language version from Berkeley. CAL-90 contains all the matrix commands from CAL-78 plus some additional ones. CAL-90 is an interactive-based program that uses a database and allows macro commands. The program implements a problem-oriented language for structural analysis. Each command in CAL performs some matrix operation. To start the version included with this book, at the prompt type

C:\ > CAL90

CAL-90 will then initialize its database, read any saved matrices, and return with the interactive prompt

Storage Remaining 3999 - Enter Command or H for Help
CAL90 >

CAL-90 uses three files: INPUT, OUTPUT, and DATABASE. The first two are formatted files and can be printed and edited. The DATABASE file is unformatted and cannot be viewed. It is used to save and transfer data between other CAL programs. The INPUT file is where CAL-90 looks to find macros defined by the user. It assumes that the file has already been created before starting CAL-90.

Commands in CAL-90 are all of the form

OPERATION M1+ M2 M3− N=N1,N2

where OPERATION is the command to be performed. These are matrix functions such as add, subtract, multiply, and so on. M1, M2, and M3 are arbitrary matrix names. There can be from zero to seven matrix names required by a command. For example:

ADD STIFA STIFB

This adds the two matrices STIFA and STIFB and puts the results into matrix STIFA. The + and − after the matrix name indicate whether the matrix is created (+) or modified (−). Note that this + and − should not be used as part of the matrix name. N1 and N2 are additional parameters required by some commands. If required, they will be explained in the help command. To input these, the form is

N= 4 , −3

Note that the letter and = must have *no* space between them. There can be any number of spaces and/or a comma between the data after the equal sign.

CAL-90 has a free-formatted input structure. It can read real, integer, and exponential formats. If the required number is real, in-line arithmetic can be performed. For example, the load command reads real matrices. It expects the number of lines of data equal to the number of rows of the matrix. To load a real matrix A:

```
LOAD A R=3 C=3
25+3*12  3.45  12.e-3
1.2,-1,0
3.5*12 -3.4e2 1
```

The final matrix looks like

$$
\begin{bmatrix}
336.000 & 3.450 & 0.012 \\
1.200 & -1.000 & 0.000 \\
42.000 & -340.000 & 1.000
\end{bmatrix}
$$

Note that arithmetic operations are sequential. In interactive mode, any missing data are prompted for. The command HELP will give information on every command available.

CAL-90 also has a batch or macro capability. Macros are contained in the file named INPUT. The commands are the same form as the interactive ones. However, all data must be specified for a macro; missing data cannot be prompted for. The file INPUT can be created with any word processor provided that the file is in ASCII format. [Some word processors require special save commands for ASCII files (e.g., Word, Ami, WordPerfect).]

Macro commands must be given a name or header. The end of the macro must also be defined. The name or header is an arbitrary one- to six-character string. The end can be defined by the RETURN or IF commands. For example,

```
DO-A
LOAD A R=2 C=2
1 2
2 1
P A
RETURN
```

is a macro named DO-A. The macro loads the matrix A and prints the matrix. The macro returns to interactive mode when complete. To use this macro, from the interactive prompt, type

CAL90 > SUBMIT DO-A

Any number of macros can be contained in an input file. They can be stacked in any order. Using the SUBMIT command, the commands can be performed in any order and as many times as desired. For example, the input file can contain

```
DO-A
LOAD A R=2 C=2
1 2
2 1
P A
RETURN
DO-B
```

```
LOAD B R=2 C=1
7.5
−3.5
P B
SOLVE A B
P B
RETURN
DO-C
LOAD C R=2 C=1
−7.5
 3.5
P C
SOLVE A C
P C
RETURN
```

Then in an interactive session, you could do

CAL90 > SUBMIT DO-A
CAL90 > SUBMIT DO-B
CAL90 > SUBMIT DO-A
CAL90 > SUBMIT DO-C

The full user's manual appears as Appendix A. As specific commands are required throughout the book, they will be explained in detail.

2.7 Problems

Use the following matrices for these problems.

$$A = \begin{bmatrix} 43.4 & 5.6 & -1.2 & 5.6 \\ -3.3 & -25.1 & 0.0 & -12.0 \\ -10.0 & 6.3 & 57.1 & -1.1 \end{bmatrix}$$

$$B = \begin{bmatrix} 21.0 & -33.1 \\ 5.7 & 2.1 \\ 44.2 & -12.0 \end{bmatrix} \qquad C = \begin{bmatrix} 3.04 & -12.0 \\ -12.0 & 60.3 \end{bmatrix}$$

$$D = \begin{bmatrix} 31.3 & 12.5 & -12.7 \\ 3.4 & -51.0 & 0.0 \\ 4.6 & -12.3 & 17.5 \end{bmatrix}$$

$$E = \begin{bmatrix} 43.2 & 8.7 & 13.1 & 10.1 & -21.0 \\ 17.0 & -100.0 & 23.0 & 11.7 & -4.5 \\ -19.9 & 8.2 & -91.4 & 33.7 & -27.1 \end{bmatrix}$$

$$\mathbf{F} = \begin{bmatrix} 1.5 \\ 2.3 \\ -4.5 \\ 1.2 \\ 7.2 \end{bmatrix} \qquad \mathbf{G} = \begin{bmatrix} 3.4 & -3.2 \\ 5.6 & -1.1 \\ -1.1 & 4.5 \end{bmatrix}$$

$$\mathbf{H} = \begin{bmatrix} 43.5 & -10.0 & 6.7 \\ -10.0 & 34.5 & -7.6 \\ 6.7 & -7.6 & 43.5 \end{bmatrix} \qquad \mathbf{J} = \begin{bmatrix} 2.3 \\ -6.5 \\ 21.0 \end{bmatrix}$$

$$\mathbf{K} = \begin{bmatrix} 0.0 \\ 57.0 \end{bmatrix} \qquad \mathbf{L} = \begin{bmatrix} 45.0 & -12.0 & 7.6 & 12.1 \\ 10.0 & 63.0 & 12.1 & 0.0 \end{bmatrix}$$

2.1. By hand, perform the following matrix operations.

   (a) B + G   **(b) B − G**   **(c) D + H**
   (d) H − D   **(e) E · F**   **(f) $\mathbf{A}^\mathrm{T} \cdot \mathbf{G}$**
   (g) $\mathbf{J}^\mathrm{T} \cdot \mathbf{E} \cdot \mathbf{F}$ **(h) $(\mathbf{J}^\mathrm{T} \cdot \mathbf{H} \cdot \mathbf{J}) + (\mathbf{J}^\mathrm{T} \cdot \mathbf{D} \cdot \mathbf{J})$**

2.2. Use CAL-90 to perform the matrix operations given in Problem 2.1.

2.3. Use the Gauss elimination method to solve the following equations for the unknowns **X**.
   (a) $\mathbf{H} \cdot \mathbf{X} = \mathbf{J}$ **(b) $(\mathbf{D}^\mathrm{T} \cdot \mathbf{H} \cdot \mathbf{D})\mathbf{X} = \mathbf{J}$**

2.4. Use CAL-90 to solve the equations in Problem 2.3 for the unknowns **X**.

2.5. Use Gauss elimination to solve the following set of simultaneous equations:

$$\begin{bmatrix} 16 & 2 & 0 & 2 & 0 & 1 \\ 2 & 16 & 0 & 0 & 2 & 1 \\ 0 & 0 & 10 & -2 & -2 & -4 \\ 2 & 0 & -2 & 16 & 2 & 1 \\ 0 & 2 & -2 & 2 & 16 & -1 \\ 1 & 1 & -4 & 1 & -1 & 10 \end{bmatrix} \begin{bmatrix} X_1 \\ X_2 \\ X_3 \\ X_4 \\ X_5 \\ X_6 \end{bmatrix} = \begin{bmatrix} 0 \\ 0 \\ 1 \\ 0 \\ 0 \\ 2 \end{bmatrix}$$

2.6. Use CAL-90 to solve the equations in Problem 2.5.

2.7. Use CAL-90 to perform the following matrix equations.
   (a) $(\mathbf{L} \cdot \mathbf{A}^\mathrm{T} \cdot \mathbf{B} - \mathbf{C}) \cdot \mathbf{G}^\mathrm{T}$ **(b) $\mathbf{H} \cdot \mathbf{E} - \mathbf{E} \cdot \mathbf{F} \cdot \mathbf{F}^\mathrm{T}$**

2.8. Use CAL-90 to solve the following set of equations for $\mathbf{X}$:

$$\mathbf{E} \cdot \mathbf{E}^\mathrm{T} \cdot \mathbf{X} = \mathbf{J}$$

2.9. Use CAL-90 to get to form the following matrix:

$$(\mathbf{A}\mathbf{A}^\mathrm{T})^{-1}$$

(*Hint:* To get an inverse, solve the equation $\mathbf{A}\mathbf{x} = \mathbf{I}$, where $\mathbf{I}$ is the identity matrix.)

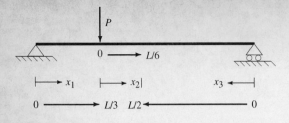

VIRTUAL WORK
AND CONSISTENT
DEFORMATIONS

3.1 Use for Derivations

Virtual work is probably the most common method for developing displacement-based formulations in structural analysis. There are other methods, such as calculus of variations, Galerkin, and potential energy. Although each has its place, virtual work will handle all of the requirements needed for the basics covered in this book. Virtual work is an energy method which is based on the assumption that the structure is at a stable equilibrium position. If a small perturbation (movement) from this position is introduced, the structure will return to its equilibrium position. In addition, the work done to move the structure (in an energy sense) is zero. This means that the work of the applied loads moving through the perturbation is stored in the structure as elastic energy. When the structure is released, the stored energy is released and pushes the structure back to its original position. *Work* is the standard force times displacement.

There are two types of work that must be considered in order for the virtual work to balance to zero. The external work is the work caused by loads acting on the structure. These loads are in equilibrium, but the small perturbation causes them to move and create work. The internal work is a potential energy type of work that is stored in beams, columns, and trusses as a result of stretching and bending. When a beam bends, it builds up an elastic energy. If the beam is released, the energy is returned when the beam returns to its initial position. It should be obvious that the energy stored in a beam (internal work) is exactly equal and opposite to the amount of work (external) required to bend the beam. Note that this assumes that there is no energy loss, such as plastic deformation, heat, and so on. The equation for the balance can be written as

$$\delta W_e = \delta W_i \tag{3.1}$$

It is assumed that the same sign convention is used to derive external and internal work, and therefore the two are equal. In reality, they should have opposite signs and their sum should be equal to zero.

Using this balance of internal and external work, we derive the needed formulations for structural analysis. There are two common forms of virtual work used in structural analysis: virtual displacements and virtual forces. *Virtual displacements* assumes that the perturbation is in the form of a small displacement applied to the structure. This principle is used to derive formulations that have displacements as unknowns. *Virtual forces* assumes that a small load is applied to the structure to cause the perturbation. This is used to derive formulations that have forces as unknowns. Both are equally useful and in reality are the same (as will be shown later).

First we need some definitions and then we derive the needed relationships for calculating both internal and external work for the types of structures we will analyze. $\mathbf{R}$ is a vector of externally applied concentrated loads. Note that loads consist of forces and moments. Any one of the individual terms in the vector $\mathbf{R}$ can be identified as R_i. In Figure 3.1 we have R_1, R_2, and R_3 as well as a distributed load.

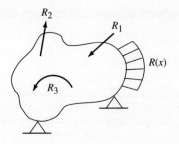

Figure 3.1 External loads applied to a structure.

$\mathbf{r}$ is a vector of global external displacements for a structure. These displacements have a one-to-one correspondence to the external loads. Note that the process of looking at a set of concentrated loads and displacements as a measure of the entire structure is a discretization process. It is the virtual work method that relates these discrete loads and displacements to the continuum of the entire structure. The external work done on a structure can be calculated by

$$W_e = \sum_j R_j r_j = \mathbf{R}^{\mathrm{T}}\mathbf{r} = \mathbf{r}^{\mathrm{T}}\mathbf{R} \tag{3.2}$$

or

$$\int_x R(x)r(x)\,dx \tag{3.3}$$

where $R(x)$ and $r(x)$ are continuous functions of load and displacement, respectively. For the discrete versions using $\mathbf{R}$ and $\mathbf{r}$, the transpose is needed to form the required scalar value.

The external virtual work can be calculated in an identical manner except that either the force or the displacement needs to be virtual (i.e., small perturbation of load or displacement). *Virtual* means small, compatible, and arbitrary. The external virtual work can be written as

$$\delta W_e = \sum_j \overline{R}_j r_j = \sum_j R_j \overline{r}_j = \overline{\mathbf{R}}^{\mathrm{T}}\mathbf{r} = \overline{\mathbf{r}}^{\mathrm{T}}\mathbf{R} \tag{3.4}$$

where the bar over the vector indicates that the quantity is a virtual quantity. Note that this also includes support reactions moving through support settlements. Next, we need some definitions for the internal forces and displacements. $\mathbf{S}$ is the vector of internal forces. They are discrete (concentrated) values at particular points on a member. For example, the end moments on a beam are internal forces (Figure 3.2).

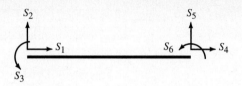

Figure 3.2 Definition of internal forces.

v is the vector of internal displacements. They are concentrated values corresponding to the internal loads: for example, the end rotations of a beam that correspond to the end moments (Figure 3.3). The method used to calculate internal work done by these concentrated forces is identical to that used to calculate external work:

$$W_i = \sum_j S_j v_j = \mathbf{S}^T \mathbf{v} = \mathbf{v}^T \mathbf{S} \tag{3.5}$$

The same can be said about the internal virtual work:

$$\delta W_i = \sum_j \overline{S}_j v_j = \sum_j S_j \overline{v}_j = \overline{\mathbf{S}}^T \mathbf{v} = \overline{\mathbf{v}}^T \mathbf{S} \tag{3.6}$$

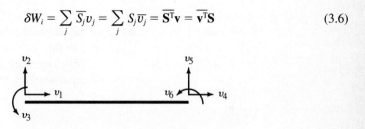

Figure 3.3 Definition of internal displacements.

Although this is a nice representation of internal work, we need some way to relate these discrete values to the actual continuous value of work or stored potential energy. In reality, the potential energy stored in a deformed beam is spread throughout every fiber in the beam. The same is true for a stretched axial member. Therefore, we need to relate this distributed work to that done by these concentrated values. This relationship is what the study of finite elements entails for more complex elements such as plates and shells. We look at the most common structural elements, trusses, and beams first. In later chapters we cover the common finite elements.

3.2 Internal Work for Trusses

The internal work for a truss is the simplest of all elements to calculate. Consider a bar subjected to an axial load S_1. It has a displacement v_1 at the end of the bar (Figure 3.4). Clearly, the work (force times displacement) is

$$W_i = S_1 v_1 \tag{3.7}$$

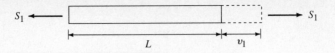

Figure 3.4 Deformed truss bar.

Remember, the force is constant throughout the bar. The displacement is an integrated value consisting of small displacements throughout the length. From mechanics, we know that the displacement of an axial member is related to the applied force by

$$S_1 = \frac{AE}{L} v_1 \tag{3.8}$$

Solving for v_1, we get

$$v_1 = \frac{S_1 L}{AE} \tag{3.9}$$

Substituting into the work equation, we have

$$W_i = \frac{S_1 S_1 L}{AE} \tag{3.10}$$

Clearly, if the virtual work is desired, either S_1 or v_1 needs to be made virtual. This can be done by making the appropriate value the virtual quantity:

$$\delta W_i = v_1 \overline{S_1} = \overline{v_1} S_1 = \frac{\overline{S_1} S_1 L}{AE} \tag{3.11}$$

3.3 Internal Work for Beam Elements

Beam work is a little more difficult, but not much. The stored energy in a beam can only come from bending (ignoring axial and shear work). No matter what the initial and final configurations are of a beam, they can be broken up into rigid-body motion (no strain) and bending. As a result, the only work that needs to be considered is that due to bending. The internal work for any continuum can be measured as *strain energy*, the product of stress times strain. This is exactly equivalent to force times distance. Since stress and strain are measured at a point, the entire strain energy stored in a body has to be arrived at by integration. If we look at the cross section of a beam, we have a stress and corresponding strain on that face. This face is designated as the *Y–Z plane*. The third direction, *X*, is a vector along the centroid of the beam (Figure 3.5).

The internal work, equal to the strain energy of the beam, is

$$W_i = \int_{\text{length}} \int_{\text{area}} \sigma\varepsilon \, dA \, dx = \int_{\text{length}} \int_{\text{width}} \int_{\text{height}} \sigma\varepsilon \, dy \, dz \, dx \tag{3.12}$$

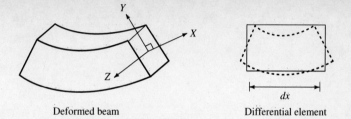

Deformed beam   Differential element

Figure 3.5 Pure bending of beam element.

Again from mechanics, we know the following:

$$\varepsilon = \frac{\sigma}{E} \tag{3.13}$$

$$\sigma = \frac{My}{I} \tag{3.14}$$

where M is the moment on the face, y the distance from the centroid, I the moment of inertia, and E is Young's modulus. Substituting equation (3.14) into (3.13) and then this result and equation (3.14) into (3.12), we get

$$W_i = \int_{\text{length}} \int_{\text{area}} \frac{My}{I} \frac{My}{EI} \, dA \, dx \tag{3.15}$$

The moment, M, moment of inertia, I, and Young's modulus, E, are all constant for a cross section and can be removed from the integration on the area, dy and dz. In general, this assumption of constant values is not necessary, and then the values need to be included in the integration. This leaves

$$W_i = \int_{\text{length}} \frac{MM}{EI^2} \int_{\text{area}} y^2 \, dy \, dz \, dx \tag{3.16}$$

But the integral of y^2 is the definition of the moment of inertia:

$$\int_{\text{area}} y^2 \, dy \, dz = I \tag{3.17}$$

This simplifies equation (3.17) to the form

$$W_i = \int_{\text{length}} \frac{MM}{EI} \, dx \tag{3.18}$$

Equation (3.18) gives the standard form for the internal work of a beam. If virtual work is needed, one of the moments needs to be made a virtual quantity:

$$\delta W_i = \int_{\text{length}} \frac{\overline{M}M}{EI} \, dx \tag{3.19}$$

Note that M/EI is defined as the curvature (ψ). Therefore, beam work can be thought of as moment times curvature integrated over the length of the beam:

$$\delta W_i = \int_{\text{length}} \overline{M}\psi\,dx \qquad (3.20)$$

This is equivalent to moment times rotation. Nowhere is it stated that the curvature must come from applied loads. As a result, the curvature can be temperature induced or could even result from initial camber. The virtual work formula for beams is good for any general curvature.

3.4 Example of Virtual Forces (Dummy Unit Load)

The first application of virtual work will be to get the deflections of a beam. We will use the virtual forces principle since the unknown is to be a deflection. The method consists of balancing the external with the internal virtual work on a structure subjected to a load. To find the required deflection, a virtual load must be applied to the beam. This load will, in turn, generate a virtual moment to be used in equation (3.19).

Given the simply supported beam with a point load at the third point (Figure 3.6), find the deflection at the centerline. The structure can be thought of as having two loadings, the real load and the virtual load. The virtual load consists of a unit point load acting on the structure at the location and in the direction of the desired displacement. We need to find the moment equations that result from each of these loadings. In using the virtual work methods, it is important to recognize that for beams, we need to perform an integration over the length. Recall that the integral of a sum is equal to the sum of the integrals. This means that we can integrate the beam by parts using whatever coordinate system we wish. In addition, we can switch coordinate systems as many times as we wish as long as we are sure to cover the entire beam. If the structure consists of more than one beam, just include the work of all of them. This is important since, as in the example, whenever the point load is encountered, the moment equation changes. Since there are two point loads (real and virtual) applied at different locations, we have three different spans to consider for the integration.

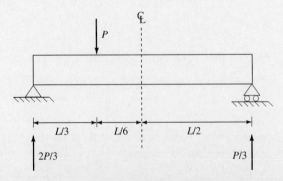

Figure 3.6 Virtual work—beam example.

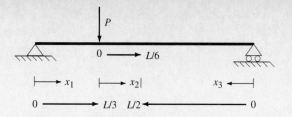

Figure 3.7 Integration coordinate system.

The easiest way to proceed with any virtual work problem is to divide the structure into spans. These divisions occur at any location where a point load is applied, a support exists, or a distributed load starts or stops. For our example the divisions and coordinate systems shown in Figure 3.7 are chosen. The next step is to write the moment equations for all of the spans for each of the two loadings. The real loading is shown in Figure 3.8.

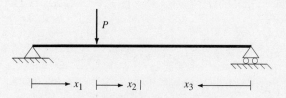

Figure 3.8 Structure with real loading.

Drawing the free bodies for each span, we can write an equation for the moment in the span as a function of x. The sign convention for the moments is not important provided that it is consistent between the different load cases (real and virtual) for each span. In this book all moments will be considered positive on the end of a member if counterclockwise. This is consistent with a right-handed coordinate system. The free bodies and moment equations for the three coordinate spans are as follows:

Free-body segment 1 (Figure 3.9):

$$M_1 = \frac{2}{3} P x_1 \tag{3.21}$$

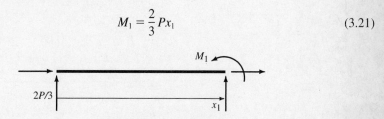

Figure 3.9 Free-body segment 1.

Free-body segment 2 (Figure 3.10):

$$M_2 = \frac{2}{3} P\left(\frac{L}{3} + x_2\right) - P x_2 \tag{3.22}$$

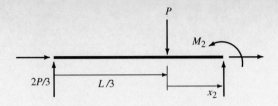

Figure 3.10 Free-body segment 2.

Free-body segment 3 (Figure 3.11):

$$M_3 = -\frac{P}{3}x_3 \qquad (3.23)$$

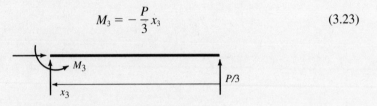

Figure 3.11 Free-body segment 3.

Now, moment equations must be written for the virtual loading. The virtual loading consists of a unit load at the center of the beam where the displacement is desired. If a rotation is the desired value, the unit load becomes a unit moment. Figure 3.12 shows the virtual loading. Again, using the required free bodies for the individual coordinate systems, we can generate the following moment equations:

$$\overline{M}_1 = \frac{x_1}{2} \qquad (3.24)$$

$$\overline{M}_2 = \frac{(L/3) + x_2}{2} \qquad (3.25)$$

$$\overline{M}_3 = -\frac{x_3}{2} \qquad (3.26)$$

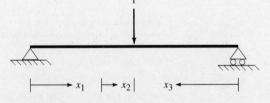

Figure 3.12 Structure with virtual load.

It is important to note that the divisions and coordinates chosen are unimportant. It is, however, important that they be the same for the real and virtual loading cases. It is usually easier to switch the coordinate so that the limits on integration will be from zero to some value. This has the tendency to cause fewer errors when evaluating the integral for the limits of integration.

The final stage is to calculate the internal and external virtual work. The internal work consists of integrating the moment equations over the limits of each span using equation (3.19). The moment of inertia as well as Young's modulus can also be a function of x. Note that there is a real and a virtual moment in the equation. The internal virtual work for this example is

$$\delta W_i = \frac{1}{EI}\left(\int_0^{L/3} M_1 \overline{M_1}\, dx + \int_0^{L/6} M_2 \overline{M_2}\, dx + \int_0^{L/2} M_3 \overline{M_3}\, dx\right) \tag{3.27}$$

Substituting for the moment equations for each span, we get

$$\delta W_i = \frac{1}{EI}\left[\int_0^{L/3} \frac{2Px_1}{3}\left(\frac{x_1}{2}\right) dx + \int_0^{L/6}\left[\frac{2P}{3}\left(\frac{L}{3}+x_2\right) - Px_2\right]\frac{(L/3)+x_2}{2}\, dx\right.$$
$$\left. + \int_0^{L/2}\left(-\frac{Px_3}{3}\right)\left(-\frac{x_3}{2}\right) dx\right] \tag{3.28}$$

$$= \frac{69PL^3}{3888EI} \tag{3.29}$$

The external work is calculated using equation (3.4), where the load is virtual and the displacement is real. In this case, the virtual load is the unit load and the displacement is the real displacement corresponding to the virtual load location. The virtual load is chosen such that the displacement is the one desired at the centerline. The external virtual work is

$$\delta W_e = 1 \times r \tag{3.30}$$

Equating internal and external work, we find that the displacement is equal to the internal virtual work [(3.29)]. This is the standard result when using a unit virtual load method.

3.5 Consistent Deformations

In Section 3.4 we reviewed virtual work, one of the most powerful methods for finding displacements for a statically determinate structure. However, most structures are indeterminate. This means that the support reactions and/or internal forces cannot be determined by statics alone. This is because in two dimensions, only three independent equations of equilibrium exist. There are six independent equations in three dimensions. If what is truly desired is the support reactions, the most straightforward method to use is one based on a flexibility approach. Flexibility methods have forces as the unknowns in the problem. Perhaps the easiest flexibility method to explain is consistent deformations. The consistent deformations method is based on the principle of compatibility. *Compatibility* states that a structure can deform in any shape provided that the displacements conform to the boundary conditions and the material restrictions (i.e., the material cannot tear apart). The correct solution is the one that causes the minimum work to be done.

Consistent deformations is a simple method in concept. It states that you remove all the redundant reactions and forces (i.e., add hinges) in a structure until it becomes determinate. Redundant reactions are extra support forces not required to make the structure stable. Redundant forces are extra internal forces, like moments at a joint, that could be made zero (released) and the structure would remain stable. After you have removed the redundants, you then apply forces at the removed redundants to put the structure back into a compatible displacement, in other words, put the structure back in a position where it meets its supports or original material conditions (no hinges). As an example, take the uniformly loaded propped cantilever beam (Figure 3.13). There is one redundant (or extra) support. If we remove the prop or end support, we have a cantilever beam. The end of the beam deflects downward under the applied loading. We now apply a force at the point where the support was removed. We need to calculate how large this force needs to be to push the structure back up until the point matches the original position of the support.

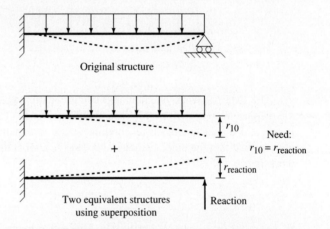

Figure 3.13 Derivation of consistent deformation method.

The calculation of the magnitude of this force is easy if we assume that the structure's response is linear and static. In that case, we need a force that will cause a displacement equal to that which occurs from removing the support. Looking again at the figure with the support removed, we see that the end deflects downward an amount

$$r_{10} = \frac{wL^4}{8EI} \tag{3.31}$$

This value can easily be determined by any number of methods, including virtual work. Next, if we apply an upward unit force to the end of the beam (Figure 3.14) and calculate the deflection, we get

$$f_{11} = \frac{L^3}{3EI} \tag{3.32}$$

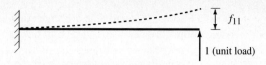

Figure 3.14 Displacement caused by unit load.

Since the structure's response is linear, we need to apply a tip force equal to the ratio of the deflection without the prop (the one we need to push back) to that of the unit force. This is true since when we apply a unit force, we need to scale the deflection up (or down) to match the required amount. Therefore, the required prop force is

$$R_{prop} = \frac{wL^4/8EI}{L^3/3EI} \tag{3.33}$$

Notice that this is a linear method. Also note that it requires the calculation of deflections. Finally, the result is a force.

We would like to formalize the method so that it can easily be extended to multiply indeterminate structures. First we need to define some notation:

r_i *is the final displacement at removed redundant i. This is usually zero but can also be a final support displacement such as a support settlement.*

r_{is} *is the displacement at redundant i on the determinate base structure due to settlement of the supports. This is only a rigid-body displacement and requires only geometry to calculate.*

r_{it} *is the displacement at redundant i on the determinate base structure due to temperature.*

r_{ie} *is the displacement at redundant i on the determinate base structure due to initial imperfections.*

r_{i0} *is the displacement at redundant i on the determinate base structure due to the applied loading.*

f_{ij} *is the displacement at redundant i on the determinate base structure due to a unit force at redundant j.*

X_j *is the unknown redundant force at redundant j. This can be a support reaction or an internal force such as a moment or an axial force.*

The second index is used to refer to the load applied that caused the deflection. For example, the applied loads are load case 0 and temperature is load case t. Using the given notation, the method of consistent deformations can be described by the following procedure.

Consistent Deformation Solution Procedure

    1. Remove all redundant forces to form a determinate base structure. This can involve removing supports, adding hinges, or cutting members. The choice of which forces to remove is arbitrary; however, clever choices can make the calculations simpler.

2. Calculate the displacements at each of the removed redundants due to all the applied loading. This should include loads, temperature, and settlement. Each type of loading is considered to be its own load case. Therefore, applied loads are considered as load case 0, temperature is considered as load case t, and so on.

3. Apply a unit force at each redundant, one at a time, and find the displacements at all of the other removed redundants. Each redundant creates a separate load case, designated as load case j, where j corresponds to the redundant at which the load is applied. This will develop the terms f_{ij}. The applied unit load is at redundant j. The calculated displacements are at all other redundants i. Each load case will fill out a column of the **F** matrix. Applying the unit load at another redundant will develop another column of the **F** matrix.

4. Apply the compatibility equation:

$$r_i = r_{is} + r_{it} + r_{ie} + r_{i0} + \overset{\text{all redundants}}{\underset{j}{\sum}} f_{ij} X_j \tag{3.34}$$

There is one compatibility equation for each removed redundant. This will create a set of simultaneous equations with X_j's as the unknowns. Solving the set of equations gives the values for the removed redundant forces.

3.5.1 Consistent Deformations Example

As an example, let's solve the $1°$ indeterminate structure of the two-span continuous beam shown in Figure 3.15. The first stage, step 1, is to create the determinate base structure

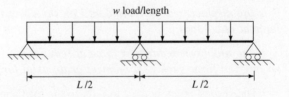

Figure 3.15 Consistent deformations example.

by removing supports. The choice is arbitrary as long as the structure remains stable. For the given structure, there are three choices for determinate base structures (Figure 3.16). Any of the three are valid choices. We will use the first since the solution for the

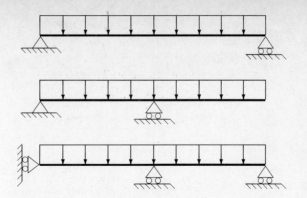

Figure 3.16 Possible determinate base structures.

displacements is easier. The structure is determinate, so the reactions can easily be found (Figure 3.17).

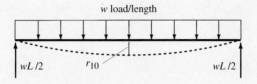

Figure 3.17 Load case 0.

Step 2 says to calculate the displacement at the center (where the redundant was removed) as a result of the applied loading. The only applied loading is the distributed load. There is only one removed redundant labeled 1. In this book we use virtual work to calculate all deflections. To calculate the displacement at the center of the structure, r_{10}, we need a virtual unit load at the center of the structure. The virtual load is depicted in Figure 3.18. If we define the coordinate system for integration as shown in Figure 3.18, we can write the moment equations for both the real and the virtual loading. The moment equations for the real loading are

$$M_1 = \frac{wL}{2}x_1 - \frac{w}{2}x_1^2 \tag{3.35}$$

$$M_2 = \frac{wL}{2}x_2 - \frac{w}{2}x_2^2 \tag{3.36}$$

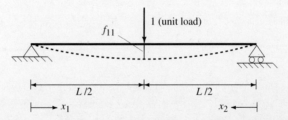

Figure 3.18 Virtual load and load case 1.

The moment equations for the virtual loading are

$$\overline{M_1} = \frac{x_1}{2} \tag{3.37}$$

$$\overline{M_2} = \frac{x_2}{2} \tag{3.38}$$

The internal virtual work is given by

$$\delta W_i = \frac{1}{EI} \int_0^{L/2} \left(\frac{wL}{2} x_1 - \frac{w}{2} x_1^2 \right) \frac{x_1}{2} dx + \frac{1}{EI} \int_0^{L/2} \left(\frac{wL}{2} x_2 - \frac{w}{2} x_2^2 \right) \frac{x_2}{2} dx \tag{3.39}$$

Notice that because of the choice of coordinates, the symmetry of both loading and structure, the two integrals are exactly the same. This means that we can integrate over only half the structure and multiply by 2. It also means that only half the number of moment equations are required. Whenever possible, symmetry should be taken advantage of. Remember that this means both symmetry of loading and structure.

The resulting displacement at the center after integration is

$$r_{10} = \frac{5wL^4}{384EI} \tag{3.40}$$

Step 3 says to calculate the displacement at each removed redundant due to a unit load applied at all of the redundants, one at a time. Since we have only one redundant, we need to calculate the displacement at the center, f_{11}, due to a unit load at the center. The load for this structure is shown in Figure 3.18. It is important to note that the applied unit load is also the virtual load required to calculate the displacements. It is also the virtual load used to calculate the r_{10} displacement. This is an important observation about consistent displacements. When using virtual work, there are considerable savings in both formulation of moment equations and integrating to get displacements.

Since we already have the moment equations for both the real load and the virtual load (one in the same), we are ready to integrate. Taking advantage of symmetry, the internal virtual work is

$$\delta W_i = \frac{2}{EI} \int_0^{L/2} \frac{x}{2} \frac{x}{2} dx \tag{3.41}$$

The result of the integration is

$$f_{11} = \frac{L^3}{48EI} \tag{3.42}$$

Step 4 is to enforce compatibility, that is, to apply equation (3.34). In this case, $r_1 = r_{1t} = r_{1e} = r_{1s} = 0$; therefore, we have

$$0 = f_{11} X_1 + r_{10} \tag{3.43}$$

Substituting the values for the terms in equation (3.43), we have

$$0 = \frac{L^3}{48EI} X_1 + \frac{5wL^4}{384EI} \tag{3.44}$$

Solving for X_1, we have

$$X_1 = -\frac{5wL}{8} \tag{3.45}$$

Note that the minus sign means that the force X_1 is in the opposite direction to that of the unit load used for load case 1. As a result, the support reaction is known and the structure is determinate. The final structure is shown in Figure 3.19.

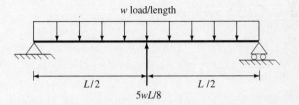

Figure 3.19 Resulting determinate structure.

Now you can use any method desired to determine displacements, rotations, moments, shears, and so on. There is one important point to mention. There is still a lot of useful information contained in all the work done in finding the unknown support reactions. The final value of any quantity is also the linear sum of the base structure with the applied loading and the unit load cases times their appropriate scale factor. For example, the final support reactions for the structure can be calculated from the equation

$$\text{reactions} = \text{reactions}_0 + X_1 \times \text{reactions}_1 \tag{3.46}$$

where the subscripts refer to the load cases. As an example, the final support reactions from the example above are

$$\text{reactions} = \frac{wL}{2} + \left(-\frac{5wL}{8}\right)\left(\frac{1}{2}\right) = \frac{3wL}{16} \tag{3.47}$$

3.6 Consistent Deformations for Trusses

The method of consistent deformations also works with truss structures. Trusses are simpler since the integration over the members can easily be related to the single displacement allowed in a truss, its axial deformation. As seen in the derivation, the internal work for a truss member is just

$$\delta W_i = \frac{\overline{S}SL}{AE} \tag{3.48}$$

The internal work for the entire structure is then the sum of all the members' internal work. This summation is adaptable to tabular form. To apply the procedure of consistent deformations, all redundants need to be removed. For a truss, the following formula can be used to help determine how many redundants exist:

$$D = M + R - 2N \qquad (3.49)$$

where M is the number of members.
        R is the number of reactions.
        N is the number of nodes or joints.

If $D = 0$, the structure is determinate.

If $D > 0$, the structure is indeterminate to the Dth degree.

If $D < 0$, the structure is a mechanism (unstable).

3.6.1   Consistent Deformation Truss Example

Given the truss shown in Figure 3.20, find the member forces. Using equation (3.49), we can find how many redundants need to be removed.

$$M = 6, \quad N = 4, \quad R = 3: \qquad 6 + 3 - 2 \times 4 = 1 \qquad (3.50)$$

Therefore, the structure is indeterminate to the first degree.

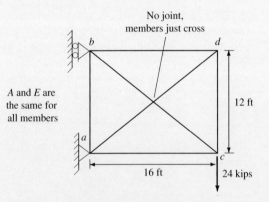

Figure 3.20 Example truss structure.

Step 1 tells us to create a determinate base structure. This means that one redundant needs to be removed. Looking at the structure, it can be seen that it is externally determinate; that is, the support reactions can be found from statics. It is an internal force that needs to be removed. This means cutting a member and considering the opening or closing of the cut to be the redundant displacement. Cutting the member and treating the gap at the cut as the redundant displacement has the same effect as removing the member. Because it is cut, the member cannot take any force. The determinate base structure shown in Figure 3.21 will be used.

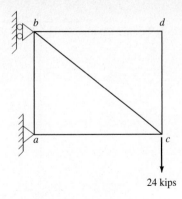

Figure 3.21 Load case 0.

Step 2 tells us to calculate the displacements at all removed redundants due to the external loads. In this problem, only internal forces exist, as opposed to removed support displacements. We will again use virtual work to solve for the displacements. Since it is a member that was cut, the required displacement is the one that the gap (or cut) goes through due to external loads. This is the same as the relative displacement of the joints along the line of the removed member. Using the dummy unit load method, this means applying a unit load along the line of action of the member at each end where it is connected. This can be thought of as a unit tension in the member that has been removed. This will calculate the opening (or closing) of the position where the member needs to be put back in. Then the amount of force required to push the truss apart so that the member can be reinserted will be calculated. Note that this includes the elongation of the removed member. The virtual load for the structure is shown in Figure 3.22.

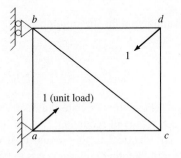

Figure 3.22 Load case 1.

The displacement, r_{10}, is the distance of the separation of the joints a and d along the line of the member connecting a and d. It can be calculated by summing the work of all members (including the removed one) using these two loadings. The forces in each member are required for this sum. The forces in the real structure are shown in Figure 3.23; the forces in the virtual structure are shown in Figure 3.24. Table 3.1 can be set up to calculate the internal work from these two loadings. Looking at the results from Table

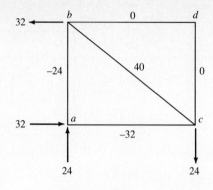

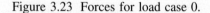

Note: + = tension, − = compression

Figure 3.23 Forces for load case 0.

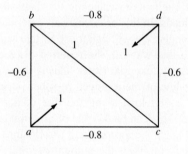

Note: + = tension, − = compression

Figure 3.24 Forces for load case 1.

3.1, the summation in the column $S_0 S_1 L$ corresponds to the r_{10} term. As a result, the resulting value for r_{10} is

$$r_{10} = \frac{1382.4}{AE} \qquad (3.51)$$

Step 3 tells us to calculate the displacement at the removed redundant due to a unit load at each removed redundant. As in the previous cases, the unit and virtual load for this case are identical to the virtual load used to determine the previous displacement. Therefore, only another column needs to be added to the table to find f_{11}, this being $S_1 S_1 L$. Summing the column, we get

$$f_{11} = \frac{69.12}{AE} \qquad (3.52)$$

The final displacement of the removed redundant must be equal to zero. This is because the cut made in the redundant member must remain unseparated. Remember, the redundant is actually the gap displacement of a cut made in the redundant member. The final gap must be equal to zero so that the member must remain connected.

Table 3.1

Member	Length, L	S_0	S_1	$S_0 S_1 L$	$S_1 S_1 L$
ab	12	−24	−0.6	172.8	4.32
ac	16	−32	−0.8	409.6	10.24
bd	16	0	−0.8	0	10.24
dc	12	0	−0.6	0	4.32
ad (removed)	20	0	1	0	20
bc	20	40	1	800	20
				$\Sigma = 1382.4$	69.12

In step 4 we substitute into the compatibility equation:

$$0 = \frac{1382.4}{AE} + \frac{69.12}{AE} X_1 \tag{3.53}$$

or

$$X_1 = -\frac{1382.4}{69.12} = -20 \tag{3.54}$$

The final force in the removed member is −20. This means that the member is in compression, since tension is defined as positive for the sign convention used. The final structural forces are shown in Figure 3.25.

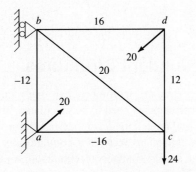

Figure 3.25 Final determinate structure.

The final forces in the structure can be found now that the redundant is known and the final structure is determinate. However, there is an easier way, using superposition. The final member forces are

$$S_{\text{final}} = S_0 + S_1 X_1 \tag{3.55}$$

This is equal to the sum of the real load forces and the virtual load forces scaled by the redundant force. Now the structure can be treated as any determinate structure for finding deflections. However, virtual work has an added advantage. The determinate base structure can be used as the virtual load case in order to determine displacements. This is because virtual work allows any compatible force or displacement pattern to be used as the virtual case. Therefore, the load and structure shown in Figure 3.26 can be used to determine the displacement under the load.

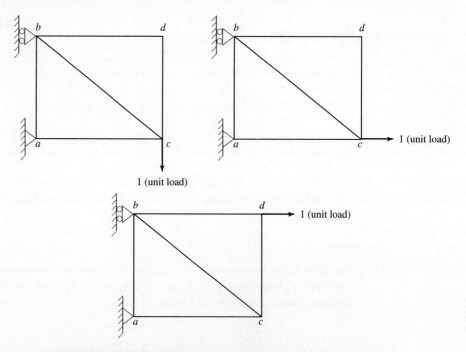

Figure 3.26 Possible virtual loads for determining displacements.

3.7 Multiple-Redundant Structures

Solving a structure with more than one redundant is just as easy as a single redundant. More displacements need to be calculated, and hence the amount of work increases. The notation and procedure developed in Section 3.6 were set up specifically for this purpose. For each redundant there is a compatibility equation and an unknown force. The end result is a set of simultaneous equations. As an example, a structure with two redundants will be solved.

3.7.1   Multiple-Redundant Structure Example

Given the structure in Figure 3.27, find the reactions. Step 1 is to form the determinate base structure. Again, there are a number of redundants to choose from. The structure

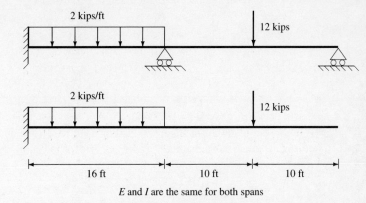

E and *I* are the same for both spans

Figure 3.27 Real and determinate base structures.

chosen is shown in Figure 3.27. As was seen in the single-redundant structures, we can label the applied loads as case 0. The unit loads applied to the removed redundants for step 3 are labeled cases 1 and 2 (Figure 3.28). The loaded structures for these cases are given below. Recall that load cases 1 and 2 are also the virtual load cases needed to find the displacements at redundants 1 and 2, respectively. Since these three loads are sufficient for finding all the displacements required, the next step is to find the moment equations for each section of the structure. We first need to divide the structure into sections. Each

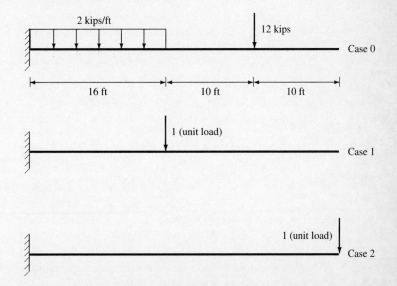

Figure 3.28 Load cases for multiple-redundant structure.

Table 3.2

Case 0	Case 1	Case 2
$M_1 = 0$	$M_1 = 0$	$M_1 = -x_1$
$M_2 = -12x_2$	$M_2 = 0$	$M_2 = -(x_2 + 10)$
$M_3 = -12(10 + x_3) - \dfrac{2x_3^2}{2}$	$M_3 = -x_3$	$M_3 = -(x_3 + 20)$

section starts or stops wherever a load starts or stops or a support exists. The coordinate system chosen is shown in Figure 3.29. For each load case and each coordinate system, the moment equations are found. Table 3.2 shows the moment equations for each load case and coordinate system.

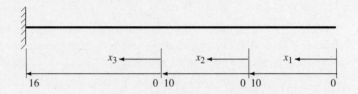

Figure 3.29 Coordinate system for multiple-redundant structure.

Step 2 says to find the displacements at the removed redundants due to the applied loads. Again since only forces exist, we need to find only r_{10} and r_{20}. Using virtual work, we simply need to integrate the moments in case 0 times the moments in case 1 for r_{10}. To find r_{20} we need to integrate case 0 times case 2. For r_{10} we get

$$r_{10} = \frac{1}{EI}\int_0^{16} -(120 + 12x + x^2)(-x) = \frac{48,128}{EI} \qquad (3.56)$$

For r_{20} we get

$$r_{20} = \frac{1}{EI}\int_0^{10} (-12x)(-x - 10) + \int_0^{16} -(120 + 12x + x^2)(-x - 20)$$

$$= \frac{154,554.66}{EI} \qquad (3.57)$$

Step 3 says to apply a unit load at each removed redundant and calculate the displacements at all of the removed redundants. This will generate the f_{ij} terms. For two unknown redundants, this means we need to find f_{11}, f_{12}, f_{22}, and f_{21}. Looking closely at what this means, to find f_{11}, we need a unit load at redundant 1 and a unit virtual load at redundant 1. This means integrate the moments of case 1 times case 1.

$$f_{11} = \frac{1}{EI} \int_0^{16} (-x)(-x) = \frac{1365.33}{EI} \tag{3.58}$$

To find f_{21}, we need to find the displacement at redundant 2 as a result of applying the load at redundant 1. This means a unit load at redundant 1 and a unit virtual load at redundant 2. Hence integrate case 1 times case 2.

$$f_{21} = \frac{1}{EI} \int_0^{16} -x(-x - 20) = \frac{3925.333}{EI} \tag{3.59}$$

To find f_{22}, we need to find the displacement at redundant 2 as a result of applying a unit load at redundant 2. This means a unit load at redundant 2 and a unit virtual load at redundant 2. For this particular load, we are going to change the coordinate system. Since both the load and the virtual load start at the end, we will let the coordinate go all the way to the support. The moment equation is the same as case 2 for the tip but is valid for the entire length of the beam.

$$f_{22} = \frac{1}{EI} \int_0^{36} (-x)(-x) = \frac{15,552}{EI} \tag{3.60}$$

To find f_{12}, we need to find the displacement at redundant 1 as a result of applying a unit load at redundant 2. This means a unit load at redundant 2 and a unit virtual load at redundant 1. Therefore, we need to integrate case 1 times case 2. However, it is important to notice that this integration is identical to finding f_{21}.

This is an important point. The flexibility matrix is symmetric. This means that $f_{ij} = f_{ji}$. Using virtual work, we have proved this to be true. This is identical to Maxwell's reciprocal theorem, which states:

For any linear structure, the displacement at point A on a structure caused by a unit load at point B is identical to the displacement at point B caused by the same unit load applied at point A.

As a result, we only have to calculate one of the terms, but not both. This symmetry reduces the work involved by almost half.

Step 4 is to apply the compatibility equation. Since there are two removed redundants (two unknown forces), we will have two equations:

$$0 = r_{10} + f_{11}X_1 + f_{12}X_2 \tag{3.61}$$
$$0 = r_{20} + f_{21}X_1 + f_{22}X_2 \tag{3.62}$$

Here we have two simultaneous equations in two unknowns. These equations represent the same thing as the one redundant structure. How much force has to be applied to redundant 1 to push it back into place on top of the support? However, pushing at redundant 1 also causes a displacement at redundant 2, and vice versa. As a result, less force is needed at both redundants and hence the interaction of the equations. Solving for X_1 and X_2, we have

$$X_1 = -24.34$$
$$X_2 = -3.79$$

This gives a final determinate structure as shown in Figure 3.30.

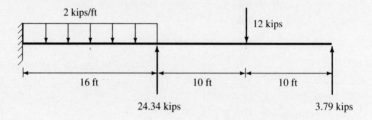

Figure 3.30 Resulting determinate structure.

To find the final support reactions (those of the determinate base structure), we can use statics or the superposition equations. The equations are

$$R_{\text{reactions}} = R_0 + X_1 R_1 + X_2 R_2 \tag{3.63}$$

For the moment at the support ($x = 0$), for the three load cases we have

$$M_0 = 568 \qquad M_1 = 16 \qquad M_2 = 36$$
$$M_{\text{final}} = 568 + (-24.34)(16) + (-3.79)(36) = 42.12 \text{ kip-ft} \tag{3.64}$$

The vertical reaction can be found the same way and is

$$V_{\text{final}} = 15.87 \text{ kips}$$

The identical method can be used for truss structures with internal, external, or both redundants. A table similar to the one used for the one redundant truss should be set up. A column needs to be set up for all permutations of the load cases. If the redundant removed is internal, the force is a unit tension in the member removed (just as in the example). If the redundant is an external support, the load is a unit load at the joint where the support was removed. In either case, element forces are generated. Then the various combinations of forces are multiplied and summed, giving all the required terms for the compatibility equations.

The method is also valid for mixed truss and beam structures. Truss members can be removed as redundants as well as supports. The displacements calculated need to include the contributions from both the beams and trusses. This is accomplished by adding the integrated moment values from the beams to the summed values from the trusses. The combinations of load cases are identical. Examples of the structure, load cases, and redundants are given in Figures 3.31 and 3.32.

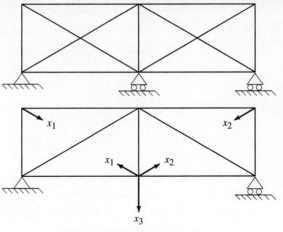

2 Internal and 1 external redundant

Figure 3.31 Example structure and determinate base structure.

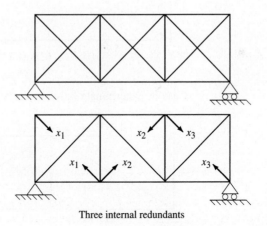

Three internal redundants

Figure 3.32 Example structure and determinate base structure.

Condensed Example: Two-Redundant Truss Structure

The truss structure shown at the top of page 68 is to be analyzed to find the member forces.

Choosing the following determinate base structure, we calculate the member forces for load cases 0, 1, and 2.

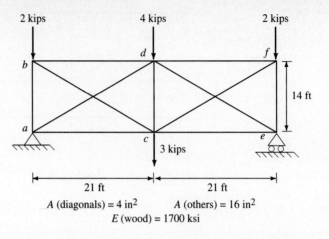

A (diagonals) = 4 in²   A (others) = 16 in²

E (wood) = 1700 ksi

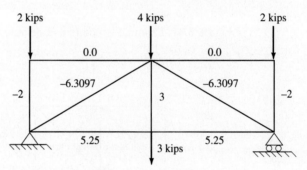

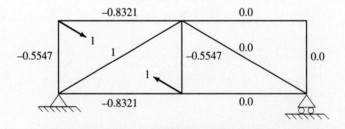

Case 0: Determinate base structure with member forces.

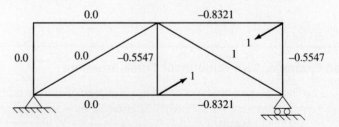

Case 1: Unit load at redundant 1, determinate base structure.

Case 2: Unit load at redundant 2, determinate base structure.

Member	Length (in.)	Area (in²)	S_0	S_1	S_2	S_0S_1L	S_0S_2L	S_1S_1L	S_1S_2L	S_2S_2L
ab	168	16	−2	−0.5547	0	11.649	0	3.2308	0	0
cd	168	16	3	−0.5547	−0.5547	−17.473	−17.473	3.2308	3.2308	3.2308
ef	168	16	−2	0	−0.5547	0	11.649	0	0	3.2308
ac	252	16	5.25	−0.8321	0	−68.804	0	10.905	0	0
ce	252	16	5.25	0	−0.8321	0	−68.804	0	0	0
bd	252	16	0	−0.8321	0	0	0	10.905	0	10.905
df	252	16	0	0	−0.8321	0	0	0	0	10.905
ad	302.86	4	−6.3097	1	0	−477.74	0	75.715	0	0
bc	302.86	4	0	1	0	0	0	75.715	0	0
de	302.86	4	−6.3097	0	1	0	−477.74	0	0	75.715
cf	302.86	4	0	0	1	0	0	0	0	75.715
					$\Sigma =$	−552.37	−552.37	179.702	3.2308	179.702

Putting the member forces, lengths, and areas into a table, we can calculate the terms required for the compatibility equation. The required compatibility equations are

$$-r_{10} = f_{11}X_1 + f_{12}X_2$$
$$-r_{20} = f_{21}X_1 + f_{22}X_2$$

From the table we get the summations, which equal the required r_{10}, r_{20}, and f_{ij}'s. Substituting these into the compatibility equation, we get the following matrix equations:

$$-552.368 = 179.702X_1 + 3.2308X_2$$
$$-552.368 = 3.2308X_1 + 179.702X_2$$

Solving the equations above, we get the results for the force in the redundants as

$$X_1 = 3.0195$$
$$X_2 = 3.0195$$

The final complete structural member force is found by the equation

$$S_{\text{final}} = S_0 + S_1X_1 + S_2X_2$$

This gives final forces of

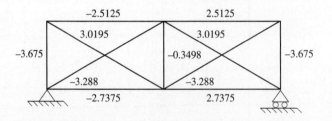

3.8 Temperature Deformations Using Virtual Work

Including temperature deformations using virtual work is as easy as including those due to external loads. Temperature deformations are caused by the fibers of the member extending or shrinking due to a change in temperature. In trusses, the change is assumed to be uniform through the thickness. This causes an axial extension. In beams, the temperature changes through the depth of the beam. This causes a curvature in the beam. In either case, the temperature can vary along the length of the member.

3.8.1 Temperature Deformations in Trusses

We need to derive the equations for finding the internal virtual work caused by a temperature change. We will start with trusses. Trusses have only axial extension. Recall that the virtual work formula was

$$\delta W_i = \frac{\overline{S}SL}{AE} \tag{3.65}$$

By grouping terms we can rearrange the formula and view it from another perspective:

$$\delta W_i = \overline{S}\,\frac{SL}{AE} = \overline{S}\,\delta L \tag{3.66}$$

This can be viewed as a virtual force times a real displacement. Since the cause of the real displacement is unimportant, we can, equally, consider it to be caused from temperature changes. Recall the displacement in an axial member due to a uniform temperature increase:

$$\delta L = \alpha\,\delta T\,L \tag{3.67}$$

where α is the coefficient of thermal expansion, δT the change in temperature, and L the member length. Therefore, for a structure we still just need to sum up the virtual work contribution from all members. For a truss structure, this means the same tabular procedure with the addition of an $\alpha\,\delta T\,L$ column.

3.8.2   One-Redundant Truss with Temperature Example

As an example, let's look at the truss structure having one redundant shown in Figure 3.20. Figure 3.33 shows the structure with the additional temperature increase in two members. To this structure we have added the fact that members bd and bc have a uniform increase in temperature of 15°, as shown in Figure 3.33. If we use the coefficient of expansion for steel (6.5×10^{-6}), we get a table of forces and virtual work (Table 3.3). From Table 3.3, the final summations give the required values for the compatibility equation. The first two summations are identical to the earlier values and give r_{10} and f_{11}. The final column gives the temperature load case displacement:

$$r_{1T} = 0.0084 \text{ in.}$$

It is important to note that when including temperature, as well as settlements, care must be taken with the units. Here, α is given in inches per inch-degree. As a result, the final r_{1T} is in inches and the length used was in inches. When A and E are plugged in for

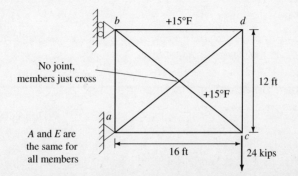

Figure 3.33 Example truss structure with temperature.

Table 3.3

Member	Length, L	S_0	S_1	δT	$S_0 S_1 L$	$S_1 S_1 L$	$\alpha\,\delta T\,S_1 L$
ab	12	-24	-0.6	0	2073.6	51.84	0
ac	16	-32	-0.8	0	4915.2	122.88	0
bd	16	0	-0.8	15	0	122.88	-0.015
dc	12	0	-0.6	0	0	51.84	0
ad (removed)	20	0	1	0	0	240	0
bc	20	40	1	15	9600	240	0.02340
					$\Sigma = 16{,}588$	829.44	0.0084

r_{10} and f, they should also be in units of inches. Substituting into the compatibility equation, we have

$$0 = 0.0084 + \frac{16{,}588}{AE} + \frac{829.44}{AE} X_1 \tag{3.68}$$

Note that since there is no temperature change in the member removed, the final displacement is zero. This means that the member extends but the cut/gap will not separate. If the member removed had a temperature change, it would be included as the final displacement for that redundant. Here we assume that A and E are constant for all members and have a value of 3.0 in^2 and 29,000 ksi, respectively. If that were not the case, we would add another column for A and/or E. Then they would be included in the summed columns for the work done. Solving for X_1 gives the final force in member ad,

$$X_1 = -30.6 \text{ kips}$$

As before, we now have a determinate structure with all the forces known and can continue to find deflections if required.

3.8.3 Temperature Deformations in Beams

Next, we will derive the equation required for solving temperature deformation problems in beam structures. For a beam to deflect, there must be a curvature. For temperature changes to cause a curvature, there must be a difference through the depth of the beam. Assuming plane sections remain plane and that the temperature change through the depth is linear, a relationship can be derived using Figure 3.34.

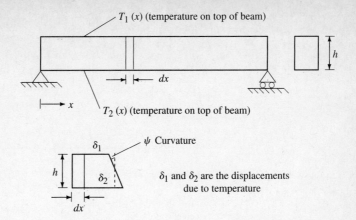

Figure 3.34 Beam subjected to temperature differences.

For the small dx segment of the beam, the rotation of the cross section is equal to

$$\psi = \frac{\delta_2 - \delta_1}{h} \tag{3.69}$$

This is defined as positive to be consistent with our positive definition of moment as being counterclockwise. Note that h is the depth of the cross section. This will cause a small rotation counterclockwise which is positive. As was defined for axial members, the extensions of the top and bottom fibers are

$$\delta_1 = \alpha T_1 \, dx \qquad \text{(top of beam)} \tag{3.70}$$

$$\delta_2 = \alpha T_2 \, dx \qquad \text{(bottom of beam)} \tag{3.71}$$

as before, the formula for internal virtual work will be regrouped to give

$$\delta W_i = \int_{\text{length}} \overline{M} \frac{M}{EI} \, dx = \int_{\text{length}} \overline{M} \psi \, dx \tag{3.72}$$

Since it is unimportant what causes the curvature, we will assume that it is caused by a change in temperature. Therefore, we are integrating the real curvature times the virtual moment. If we substitute equations (3.70) and (3.71) into (3.69) and then the result into (3.72), we get

$$\delta W_i = \int_{\text{length}} \overline{M} \frac{\delta_2 - \delta_1}{h} \, dx = \int_{\text{length}} \overline{M} \frac{T_2 - T_1}{h} \alpha \, dx \tag{3.73}$$

Usually, both h and α are constant along the length of the beam. If this is true, they can be removed from the integration. In addition, if the temperature difference between the top and the bottom are constant along the length, it can also be removed, leaving

$$\delta W_i = \frac{T_2 - T_1}{h} \alpha \int_{\text{length}} \overline{M} \, dx \tag{3.74}$$

3.8.4   One-Redundant Beam with Temperature Example

As an example, let's solve the problem shown in Figure 3.35. The temperatures at the top and bottom are 20 and 50°F, respectively. The structure is a propped cantilever.

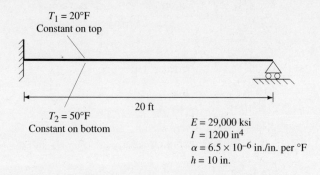

$T_1 = 20°F$
Constant on top

20 ft

$T_2 = 50°F$
Constant on bottom

$E = 29{,}000$ ksi
$I = 1200$ in^4
$\alpha = 6.5 \times 10^{-6}$ in./in. per °F
$h = 10$ in.

Figure 3.35 Example indeterminate beam with temperature.

There are no other loads except temperature. Following the procedure for indeterminate structures, the determinate base is defined as removing the prop. Next, the load cases

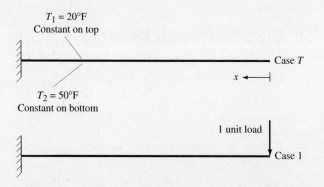

$T_1 = 20°F$
Constant on top

Case T

x

$T_2 = 50°F$
Constant on bottom

1 unit load

Case 1

Figure 3.36 Determinate base structure and load cases.

required for steps 2 and 3 are defined (temperature and unit load). Using the coordinate system shown in Figure 3.36, we obtain the moment equation for case 1:

$$\overline{M} = -x \qquad (3.75)$$

Solving for r_{1T} using equation (3.74) gives

$$r_{1T} = \frac{50 - 20}{10}(6.5 \times 10^{-6}) \int_0^{240} -x \, dx = -0.5616 \text{ in.} \qquad (3.76)$$

Using the standard formula for internal work due to loads for f_{11}, we have

$$f_{11} = \frac{1}{EI} \int_0^{240} (-x)(-x) \, dx = 0.1324 \text{ in./kip} \qquad (3.77)$$

Substituting into the compatibility equation, we have

$$0 = -0.5616 + 0.1324X_1 \tag{3.78}$$

The final support reaction due to temperature is

$$X_1 = 4.24 \text{ kips}$$

Figure 3.37 shows the final determinate structure with all external loads.

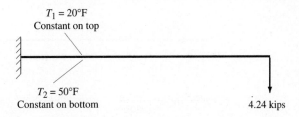

Figure 3.37 Resulting determinate structure.

3.9 Support Displacements Using Consistent Deformations

Consistent deformations also handle support displacements very easily. This step is simpler than any of the other portions of the procedure because the required displacements are either given or are calculated by geometry alone. The calculated displacements are always for a determinate structure. As an example, the structure shown in Figure 3.38 with the given support displacements will be solved. This is the identical single-redundant structure used in explaining the method of consistent deformations. The only difference is the addition of the support settlements. Since the structure was already solved, we have the r_{10} and f_{11} terms [equations (3.40) and (3.42)]. The only new terms to find are r_{1S} and r_1. Let's start with the easiest one, r_1.

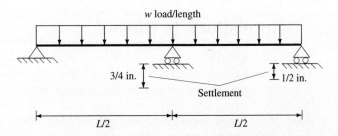

Figure 3.38 Settlement example structure.

By definition, r_1 is the final displacement for redundant 1. Since this is a support, the known settlement is exactly the value needed. Normally, the settlement is assumed to be zero ($r_1 = 0$). For this example, we have

$$r_1 = 0.75 \text{ in.}$$

Next we need r_{1S}. This is defined as the displacement at redundant 1 due to support settlement of the determinate base structure. Since the structure is determinate, settlements will not cause any stresses but only rigid-body displacements. As a result, the required displacement can be calculated by geometry alone. Figure 3.39 shows the deformed shape

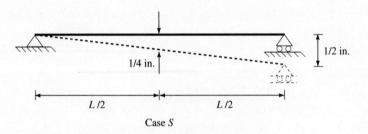

Case S

Figure 3.39 Settlement for determinate base structure.

of the structure due to allowing the supports to settle. By similar triangles, the settlement at redundant 1 for load case S is

$$r_{1S} = 0.25 \text{ in.}$$

Using the previously calculated values for r_{10} and f_{11}, we can substitute into the compatibility equation:

$$0.75 = 0.25 + \frac{5wL^4}{384EI} + \frac{L^3}{48EI}X_1 \tag{3.79}$$

Given values for w, E, L, and I, you can easily solve for X_1, the support reaction. It is important to remember that as with temperature, units are very important since E and I do not cancel out. Also notice that in indeterminate structures, stresses are induced by support settlements as well as temperature changes.

3.10 Shear Deformations in Beams

So far, the beam theory that we have been using includes only bending deformations. This is because in most cases, bending is the dominant cause of deformation for the types of members commonly used in structural practice. However, in beams and girders where the depth-to-span ratio is high and in some wide-flange sections, the deformation due to shear strains can be considerable. Virtual work can also be used to determine deflections due to shear strain. If we assume a rectangular cross section with uniform shear across the face, we can derive the internal virtual work due to shear as the virtual shear stress times the shear strain. From mechanics, we know that the shear strain is

$$\gamma = \frac{V}{A_sG}$$

where V is the shear stress.
       A_s is the shear area.
       G is the shear modulus.

Using Figure 3.40, we see that if we integrate the strain times stress for the infinitesimal segment over the entire length, the internal virtual work is

$$\delta W_i = \int_{\text{length}} \frac{\overline{V}V}{A_sG}\,dx \tag{3.80}$$

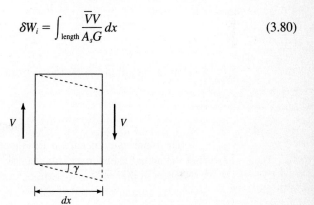

Figure 3.40 Derivation of shear deformation work.

As a simple example, we can find the amount of additional displacement due to shear strain in the example shown in Figure 3.6. The solution follows the same coordinate systems used for the moment equations. Therefore, the required shear equations for the real load in the three spans can be calculated from the free bodies shown in Figures 3.9, 3.10, and 3.11. The shear forces for the real load are

$$V_1 = \frac{2P}{3} \tag{3.81}$$

$$V_2 = -\frac{P}{3} \tag{3.82}$$

$$V_3 = -\frac{P}{3} \tag{3.83}$$

The equations for the virtual load can be calculated from Figure 3.12. The required equations are

$$\overline{V}_1 = \frac{1}{2} \tag{3.84}$$

$$\overline{V}_2 = \frac{1}{2} \tag{3.85}$$

$$\overline{V}_3 = -\frac{1}{2} \tag{3.86}$$

Substituting these shear equations into equation (3.80), we get the internal virtual work due to shear:

$$\delta W_i = \frac{1}{A_s G}\left[\int_0^{L/3} \frac{2P}{3}\left(\frac{1}{2}\right) dx + \int_0^{L/6}\left(-\frac{P}{3}\right)\left(\frac{1}{2}\right) dx + \int_0^{L/2}\left(-\frac{P}{3}\right)\left(-\frac{1}{2}\right) dx\right] \tag{3.87}$$

After integrating, the final virtual shear work (or displacement at the centerline due to shear) is

$$\delta W_i = \frac{PL}{6A_s G} \tag{3.88}$$

The shear modulus can be found from solid mechanics as the following equation:

$$G = \frac{E}{2(1 + \nu)} \tag{3.89}$$

The other value needed is the shear area. These shear areas are derived from the fact that the actual shear in a cross section is not constant (as assumed). The shear area is a correction in the true cross-sectional area to account for the nonconstant shear stress. The two most commonly needed values are for rectangular sections and wide-flange sections (I-beams). For the rectangular cross section, we have

$$A_s(\text{rectangular}) = \tfrac{5}{6}A = \tfrac{5}{6}bh \tag{3.90}$$

For wide-flange sections, the shear area is taken as the area of the web.

$$A_s(\text{wide-flange section}) = A_w(\text{area of the web}) \tag{3.91}$$

If we pick an average wide-flange section, we can compare the values of displacement from bending and shear. The equation for deflection is the sum of the shear and bending parts:

$$\delta_{\text{centerline}} = \frac{69PL^3}{3888EI} + \frac{PL}{6A_s G} \tag{3.29}$$

If we assume steel and use a W24 × 160, the required values are $A_s = 16.2$, $I = 5110$, and $\nu = 0.3$. Using a load of 10 kips and a span of 40 ft, we have the center displacement for the example as

$$\delta_{\text{centerline}} = \frac{69(10)(40 \times 12)^3}{3888(29,000)(5110)} + \frac{10(40 \times 12)}{6(16.2)(11,153.8)} = 0.13244 + 0.00442 \tag{3.92}$$

As seen above, the shear contribution to the deflection is on the order of 3 percent and in most cases is not significant.

3.11 Problems

3.1. Given the following structure, find the deflection under the load and at the centerline.

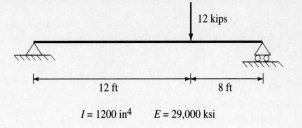

12 kips

12 ft   8 ft

$I = 1200$ in^4 $E = 29,000$ ksi

3.2. For the following structure, find the deflection at the centerline.

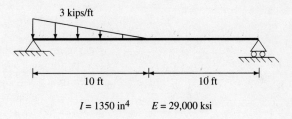

3 kips/ft

10 ft   10 ft

$I = 1350$ in^4 $E = 29,000$ ksi

3.3. Given the following structure, find the support reactions.

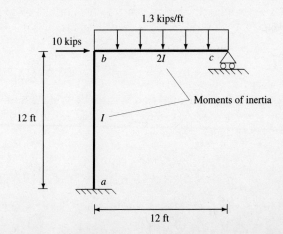

1.3 kips/ft

10 kips

b $2I$ c

Moments of inertia

12 ft I

a

12 ft

3.4. For the following truss, find the member forces.

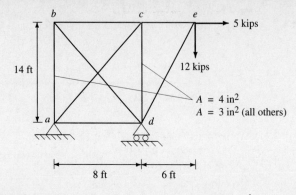

The diagonal members *ab* and *de* have an area of 4 in²; the other members have an area of 3 in².

3.5. For the following multiple-redundant structure, find the support reactions.

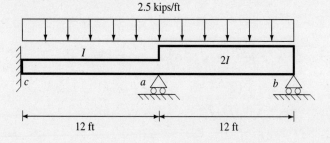

3.6. The following structure is subjected to a temperature differential in addition to the loads.

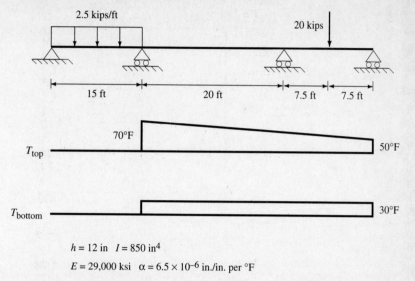

$h = 12$ in $I = 850$ in^4

$E = 29{,}000$ ksi $\alpha = 6.5 \times 10^{-6}$ in./in. per °F

The properties are continuous. Find the support reactions.

3.7. Find the member forces in the following multiple-redundant truss.

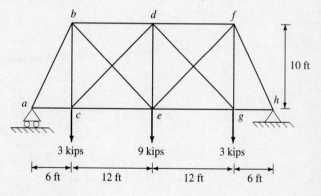

3.8. Find the displacement δ at the upper right-hand joint of the following structure.

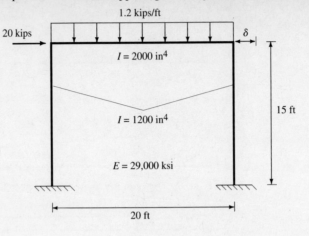

3.9. Find the fixed-end moments for the following structure.

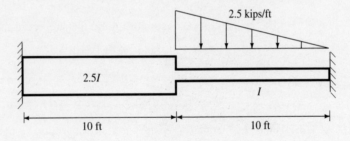

3.10. Find the support reactions for the following structure.

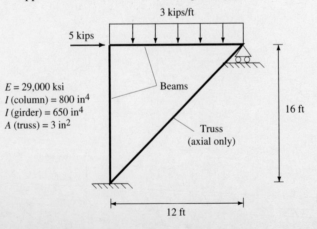

3.11. Find the support reactions and the force in the truss element for the following structure.

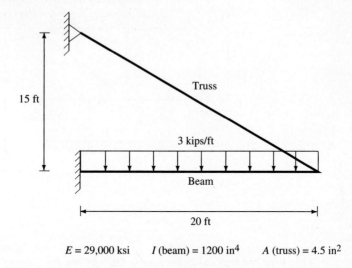

$E = 29,000$ ksi I (beam) = 1200 in^4 A (truss) = 4.5 in^2

3.12. Find the fixed-end forces for the following structure.

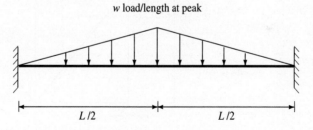

3.13. Find the deflection at the center on the bottom chord of the following structure.

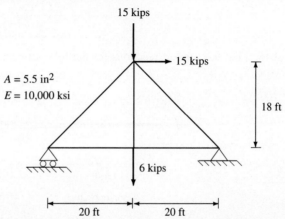

3.14. Find the vertical deflection under the load for the following structure.

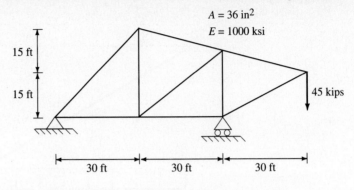

$A = 36$ in^2
$E = 1000$ ksi

15 ft

15 ft

45 kips

30 ft   30 ft   30 ft

3.15. Find the vertical deflection at the load for the following structure.

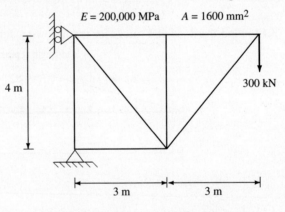

$E = 200,000$ MPa $A = 1600$ mm^2

4 m

300 kN

3 m   3 m

3.16. Find the support reactions for the following structure.

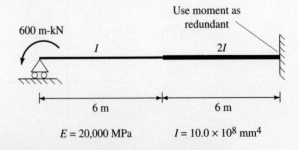

Use moment as redundant

600 m-kN

I $2I$

6 m   6 m

$E = 20,000$ MPa $I = 10.0 \times 10^8$ mm^4

Use the moment at the fixed end as the redundant.

3.17. Find the support reactions and the moment at the joint for the following structure.

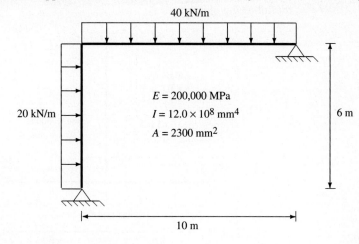

$E = 200,000$ MPa
$I = 12.0 \times 10^8$ mm^4
$A = 2300$ mm^2

40 kN/m

20 kN/m

6 m

10 m

3.18. Find the support reactions and the vertical deflection at the load for the following structure.

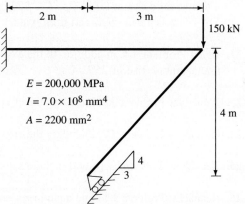

2 m

3 m

150 kN

$E = 200,000$ MPa
$I = 7.0 \times 10^8$ mm^4
$A = 2200$ mm^2

4 m

4

3

3.19. Find the support reactions and the member forces for the following structure.

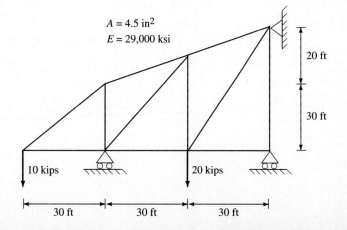

$A = 4.5$ in^2
$E = 29,000$ ksi

20 ft

30 ft

10 kips

20 kips

30 ft

30 ft

30 ft

3.20. Find the support reactions and the moment in the beam at the truss connection for the following structure.

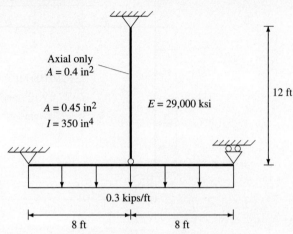

3.21. Using Problem 3.3, in addition to the existing load, the top of the beam experiences a temperature rise of 20°F. Use $I = 850$ in⁴, $E = 29,000$ ksi, and $\alpha = 6.5 \times 10^{-6}$. The beam is 12 in. high and 6 in. wide.

3.22. Using Problem 3.4, in addition to the existing load, members ac and cd experience a temperature rise of 25°F. Use $E = 29,000$ ksi and $\alpha = 6.5 \times 10^{-6}$.

3.23. Reanalyze Problem 3.1, including the effect of shear deformations. Use $A_s = 20$ in².

3.24. Reanalyze Problem 3.2, including the effect of shear deformations. Use $A_s = 23$ in².

3.25. Reanalyze Problem 3.10 and include settlement at the roller of 2 in.

3.26. Reanalyze Problem 3.17 and include settlement at the upper right-hand pin of 32 mm.

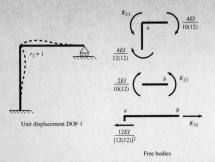

K_{11}

$\dfrac{4EI}{10(12)}$

$\dfrac{4EI}{12(12)}$

$\dfrac{2EI}{10(12)}$ $\quad a \quad$

$\dfrac{2EI}{10(12)}$ $\quad b \quad K_{21}$

a $\qquad\qquad b$

K_{31}

$\dfrac{12EI}{[12(12)]^2}$

$r_1 = 1$

Unit displacement DOF 1

Free bodies

STIFFNESS METHODS

Stiffness methods are the counterparts to flexibility methods. Whereas flexibility has forces as unknowns, stiffness has displacements. To derive a flexibility matrix, you calculate displacements; for a stiffness matrix you calculate forces. Flexibility solutions are based on compatibility, stiffness is based on equilibrium. This association can be carried further to show that the inverse of a flexibility matrix is the corresponding stiffness matrix. The reverse is not always true. The inverse of stiffness is flexibility—provided that the inverse exists. Stiffness matrices for unstable (insufficient boundary conditions) structures cannot be inverted. A stiffness method can thus be described as an analysis method that has displacements as the unknowns to the problem. The simplest method of this type is slope deflection.

4.1 Slope Deflection

Slope deflection is the first stiffness method we will discuss. The reason is that it can easily be derived using the methods in Chapter 3 on virtual work and consistent deformations. We will use compatibility to find all the required terms in the slope deflection formulation.

Slope deflection is based on the superposition of moments for a member. The moment at the end of any member can be broken up into four parts (Figure 4.1). The total moment at any end of a beam (M_a for the a end) is the sum of (1) the moment caused by the external load assuming that the ends of the element are fixed, called the fixed-end moments (FEMs), (2) the moment due to a rotation of the end in consideration (θ_a), (3) the moment due to a rotation of the opposite end of the elements (θ_b), and (4) the moment due to a relative displacement of the two ends (δ). From the figure you can see that the moment at end a is

$$M_a = M_{a_{\text{FEM}}} + M_{a_{\theta_a}} + M_{a_{\theta_b}} + M_{a_\delta} \qquad (4.1)$$

This summation is true for any element in any type of structural configuration. Notice that axial deformations are ignored.

Let's look at each of these contributions separately to develop the required equations. First, we look at the moments due to applied loads within the span of a fixed beam. These are called *fixed-end moments* (FEMs). Fixed-end moments can be found for any type of loading using the previous method of compatible displacements. For example, fixed-end moments for a uniform distributed load can be found using the structure in Figure 4.2. Using consistent deformations, we remove the right-hand support entirely, leaving a

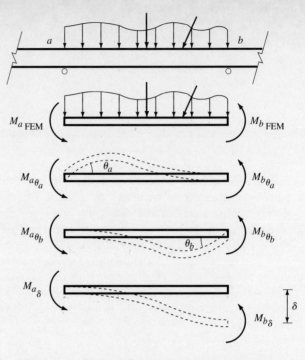

Figure 4.1 Slope deflection moment components.

cantilever base structure. Although we removed three reactions, only the vertical and the moment need to be considered since there are no horizontal loads. If we apply the unit forces and do the integration, we generate two equations for the two unknowns. Solving using the method of consistent deformations from Chapter 3, we get the forces shown in Figure 4.3.

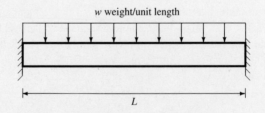

Figure 4.2 Example for fixed-end forces.

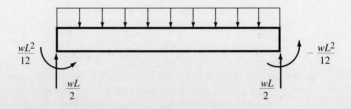

Figure 4.3 Resulting fixed-end forces.

This process can be done for any loading. It must also be done for nonprismatic (nonconstant) sections since the fixed-end forces are different. Remember, the solution process involves an integral over the length including E and I. This means that the fixed-end moments (FEMs) are different for a nonconstant cross section. A list of the most common FEMs are given in Table 4.1. The table gives only the moments. The vertical forces must be calculated by statics.

Next, we want to find the moment, M_θ, due to rotation. This means that we need a relationship between moment and rotation. Again, we can use flexibility to solve for the required equation. Note that we need to find the moment that results from a rotation θ. This can be done through Figure 4.4. Given the load of $M_{a_{\theta_a}}$, we need to find the angle θ_a. As can be seen, there is one redundant, the prop. The prop can be removed to form a determinate base structure. Again using compatibility, we can find the support reaction. The virtual load case is shown in Figure 4.5. Integrating for the r_{10} and f_{11} terms and then substituting into the compatibility equation and solving, we get the final support reaction and resulting determinate structure (Figure 4.6).

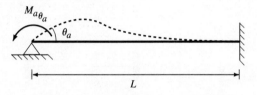

Figure 4.4 Moment due to rotation.

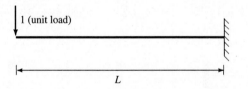

Figure 4.5 Virtual load to find support reaction.

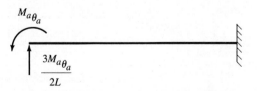

Figure 4.6 Resulting determinate structure.

Now we can solve for the angle θ_a due to the moment M_a since the structure is determinate with two applied loads. The virtual load case needed for finding the rotation is shown in Figure 4.7. Using virtual work and integrating, we find that

$$M_{a_{\theta_a}} = \frac{4EI}{L}\theta_a \qquad (4.2)$$

Table 4.1

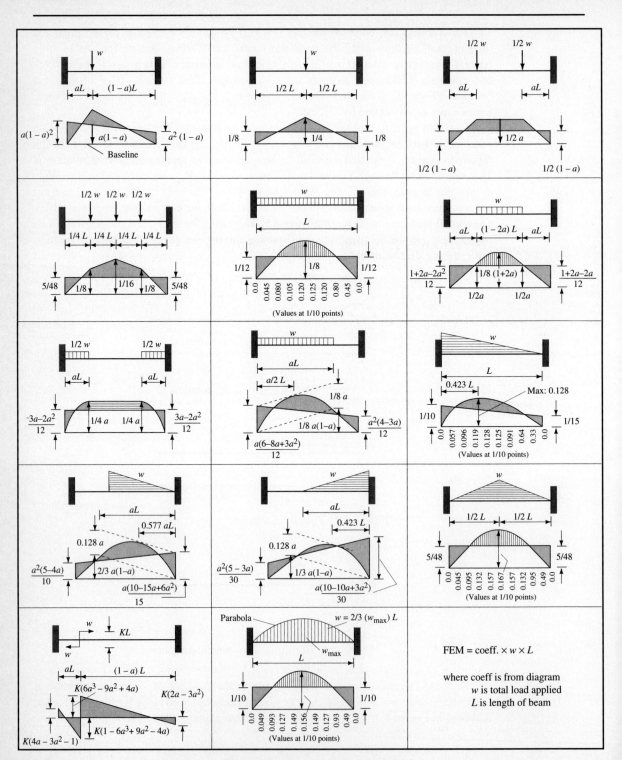

FEM = coeff. $\times w \times L$

where coeff is from diagram
w is total load applied
L is length of beam

92

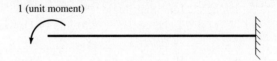

1 (unit moment)

Figure 4.7 Virtual load for moment-rotation equation.

We can now find $M_{b_{\theta_a}}$ from statics. Summing moments about point b and substituting for $M_{a_{\theta_a}}$ using equation (4.2), we get

$$M_{b_{\theta_a}} = \frac{2EI}{L}\,\theta_a \qquad\qquad (4.3)$$

Equation (4.3) says that the moment at the opposite end of a rotation is equal to half the moment at the rotated end. This, of course, is assuming that the beam is prismatic (constant cross section). Clearly, the same values hold true for a rotation at the b end, θ_b.

The only term left is the moment due to a relative displacement between the two ends. It is important to remember that it is the integration of the curvature that causes work in a beam. This is because only the strain energy due to bending is considered. Looking carefully at the figure for the definition of relative displacement, it can be seen that the relative displacement is composed of three parts (Figure 4.8). The first two parts are a rigid-body displacement and a rigid-body rotation. The third part is an equal rotation at both ends. By definition, rigid-body motion causes no strain. Thus only the third part is considered in virtual work calculations. If small angles are considered, the angle that each end rotates is

$$\theta_a = \theta_b = \frac{\delta}{L} \qquad\qquad (4.4)$$

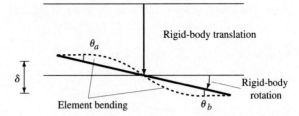

Figure 4.8 Bending deformation due to element displacement.

Substituting this angle into equation (4.2), we get the relationship for moment as a result of relative displacement:

$$M_\delta = \frac{6EI}{L^2}\,\delta \qquad\qquad (4.5)$$

Summing the four contributions to get the total, we substitute equation (4.2) [equivalent of (4.3) for θ_b] and equation (4.5) into (4.1), which gives

$$M_a = M_{\text{FEM}_a} + \frac{2EI}{L}\left(2\theta_a + \theta_b + \frac{3\delta}{L}\right) \tag{4.6}$$

The sign convention followed is that all rotations are positive counterclockwise (this follows the right-hand rule). The convention for positive and negative displacements is shown in Figure 4.9. Note that this shape can be rotated in the plane to compare to any displaced shape. If the first pattern matches, the displacement is positive; if the second pattern matches, it is negative.

Positive displacement   Negative displacement

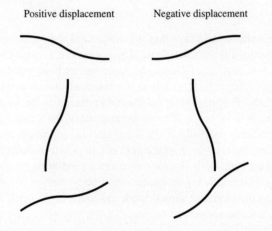

Figure 4.9 Displacement sign convention examples.

A helpful method to use for determining the sign of the displacement for an element is shown in Figure 4.10. Take any deformed element and draw a chord between the two ends. Then look at the rotation of the ends. If the ends rotate counterclockwise from the chord, the displacement is defined as positive.

Chord between ends

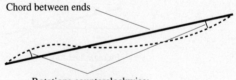

Rotations counterclockwise;
displacement is positive

Figure 4.10 Determination of displacement sign by rotation.

4.2 Application of Slope Deflection

A procedure for solving structures using slope deflection can be developed similar to that for consistent deformations. The procedure consists of the following:

1. Define the unknown displacements. This is a little more difficult than finding unknown redundants because redundants deal with equilibrium, which is more physical. These unknown displacements are generally called *degrees of freedom* (DOFs).

2. Develop an equilibrium equation for every unknown displacement. For rotations, this is a summation of moments about a joint. For translations this is some form of horizontal, vertical, and/or moment equilibrium using a part of or the entire structure.

3. Substitute the moment equation from slope deflection for each of the forces into the equilibrium equations. This will generate a set of equations in the unknown displacements.

4. Solve the set of equations for the displacements. Then substitute the displacements into the slope deflection equations to recover the member end moments. Shears can be recovered from statics.

This procedure is common to all displacement approaches. The unknowns are displacements; however, the result generally needed is forces. An additional step is required to recover the forces from the displacements. Although this seems like additional work, in reality it is comparable to flexibility methods. The advantage is that displacement methods are easier to automate.

The first step requires definition of the DOFs. This is a very crucial step. It is important to be consistent with the assumptions of the method. Only unique displacements must be chosen. Since axial deformations are ignored, displacements that allow this must not be permitted. See, for example, the unique DOFs for the structure shown in Figure 4.11. The vertical displacements at each column are not allowed because this would imply axial deformations of the columns. Only one horizontal displacement is allowed, since two would imply axial deformation of the beam. Finally, rotations at each joint are allowed. Generally, all rotations can be allowed. This would not be the case if members were connected by rigid blocks such as a rigid floor diaphragm in a building. As an example, see the structures shown in Figure 4.12 with their DOFs. DOFs and structural displaced shapes are discussed further in Section 4.4.

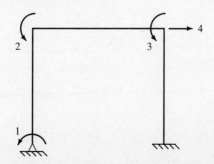

Figure 4.11 Definition of unique DOFs.

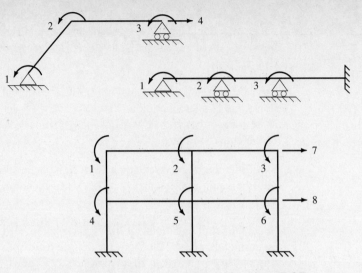

Figure 4.12 Example structures with unique DOFs.

4.2.1 Simple Slope Deflection Example

A classic example for slope deflection is the structure shown in Figure 4.13. Using the solution procedure given earlier, step 1 is to define the unknowns for the analysis, that is, the unique displacements that will define the response of the structure. For the structure of Figure 4.13, it is clear that the only choice is the rotation at the center joint (Figure 4.14). Therefore, this is a single DOF structure for a displacement method solution. Recall that for flexibility we would have to remove four reactions, thus four unknowns. This makes the solution by slope deflection much simpler.

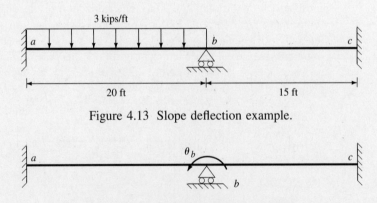

Figure 4.13 Slope deflection example.

Figure 4.14 DOFs defined for example.

Step 2 says to develop an equilibrium equation for each unknown. For rotations, this is generally the summation of moments about the joint. For the rotation given, a free body can be drawn for the joint (Figure 4.15). The moments on the ends of the members are those from the slope deflection equations. The notation for the member end moments

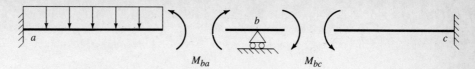

Figure 4.15 Free body for moment equilibrium at b.

is M_{ba}, the moment at the b end of member ab. The equal and opposite moments are shown on the joint. Note that the sign convention is positive counterclockwise *on the end of the member*. Summing moments about joint b, from the free body of the joint, we get

$$M_{ba} + M_{bc} = 0 \qquad (4.7)$$

The slope deflection equations for the moments M_{ba} and M_{bc} are

$$M_{ba} = M_{\text{FEM}_{ba}} + \frac{2EI}{L_{ba}}\left(2\theta_b + \theta_a + \frac{3\delta}{L_{ba}}\right) \qquad (4.8)$$

$$M_{bc} = M_{\text{FEM}_{bc}} + \frac{2EI}{L_{bc}}\left(2\theta_b + \theta_c + \frac{3\delta}{L_{bc}}\right) \qquad (4.9)$$

where L_{ba} is the length of member ba.
         L_{bc} is the length of member bc.

The fixed-end moments for each span can be calculated from the formulas given earlier. For span ab there is a uniform load; therefore, the fixed-end moments are

$$M_{\text{FEM}_{ba}} = -\frac{wL^2}{12} = \frac{(-3 \text{ kips/ft/12 in./ft})[(20 \text{ ft}) (12 \text{ in./ft})]^2}{12} = -1200 \text{ kip-in.} \qquad (4.10)$$

$$M_{\text{FEM}_{ab}} = \frac{wL^2}{12} = 1200 \text{ kip-in.} \qquad (4.11)$$

There is no load on span bc; therefore, there are no fixed-end moments. Remember that FEMs exist only if there is a load within the span. Looking at equations (4.8) and (4.9), we can see that θ_a, θ_c, and the relative end displacement δ drop out since according to the boundary conditions, they are zero.

Step 3 says to substitute equations (4.8) and (4.9) into the equilibrium equation (4.7). Removing θ_a and θ_c, we have

$$-1200 + \frac{4EI}{20(12)}\theta_b + \frac{4EI}{15(12)}\theta_b = 0 \qquad (4.12)$$

The final step, 4, is to solve for θ_b and recover the forces.

$$\theta_b = \frac{30,857}{EI} \qquad (4.13)$$

To recover the forces, we first recover the member end moments by using the slope deflection equations (4.8) and (4.9). The two additional slope deflection equations that were not used are for the moments at the fixed ends of the members. They are

$$M_{ab} = M_{\text{FEM}_{ab}} + \frac{2EI}{L_{ba}}\left(2\theta_a + \theta_b + \frac{3\delta}{L_{ba}}\right) \tag{4.14}$$

$$M_{cb} = M_{\text{FEM}_{cb}} + \frac{2EI}{L_{bc}}\left(2\theta_c + \theta_b + \frac{3\delta}{L_{bc}}\right) \tag{4.15}$$

Substituting θ_b and noting that θ_a, θ_c, and δ are zero, we can find M_{ab} and M_{cb}:

$$M_{ab} = 1200 + \frac{2EI}{L_{ab}}\frac{30{,}857}{EI} = 1200 + \frac{2(30{,}857)}{20(12)} = 1457.1 \text{ kip-in.} \tag{4.16}$$

$$M_{cb} = 0 + \frac{2EI}{L_{cb}}\frac{30{,}857}{EI} = \frac{2(30{,}857)}{15(12)} = 342.9 \text{ kip-in.} \tag{4.17}$$

Note that since E and I are constant, they drop out of the calculation of forces. Actually, it is only the relative values of E and I between members that affect the member forces. If we also find the moments at the ends connected to joint b, we have a sufficient number of forces to use statics to solve for the other support forces.

$$M_{ba} = -1200 + \frac{4EI}{L_{ab}}\frac{30{,}857}{EI} = -1200 + \frac{4(30{,}857)}{20(12)} = -685.7 \text{ kip-in.} \tag{4.18}$$

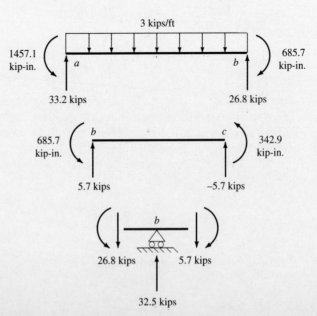

Figure 4.16 Final free bodies for slope deflection example.

$$M_{bc} = 0 + \frac{4EI}{L_{cb}}\frac{30{,}857}{EI} = \frac{4(30{,}857)}{15(12)} = 685.7 \text{ kip-in.} \tag{4.19}$$

Now, from the free-body diagrams and statics, we can find the final equilibrium forces given in Figure 4.16.

4.3 Modified Slope Deflection

The slope deflection equations developed so far have an implied assumption that the ends of all beams have either an unknown rotation or a zero rotation. If the displacement is unknown, its value is one of the unknowns in the problem. The zero-displacement assumption is used for all fixed supports with no rotation. Therefore, using these equations, the structure shown in Figure 4.17 requires two unknown rotations. There is, however, some additional information that can be used to simplify the solution process. The fact that the moment at joint a is known to be zero can be included in the formulation. By using this fact, the rotation at that joint need not be included in the formulation. Note that including all rotations will always give the correct solution. This reduced solution is included since it is used later to develop the additional terms required for the development of a pinned-end stiffness matrix.

Figure 4.17 Structure for modified formula derivation.

To take advantage of the known value for the moment at a, we can manipulate the equations to generate a modified form of the slope deflection equations. The equations for the moments of member ab are

$$M_{ab} = M_{\text{FEM}_{ab}} + \frac{2EI}{L_{ab}}\left(2\theta_a + \theta_b + \frac{3\delta}{L_{ab}}\right) \tag{4.20}$$

$$M_{ba} = M_{\text{FEM}_{ba}} + \frac{2EI}{L_{ab}}\left(2\theta_b + \theta_a + \frac{3\delta}{L_{ab}}\right) \tag{4.21}$$

Setting equation (4.20) equal to zero and solving for the rotation θ_a, we get

$$\frac{2EI}{L_{ab}}\theta_a = -\frac{M_{\text{FEM}_{ab}}}{2} - \frac{EI}{L_{ab}}\left(\theta_b + \frac{3\delta}{L_{ab}}\right) \tag{4.22}$$

Rearranging and substituting this into equation (4.20) and again rearranging, we get a modified form for the slope deflection equation:

$$M_{ba}^{*} = M_{\text{FEM}_{ba}} - \frac{M_{\text{FEM}_{ab}}}{2} + \frac{3EI}{L_{ab}}\left(\theta_b + \frac{\delta}{L_{ab}}\right) \tag{4.23}$$

This modified form can be used for the equation of a moment at the end of a beam when the opposite end is pinned and has a moment of zero. It also assumes that the rotation at the opposite end is nonzero but not one of the unknowns.

4.3.1   Modified Slope Deflection Example

As an example of how to use the modified equation, let's take the frame shown in Figure 4.18. If all possible DOFs are defined, there would be three rotations, θ_a, θ_b, and θ_c. Then using the summation of moments about a, b, and c, you could generate three equations in the three unknowns. Solving would give the three rotations shown in Figure 4.19.

An easier way for hand calculations is to use the modified formula. According to the assumptions for the modified formula, the rotations θ_a and θ_c can be left out of the formulation since it is known that the moment is zero. As a result, only one DOF is

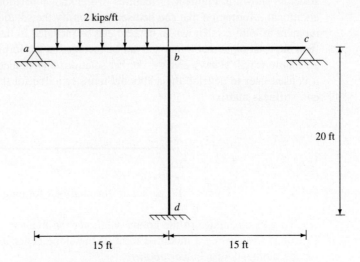

Figure 4.18 Modified slope deflection example.

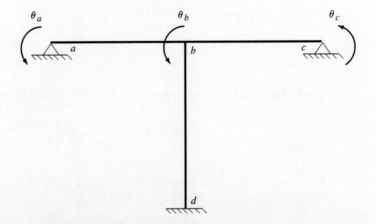

Figure 4.19 Unknowns for full slope deflection method.

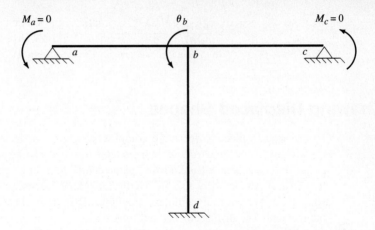

Figure 4.20 Unknowns and assumptions for modified method.

required, θ_b (Figure 4.20). The summation of moments about joint b will be used to generate the required equilibrium equation. Summing moments, we get

$$M^*_{ba} + M^*_{bc} + M_{bd} = 0 \qquad (4.24)$$

The equations for the moments come directly from the formulas

$$M^*_{ba} = \frac{wL^2_{ba}}{12} - \frac{1}{2}\frac{wL^2_{ba}}{12} + \frac{3EI}{L_{ba}}\theta_b \qquad (4.25)$$

$$M^*_{bc} = \frac{3EI}{L_{bc}}\theta_b \qquad (4.26)$$

$$M_{bd} = \frac{2EI}{L_{bd}}2\theta_b \qquad (4.27)$$

Substituting and solving for θ_b, we get

$$\theta_b = \frac{-4500}{EI} \qquad (4.28)$$

Notice that when substituting into the equations, we converted all lengths and loads to units of inches and kips to arrive at the value above. To recover the moments, you would plug into the appropriate equations, modified or regular. For example:

$$M_{db} = \frac{2EI}{L_{bd}}\theta_b = -37.5 \text{ kip-in.} \qquad (4.29)$$

Again, all other forces can be recovered from free-body diagrams by including the moments and the external loads. The rotation at a or c can also be recovered using the rearranged equation (4.22).

$$\theta_a = \frac{L_{ab}}{2EI}\left(-\frac{M_{\text{FEM}_{ab}}}{2} - \frac{EI}{L_{ab}}\theta_b\right) = -\frac{(15 \times 12)^3 w}{4EI \times 12} + \frac{4500}{2EI} \tag{4.30}$$

4.4 Drawing Displaced Shapes

To use slope deflection for more complex structures, deflected shapes must be drawn. These shapes must be used to define the DOFs for complex geometry. It is very important to use only independent unknowns. Slanted members in a structure cause the DOFs to be linked. Axial members also link DOFs. Displaced shapes are also important for verification of the behavior of complex structures. Displaced shapes are *always* drawn very exaggerated. The displacements and rotations are shown orders of magnitude larger than would happen in a real structure. This is to help clarify the movements and rotations. Remember, we are still operating under the small-deformation and small-angle assumptions.

Drawing displaced shapes can be broken down into some basic displaced shape components. There are two assumptions that must be stated: (1) no axial deformations are allowed for displaced shapes, and (2) displaced shapes are independent of loadings and loads within a span are ignored.

The first basic component needed for drawing displaced shapes is the displaced shape of a cantilever. Shown in Figure 4.21 are the three basic shapes of the cantilever: the first for a free end, the next for an end free to translate but no rotation is allowed, and the final shape somewhere in between the first two. The end is free to translate but rotation is restrained by the attached spring. Note that the spring cannot fully prevent rotation, and the rotation is between zero and that of the free cantilever.

The second set of basic shapes is for propped cantilevers, shown in Figure 4.22. The first shape is for a moment-free propped end. The second is for a propped cantilever with

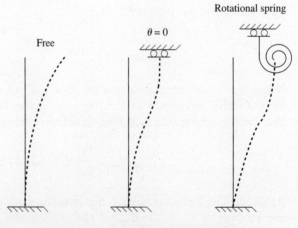

Figure 4.21 Three basic bending shapes.

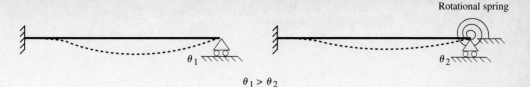

$$\theta_1 > \theta_2$$

Figure 4.22 Propped cantilever basic shapes.

partial restraint. Again the spring offers some resistance but cannot fix the rotation fully. Therefore, the end rotation is greater than zero but less than the no-spring case.

By putting these basic parts together and adding rigid-body rotations you can draw the displaced shape of a structure. Note that the spring restraint is the exact effect of attaching another beam to the end of an element. For example, a structure that exactly represents the propped cantilever with a spring is shown in Figure 4.23.

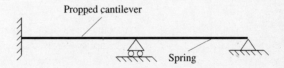

Figure 4.23 Beam acting as spring for propped cantilever.

One of the assumptions made is that axial displacements are ignored. As a result, only bending or rotational displacements are allowed. This has special meaning for our structures in that it states that only displacements perpendicular to a member are allowed. This is because in the small-displacement scenario, a beam would be allowed to rotate about its support. The arc of possible displacement is shown in Figure 4.24. For small displacements, only the initial, small portion of the curve is allowed. This curve is essentially flat for the amount of displacement allowed. Therefore, the displacement is essentially perpendicular to the member.

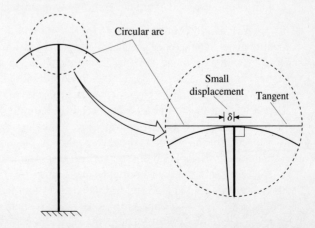

Figure 4.24 Small-deformation theory perpendicular displacement.

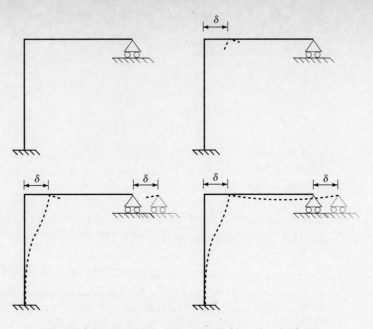

Figure 4.25 Drawing deflected shapes: simple frame.

Using these basic shapes, let's draw a displaced shape for a structure. Figure 4.25 shows the structure and the steps used to draw the final displaced shape. The general procedure is as follows:

1. To draw displaced shapes, we start at a support.

2. Let the opposite ends of the members that are connected to that support displace. This means displacing and rotating the joint.

3. Allow any other portion of the structure to follow as a rigid body.

4. Move to the next joint and continue with other members.

This process should continue until you reach the end. For example, starting at the fixed support, go to the member's end and displace the joint. Since this is acting like the cantilever with a spring at the end, it will displace and rotate. Now connect the vertical member between the fixed end and the displaced joint. The next member has to move as a rigid body since axial displacements are neglected. Therefore, the roller will translate an equal amount to the right. Since there is nothing to prevent the end from rotating, it will do so because of the left-end rotation. Now the member can be drawn to complete the displaced shape. Notice that no vertical displacements occurred because of no axial deformations in the column and no vertical displacements at the roller.

The next shape to try is a portal frame (Figure 4.26). Again starting at the lower left, the upper left joint is rotated and translated. The right joint must also translate due to the beam. However, this translation is also pushing the right-hand column like a spring-ended cantilever. Therefore, it will rotate in the same direction as the left end. The shape of the

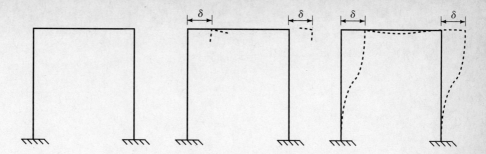

Figure 4.26 Drawing deflected shapes: portal frame.

right column will be identical to that of the left. The beam is drawn to match the two ends connected to the columns.

The final structure considered is one with a slanted member (Figure 4.27). As before, the left joint is considered first. It is again translated and rotated. Now we move to the right joint. It also has to move to the right. However, the slanted member will not allow just a rightward movement. Remember, displacements must be perpendicular to all members. Therefore, displacements must be along the lines shown, perpendicular to both members. Therefore, the point on the line where the translation to the right is equal to that of the left joint is the final position. Note that this position can be calculated by geometry. As for the rotation, the slanted member acts like a cantilever pushed along the

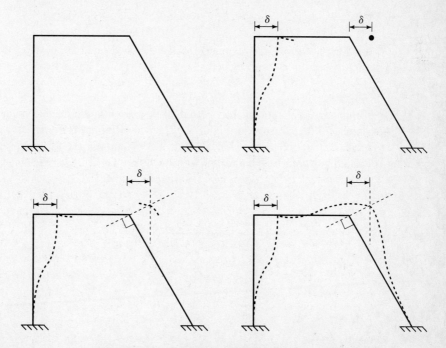

Figure 4.27 Drawing deflected shapes: slanted frame.

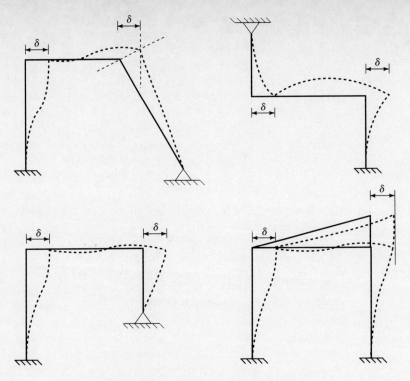

Figure 4.28 Examples of deformed shapes.

drawn line. Therefore, it will rotate clockwise. Now the members can be drawn. Additional illustrative examples of displaced shapes are given in Figure 4.28.

A helpful hint about drawing displaced shapes is the following: If you think a displacement is independent, try applying its displacement and seeing if the axial or perpendicular assumptions are violated. If they are not, the displacement is valid. As an example, for the structure in Figure 4.27, try allowing a horizontal displacement but no vertical. Drawing this shape clearly shows a violation in the assumptions since an axial deformation would be required in the slanted member (Figure 4.29).

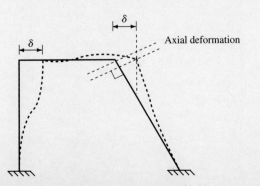

Figure 4.29 Attempted shape: axial deformation violation.

4.5 Frames by Slope Deflection

Now that we can draw the deflected shapes of structures, we can determine the required independent DOFs of a structure. Using these independent DOFs we can solve structures using slope deflection. Frames are a little more difficult to solve than beams since an equilibrium equation must be developed for the displacements in addition to rotations, as in beams. Recall that rotations are handled by summation of moments about the joint. Displacements require the use of more of the structure than just a joint or a member. It usually involves taking a free body of the entire structure. Then the summation of vertical or horizontal forces or moments for the entire structure needs to be taken. Next, the substitution of the moment equations can be performed as before.

4.5.1 Frame Example by Slope Deflection

This first frame example will demonstrate the ideas of using global equilibrium. The structure shown in Figure 4.30 will be solved. If we use the modified equation for moment M_{bc}, there are only two DOFs, θ_b and δ. The displaced shape is drawn to verify that the two displacements are the required independent DOFs (Figure 4.31). As before, the first equation will be generated by summing moments about joint b. Taking a free body of joint b, we have the moments M_{ba} and M_{bc}^* acting. The resulting equation is

$$M_{ba} + M_{bc}^* = 0 \tag{4.31}$$

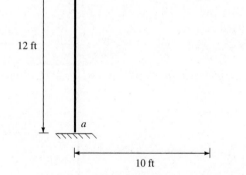

Figure 4.30 Slope deflection: frame example.

The equations for these moments are (units are kips and inches)

$$M_{ba} = \frac{2EI}{L_{ab}}\left(2\theta_b + \frac{3\delta}{L_{ab}}\right) = \frac{4EI\theta_b}{144} + \frac{6EI\delta}{144^2} \tag{4.32}$$

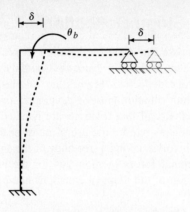

Figure 4.31 DOF for frame example.

$$M_{bc}^* = \frac{wL_{bc}^2}{12} - \frac{wL_{bc}^2}{2(12)} + \frac{3EI}{L_{bc}}\theta_b = \frac{3(3/12)(120)^2}{2(12)} + \frac{3EI}{120}\theta_b \qquad (4.33)$$

Substituting equations (4.32) and (4.33) into (4.31) gives

$$0.052777EI\theta_b + 0.000289EI\delta + 450 = 0 \qquad (4.34)$$

The second equilibrium equation is generated using a free body of the entire structure and summation of horizontal forces (Figure 4.32). It is important to include all external loads as well as internal forces when taking the free body. For the case shown in Figure 4.32, summing forces in the x direction, we get

$$10 - H_a = 0 \qquad (4.35)$$

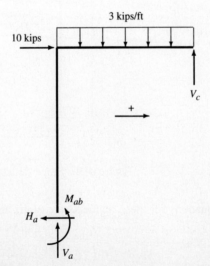

Figure 4.32 Structural free body: equilibrium.

Next, we need to substitute into (4.35) the equations that contain the unknowns, as we did with the moment equations. However, we only have the horizontal force at a in the equation. Therefore, we need to find an equation for H_a that contains the unknowns or the moments. The typical method is to take a free body of the member that has the needed force as a reaction, in this case, member ab (Figure 4.33). Now, by summing moments about point b from Figure 4.33, we generate an equation with H_a and the moments:

$$M_{ba} + M_{ab} - H_a(144) = 0 \tag{4.36}$$

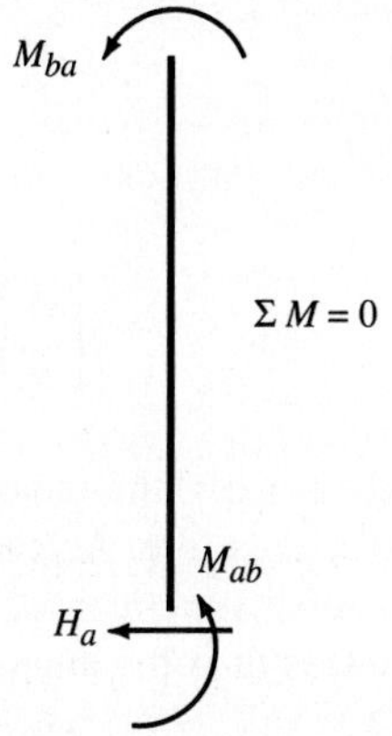

Figure 4.33 Vertical member: moment equilibrium.

Solving for H_a, we get

$$H_a = \frac{M_{ba} + M_{ab}}{144} \tag{4.37}$$

We then substitute this back into the horizontal equilibrium equation (4.35). This generates a horizontal equilibrium equation based on the moments in the structure:

$$10 - \frac{M_{ba} + M_{ab}}{144} = 0 \tag{4.38}$$

Now the moment equations for M_{ab} and M_{ba} need to be substituted into equation (4.38) to generate an equation containing the unknowns. The equation for M_{ab} is

$$M_{ab} = \frac{2EI}{L_{ab}}\left(\theta_b + \frac{3\delta}{L_{ab}}\right) = \frac{2EI}{144}\theta_b + \frac{6EI}{144^2}\delta \tag{4.39}$$

Substituting equations (4.32) and (4.39) into equation (4.38) and factoring out the denominator, we get the following equilibrium equation:

$$144 \times 10 - \frac{2EI}{144}\theta_b - \frac{6EI}{144^2}\delta - \frac{4EI}{144}\theta_b - \frac{6EI}{144^2}\delta = 0 \tag{4.40}$$

Reducing the expressions, we get the final form of the equilibrium equation:

$$1440 - 0.04166EI\theta_b - 0.0005787EI\delta = 0 \tag{4.41}$$

If we cast the two equilibrium equations into matrix form, we have

$$\begin{bmatrix} 0.05277 & 0.0002893 \\ 0.04166 & 0.0005787 \end{bmatrix} \begin{bmatrix} \theta_b \\ \delta \end{bmatrix} = \begin{bmatrix} -\dfrac{450}{EI} \\ \dfrac{1440}{EI} \end{bmatrix} \tag{4.42}$$

Using values of $E = 29,000$ ksi and $I = 400$ in^4 and solving, we get

$$\begin{bmatrix} \theta_b \\ \delta \end{bmatrix} = \begin{bmatrix} -0.00316 \text{ rad} \\ 0.44179 \text{ in.} \end{bmatrix} \tag{4.43}$$

Using these displacements, the moments and reactions can be recovered using the slope deflection formulas as before. As can be observed, there is a significant amount of extra work required to solve structures where the translations are considered. However, this is still considerably less than the amount required using a flexibility approach.

4.6 Frame with Slanted Member

A structure with an angled member will be solved using slope deflection to show the additional complexity of the method. The structure that will be solved is shown in Figure 4.34. As before, we need the deflected shape to help find the independent DOFs. This shape was discussed in Chapter 3 and is shown in Figure 4.35. Notice here that the

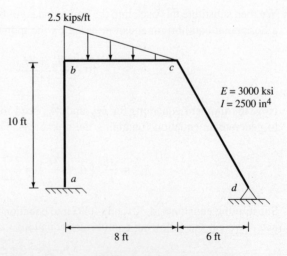

Figure 4.34 Slanted frame example.

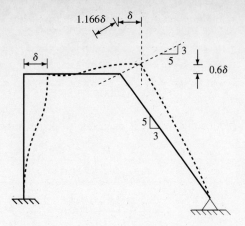

Figure 4.35 Deformed shape: active DOF.

geometry of the relationship for displacements of each member is calculated. Since the slope of the slanted member is known, we know that the line of perpendicular displacement has the inverse slope. From this, the relative displacements are calculated for each member in terms of the single unknown, δ. The results are shown in the figure. From the shape we can see that there are only four independent DOFs: θ_b, θ_c, θ_d, and δ. If we choose to use the modified equation at support c because of the pin support d, we can reduce this to the three DOFs shown in Figure 4.36. Note that instead of the horizontal DOF, we could just as easily have chosen the vertical displacement at joint c as the third DOF. This is because the horizontal and vertical displacements are linked by the fact that the displacement must follow along the line perpendicular to the slanted member.

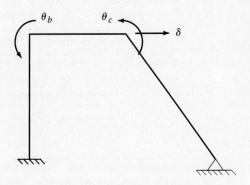

Figure 4.36 Required DOF.

We now need to generate the three equations of equilibrium. As usual, the summation of moments about a joint will be used for the two rotations. The first equation is the summation about joint b.

$$M_{ba} + M_{bc} = 0 \qquad (4.44)$$

The individual moment equations are (units are kips and inches)

$$M_{ba} = \frac{2EI}{L_{ab}}\left(2\theta_b + \frac{3\delta}{L_{ab}}\right) = \frac{2(3000)(2500)}{10(12)}\left[2\theta_b + \frac{3\delta}{10(12)}\right] \tag{4.45}$$

$$= 250{,}000\theta_b + 3125\delta \tag{4.46}$$

and (notice that -0.6δ is used due to negative bending shape)

$$M_{bc} = \frac{wL_{bc}^2}{20} + \frac{2EI}{L_{bc}}\left[2\theta_b + \theta_c + \frac{3(-0.6\delta)}{L_{bc}}\right] \tag{4.47}$$

$$= \frac{(2.5/12)[8(12)]^2}{20} + \frac{2(3000)(2500)}{8(12)}\left[2\theta_b + \theta_c - \frac{1.8\delta}{8(12)}\right] \tag{4.48}$$

$$= 96 + 312{,}500\theta_b + 156{,}250\theta_c - 2929.69\delta \tag{4.49}$$

Summing the two equations (4.46) and (4.49), we get

$$96 + 562{,}500\theta_b + 156{,}250\theta_c + 195.31\delta = 0 \tag{4.50}$$

Notice that the δ used for M_{bc} was equal to (-0.6δ). This is due to the fact that the δ values of each member are related to a single unknown δ by geometry. The geometric relationship is what gives the -0.6 factor. The negative sign comes from the shape of the displaced element. Looking at the displaced shape, we can see that it is a negative displacement.

The second equation is generated from the summation of moments about joint c:

$$M_{cb} + M_{cd}^* = 0 \tag{4.51}$$

The individual moment equations are

$$M_{cb} = -\frac{wL_{cb}^2}{30} + \frac{2EI}{L_{cb}}\left[2\theta_c + \theta_b + \frac{3(-0.6\delta)}{L_{cb}}\right] \tag{4.52}$$

$$= -\frac{(2.5/12)[8(12)]^2}{30} + \frac{2(3000)(2500)}{8(12)}\left[2\theta_c + \theta_b - \frac{1.8\delta}{8(12)}\right] \tag{4.53}$$

$$= -64 + 312{,}500\theta_c + 156{,}250\theta_b - 2929.69\delta \tag{4.54}$$

and using the modified equation, we have

$$M_{cd}^* = \frac{3EI}{L_{cd}}\left(\theta_c + \frac{1.166\delta}{L_{cd}}\right) \tag{4.55}$$

$$= \frac{3(3000)(2500)}{11.66(12)}\left[\theta_c + \frac{1.166\delta}{11.66(12)}\right] \tag{4.56}$$

$$= 160{,}806.2\theta_c + 1340.05\delta \tag{4.57}$$

Summing equations (4.54) and (4.57), we get

$$-64 + 473,306.2\theta_c + 156,250\theta_b - 1589.6\delta = 0 \qquad (4.58)$$

The third equilibrium equation is generated using some type of global equilibrium. There are many ways to generate the last equation. For example, we could sum forces in the horizontal direction for the entire structure (Figure 4.37). This gives the equation

$$H_a + H_d = 0 \qquad (4.59)$$

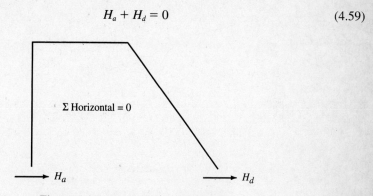

Figure 4.37 Horizontal equilibrium.

Now, taking the free body of the members with these forces as end shears, we can find the horizontal forces as functions of the end moments, as we did in the preceding example. However, for member cd and summing about joint c we have the free body shown in Figure 4.38.

Summation of moments about point c gives the equation

$$M_{cd}^* + H_d(10)(12) + V_d(6)(12) = 0 \qquad (4.60)$$

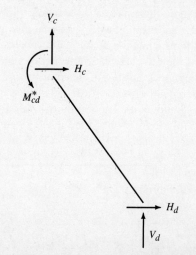

Figure 4.38 Free body of member cd.

Notice that the vertical force at end d comes into the equation. This is always true for a slanted member. The vertical and horizontal end forces both have components perpendicular to the member and contribute to the moment summation. We now have to find an additional equation for the vertical force as a function of end moments. One way to do this is using the free body of the entire structure and summing moments about joint a (Figure 4.39). The forces at joint a as well as the horizontal force at d do not enter the equation. This generates the equation

$$M_{ab} + V_d(14)(12) - \frac{wL_{cb}}{2}\frac{L_{cb}}{3} = 0 \tag{4.61}$$

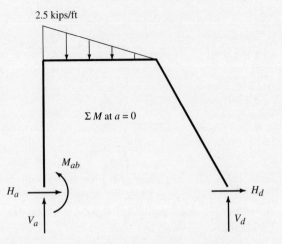

Figure 4.39 Moment summation at point a.

Note that the load now enters into the equation since the free body must include the load. We could now substitute using equations (4.59), (4.60), and (4.61) and generate the final equilibrium equation. The same amount of work would be involved if the global equation used was the summation of vertical forces instead of horizontal forces in equation (4.59). Neither of these is the easiest method of generating the final equilibrium equation.

The easiest equilibrium equation to use is the summation of moments about the intersection point of the vertical and slanted members. The free body for this equation is shown in Figure 4.40. The reason that this is the easiest is due to the fact that the only member end forces involved are those perpendicular to the members. This includes the slanted member. The axial forces in these members go through the summation point and create no moments. The perpendicular forces are easy to get from the free body of the member.

The equilibrium equation we will use is obtained by summing moments about point e. This gives

$$M_{ab} + H_a(23.333)(12) + P_d(27.211)(12) - \frac{(2.5/12)[8(12)]^2}{2(3)} = 0 \tag{4.62}$$

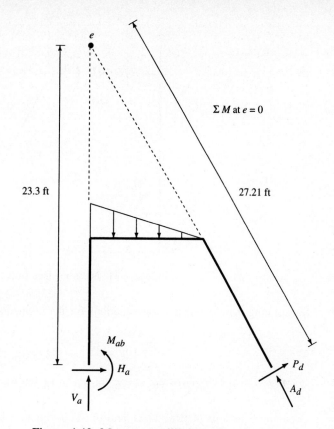

Figure 4.40 Moment equilibrium about intersection.

Note that the load needs to be included since it is part of the free body. Taking out the two members, we can generate the equations for the two perpendicular end forces. The free bodies are shown in Figure 4.41.

The equations required are for H_a and P_d in terms of the moments. These can be generated by summing the moments about points b and c, respectively. The resulting equations are

$$P_d(11.66)(12) + M_{cd}^* = 0 \tag{4.63}$$

$$M_{ba} + M_{ab} + H_a(10)(12) = 0 \tag{4.64}$$

Solving for the end forces and substituting into equation (4.62), we get

$$0 = M_{ab} - 23.33(12)\frac{M_{ba} + M_{ab}}{120} + 27.21(12)\frac{-M_{cd}^*}{11.66(12)} - 320 \tag{4.65}$$

$$0 = -1.333M_{ab} - 2.333M_{ba} - 2.333M_{cd}^* - 320 \tag{4.66}$$

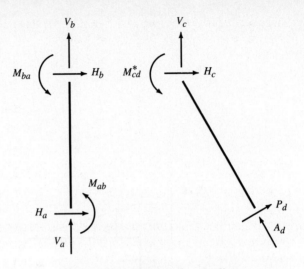

Figure 4.41 Member free bodies.

Substituting the slope deflection equations for the moments, we get the final equation:

$$M_{ab} = \frac{2EI}{L_{ab}}\left(\theta_b + \frac{3\delta}{L_{ab}}\right) = 125,000\theta_b + 3125\delta \qquad (4.67)$$

$$749,912.5\theta_b + 375,209.1\theta_c + 14,584.8\delta + 920 = 0 \qquad (4.68)$$

Grouping equations (4.50), (4.58), and (4.68) into matrix form, we have

$$\begin{bmatrix} 562,500 & 156,250 & 195.3 \\ 156,250 & 473,306.2 & -1,589.6 \\ 749,912.5 & 375,209.1 & 14,584.8 \end{bmatrix} \begin{bmatrix} \theta_b \\ \theta_c \\ \delta \end{bmatrix} = \begin{bmatrix} -96 \\ 64 \\ -320 \end{bmatrix} \qquad (4.69)$$

This matrix form represents the set of equilibrium equations required to solve for the unknown displacements. As in the preceding example, we need to solve for the unknown displacements and then substitute these back into the moment equations to recover the moments and forces. One method of solving is to use the CAL-90 solve command. Doing this, the final displacements for the structure are

$$\begin{bmatrix} \theta_b \\ \theta_c \\ \delta \end{bmatrix} = \begin{bmatrix} -0.00249 \\ 0.00183 \\ -0.18223 \end{bmatrix} \qquad (4.70)$$

4.7 Stiffness Matrices by Definition

In the preceding section we developed a set of simultaneous equations that represent the behavior of a structure. The unknowns of the equations were the displacements. The coefficient matrix was generated by equilibrium equations consisting of the summation of forces and moments. The generic matrix form of the problem after all equations are generated is

$$
\begin{bmatrix} K_{11} & K_{12} & K_{13} \\ K_{21} & K_{22} & K_{23} \\ K_{31} & K_{32} & K_{33} \end{bmatrix} \begin{bmatrix} r_1 \\ r_2 \\ r_3 \end{bmatrix} = \begin{bmatrix} R_1 \\ R_2 \\ R_3 \end{bmatrix}
\tag{4.71}
$$

The coefficient matrix, with K_{ij} terms, is called the *stiffness matrix*, **K**. The global displacements, with r_i terms, are the unknown displacements and/or rotations, referred to as the *displacement matrix*, **r**. The global loads, with R_i terms, are the applied loads (either forces or moments), referred to as the *load matrix*, **R**.

Let's look at this set of equations and see what the stiffness matrix represents. Using the structure from the preceding example and assigning the r_i's, we have the structure shown in Figure 4.42. There are three DOFs. If we set $r_1 = 1$ and $r_2 = r_3 = 0$, we can draw the deflected shape for these imposed displacements (Figure 4.43). Note that the unit displacement is small. Plugging these values into the **r** vector and multiplying by the stiffness matrix, we get the results equal to the first column of the stiffness:

$$
\begin{bmatrix} K_{11} & K_{12} & K_{13} \\ K_{21} & K_{22} & K_{23} \\ K_{31} & K_{32} & K_{33} \end{bmatrix} \begin{bmatrix} 1 \\ 0 \\ 0 \end{bmatrix} = \begin{bmatrix} K_{11} \\ K_{21} \\ K_{31} \end{bmatrix}
\tag{4.72}
$$

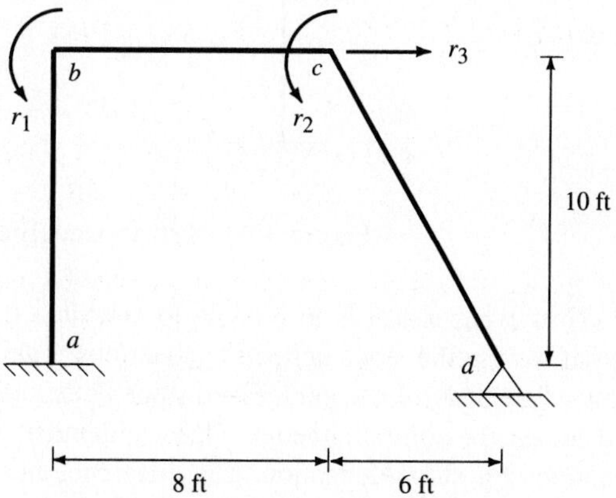

Figure 4.42 Structure for stiffness example.

This says that the loads required to generate the displaced shape given in Figure 4.43 are equal to the first column of the stiffness matrix. The same can be said about any column of a stiffness matrix. Column *i* of a stiffness matrix contains the forces required to cause a unit displacement at DOF *i* and maintain zero displacements at all other DOFs. As before, the terms of the stiffness matrix are forces. This relationship of unit displacements and forces gives the formal definition of the stiffness matrix as follows:

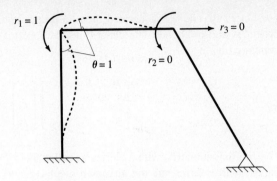

Figure 4.43 Unit displacement at DOF 1.

Any term K_{ij} of a stiffness matrix is the force required at DOF i to hold a unit displacement at DOF j with the displacements at all other DOFs equal to zero.

The unit displacement is assumed to be small. With this definition we can show the stiffness terms as forces on the displaced shape of the structure (Figure 4.44).

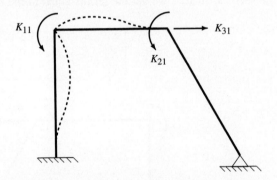

Figure 4.44 Stiffness coefficients for DOF 1.

Now what is left is to be able to calculate the value of the stiffness terms. This is easy if we use the slope deflection equations. Taking the left joint as an example, we can draw a free body of the joint. From slope deflection we can get the member end moments and hence the joint moments. This is identical to how we generated the equilibrium equations for slope deflection. The difference here is that we apply a unit displacement to a single DOF to get the displaced shape required. Recall that slope deflection gives the moment for a given complete set of displacements. In this special case we know the displacements to be unity and only at one DOF at a time. Therefore, the value of the end moment is known directly. Since the moment is derived only from displacements, no applied loads are considered and there are no fixed-end moments.

Using the familiar summation of moments about a joint, we can develop an equation for the value of K_{11}. When the stiffness term is a force instead of a moment, we will need to use vertical or horizontal equilibrium as in the previous examples. This method of stiffness derivation is called *stiffness by definition*. Therefore, to develop the stiffness matrix for any structure, you need to follow a two-step procedure:

1. Define the displacement DOF.

2. For each DOF, apply a unit displacement and calculate the forces required at all other DOFs to maintain the displaced shape.

4.7.1   Stiffness by Definition Example

As an example, let's derive the stiffness matrix for the preceding example. The displacement DOFs are defined in Figure 4.42. Applying a unit displacement to DOF 1, we get the displaced shape shown in Figure 4.44. Taking the free body of joint b, we have equilibrium as shown in Figure 4.45. From the free body of joint b, taking the summation of moments, we get equation (4.73). Notice that the stiffness term K_{11}, which is a force, must be included in the free-body diagram and the moment summation.

$$K_{11} - M_{ba} - M_{bc} = 0 \tag{4.73}$$

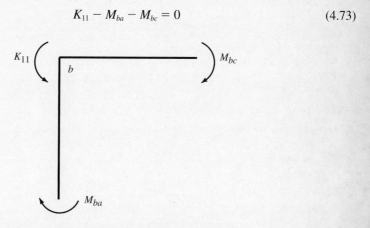

Figure 4.45 Moment equilibrium at joint b.

As in slope deflection, the required moment equations are

$$M_{ba} = \frac{2EI}{L_{ab}}\left(2r_1 + \frac{3r_3}{L_{ab}}\right) = \frac{2EI}{L_{ab}}(2) = \frac{4EI}{120} \tag{4.74}$$

$$M_{bc} = \frac{2EI}{L_{bc}}\left[2r_1 + r_2 + \frac{3(-0.6r_3)}{L_{bc}}\right] = \frac{2EI}{L_{bc}}(2) = \frac{4EI}{96} \tag{4.75}$$

In developing a stiffness matrix, the equations simplify greatly. This is because by definition, $r_1 = 1$ and $r_2 = r_3 = 0$ in order to find the first column of the stiffness matrix. Substituting equations (4.74) and (4.75) into (4.73), we get

$$K_{11} = 4EI\left(\frac{1}{96} + \frac{1}{120}\right) = 562{,}500 \text{ kip-in./rad} \tag{4.76}$$

This process needs to be continued to get all of the stiffness terms in column 1. To get the next term, we draw the free body of joint c (Figure 4.46). From the free body of joint c, we get

$$K_{21} - M_{cb} - M_{cd}^* = 0 \tag{4.77}$$

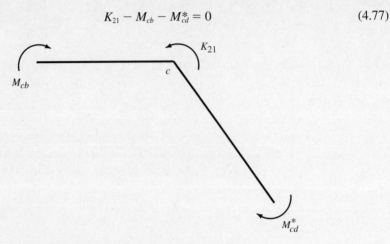

Figure 4.46 Moment equilibrium for joint c.

The required moment equations are

$$M_{cb} = \frac{2EI}{L_{cb}}\left[2r_2 + r_1 + \frac{3(-0.6r_3)}{L_{cb}}\right] = \frac{2EI}{L_{cb}} = \frac{2EI}{96} \tag{4.78}$$

$$M_{cd}^* = \frac{3EI}{L_{cd}}\left(r_2 + \frac{1.166r_3}{L_{cd}}\right) = 0 \tag{4.79}$$

Substituting equations (4.78) and (4.79) into (4.77), we get

$$K_{21} = \frac{2EI}{96} = 156{,}250 \text{ kip-in./rad} \tag{4.80}$$

The final term in the column is a force, and something other than summation of moments needs to be used. This is identical to the slope deflection problem. For stiffness we will use a similar approach of using the intersection point for the slanted and straight members. If we take a free body of the beam and the joints b and c, we get Figure 4.47. For this free body we have to include all the stiffness terms as well as the perpendicular forces for members ab and cd, since both the beam and the joints are included in the free body. Notice that in the free body, the forces on the joints (b and c) are due to the connected members ab and cd, respectively. These forces come from the ends of the member and are transferred to the joint as shown in Figure 4.48.

As in the slope deflection solution, we resolve the forces at the cuts into shear and axial forces for the cut members, then take the summation of moments about the intersection of the two points as before (Figure 4.47). Now what is needed is a term for the shear at

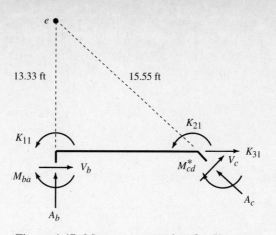

Figure 4.47 Moment summation for K_{31} term.

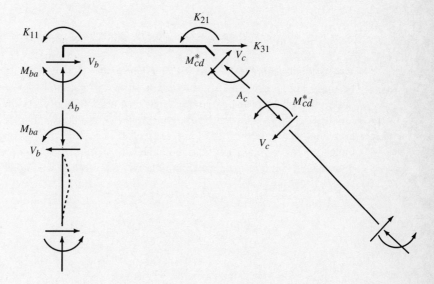

Figure 4.48 Member force transfer to joint and beam.

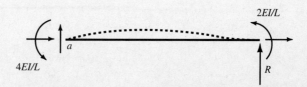

Figure 4.49 Moment summation to get shear.

the end of a member that has a unit rotation. From the free-body equilibrium in Figure 4.49 (*previous page*), with a unit rotation at one end we can derive the force needed. We already know the end moments from the slope deflection formulas. The unknown shear force R can be calculated by statics. Summing moments about the end with no displacement, we get

$$\frac{4EI}{L} + \frac{2EI}{L} + RL = 0 \tag{4.81}$$

Solving for R, we get

$$R = -\frac{6EI}{L^2} \tag{4.82}$$

Therefore, the shear at the end of a member due to a unit rotation is $6EI/L^2$.

Since we are summing moments about the intersection of the members, the axial forces have no effect on the sum and need not be calculated. The final summation becomes

$$K_{31}(13.333)(12) + K_{11} - M_{ba} + K_{21} - M_{cd}^* - \frac{6EI}{L_{ab}^2}(13.333)(12) = 0 \tag{4.83}$$

To use this equation, we need to substitute in the values K_{11} and K_{12}. This process can be made simpler by choosing a slightly different free body. Note that the moments at each joint consist of the external moment K_{11}, the moment from the cut member, and the moment on the end of the beam. Our previous free body included the column moment and the stiffness term at the joint. Both of these can be represented by the moment on the end of the beam. The shear and axial forces are the same on the end of the beam as those of the column—just switching places. Therefore, it is simpler to take the free body shown in Figure 4.50.

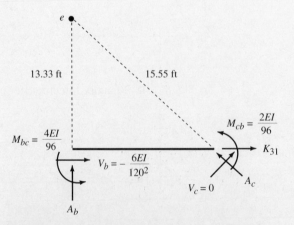

Figure 4.50 Improved moment sum for K_{31} term.

Now we just use the forces resolved in the directions of the connected members. The summation equation becomes

$$K_{31}(13.333)(12) + M_{bc} + M_{cb} - \frac{6EI}{L_{ab}^2}(13.333)(12) = 0 \qquad (4.84)$$

It would seem much simpler just to take horizontal equilibrium of the section and solve for K_{31} directly. This cannot be done unless you consider the horizontal component of the axial force in member cd. To find the horizontal component, you would need to find the axial force by summing moments on the free body of the beam to get the end shear, and then find the value of the horizontal force required to make the force axial. This horizontal component takes more computation than the summation given above but is equally valid.

The moments required for equation (4.84) are

$$M_{bc} = \frac{2EI}{L_{bc}} 2r_1 = \frac{4EI}{96} \qquad (4.85)$$

$$M_{cb} = \frac{2EI}{L_{bc}} r_1 = \frac{2EI}{96} \qquad (4.86)$$

Substituting equations (4.85) and (4.86) into (4.84) and solving, we get

$$K_{31} = 195.3 \text{ kips/rad} \qquad (4.87)$$

This completes the calculation of the stiffness terms for the first column. Next, we need to find column 2 of the stiffness matrix. This is done in the same manner as for the first column. Applying a unit displacement to DOF 2, we have the displaced shape shown in Figure 4.51. Note that the right support is pinned but the rotation is not considered to be

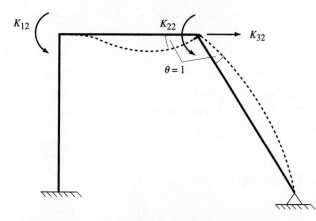

Figure 4.51 Stiffness coefficients for DOF 2.

one of the unknowns. Similar to slope deflection, we can use the modified terms to generate the equilibrium equations. It is important here that this rotation be allowed for the displaced shape. If the rotation were not allowed, it would mean another unknown for the problem and its displacement would have to remain zero for this shape.

Drawing the free body of joint b (Figure 4.52) and summing moments, we get

$$K_{12} - M_{ba} - M_{bc} = 0 \tag{4.88}$$

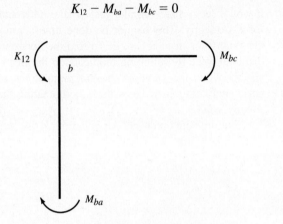

Figure 4.52   Moment equilibrium at joint b.

The required moment equations are

$$M_{ba} = \frac{2EI}{L_{ab}}\left(2r_1 + \frac{3r_3}{L_{ab}}\right) = 0 \tag{4.89}$$

$$M_{bc} = \frac{2EI}{L_{bc}}\left[2r_1 + r_2 + \frac{3(-0.6r_3)}{L_{bc}}\right] = \frac{2EI}{L_{bc}}(1) = \frac{2EI}{96} \tag{4.90}$$

Substituting equations (4.89) and (4.90) into (4.88), we get

$$K_{12} = \frac{2EI}{96} = 156{,}250 \text{ kip-in./rad} \tag{4.91}$$

Note that this is the same as equation (4.80), due to symmetry of the stiffness matrix.

Drawing the free body of joint c (Figure 4.53) and summing moments, we get

$$K_{22} - M_{cb} - M_{cd}^* = 0 \tag{4.92}$$

The required moment equations are

$$M_{cb} = \frac{2EI}{L_{cb}}\left[2r_2 + r_1 + \frac{3(-0.6r_3)}{L_{cb}}\right] = \frac{4EI}{L_{cb}} = \frac{4EI}{96} \tag{4.93}$$

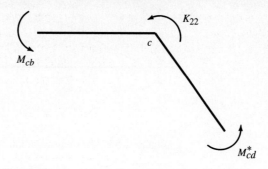

Figure 4.53 Moment equilibrium for joint c.

$$M_{cd}^* = \frac{3EI}{L_{cd}}\left(r_2 + \frac{1.166r_3}{L_{cd}}\right) = \frac{3EI}{L_{cd}} = \frac{3EI}{139.9} \tag{4.94}$$

Substituting equations (4.93) and (4.94) into (4.92), we get

$$K_{22} = \frac{4EI}{96} + \frac{3EI}{139.9} = 473{,}329.2 \text{ kip-in./rad} \tag{4.95}$$

Using the same free body and moments as before, we can draw the free body for finding the horizontal force (K_{32}) (Figure 4.54). As can be seen, we need the perpendicular force (shear force) on the end of a slanted pinned member. This is similar to the force needed when finding the K_{31} term. The difference is that while there is a unit rotation, the opposite end is pinned and free to move. The same method for finding the force can be used by substituting the modified moment equation.

Again we have a statics problem with known moments and unknown shears. From the free body of the member, we can sum moments about point a and find the shear force

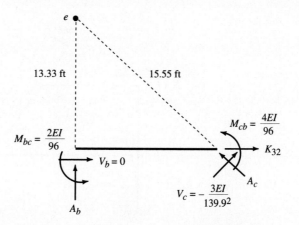

Figure 4.54 Improved moment sum for K_{32} term.

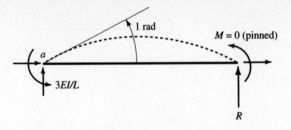

Figure 4.55 Moment summation for end shear.

(Figure 4.55). Solving for the force R, we get

$$R = -\frac{3EI}{L^2} \tag{4.96}$$

Including this force in the free body in Figure 4.53 and summing moments as before, we get

$$K_{32}(13.333)(12) + M_{bc} + M_{cb} - \frac{3EI}{L^2_{cd}}(15.55)(12) = 0 \tag{4.97}$$

Substituting for the moments and solving for the stiffness term, we get

$$K_{32} = -1589.1 \text{ kips/rad} \tag{4.98}$$

Note that the shear force on member cd is perpendicular to the member.

The final column is attained by applying a unit displacement to DOF r_3 and continuing in the same manner as the terms computed previously. The displaced shape is shown in Figure 4.56. Drawing the free body of joint b and summing moments, we get equilibrium

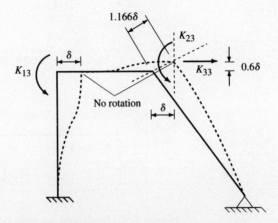

Figure 4.56 Stiffness coefficients for DOF 3.

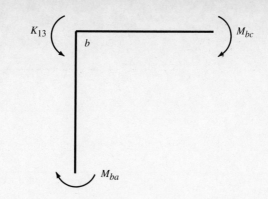

Figure 4.57 Moment equilibrium at joint b.

as shown in Figure 4.57. Moment summation about joint b gives the equation

$$K_{13} - M_{ba} - M_{bc} = 0 \tag{4.99}$$

The required moment equations are

$$M_{ba} = \frac{2EI}{L_{ab}}\left(2r_1 + \frac{3r_3}{L_{ab}}\right) = \frac{6EI}{L_{ab}^2} = \frac{6EI}{120^2} \tag{4.100}$$

$$M_{bc} = \frac{2EI}{L_{bc}}\left[2r_1 + r_2 + \frac{3(-0.6r_3)}{L_{bc}}\right] = \frac{6EI(-0.6)}{L_{bc}^2} = \frac{-6EI(0.6)}{96^2} \tag{4.101}$$

Substituting equations (4.100) and (4.101) into (4.99), we get

$$K_{13} = 6EI\left(\frac{1}{120^2} - \frac{0.6}{96^2}\right) = 195.31 \text{ kip-in./in.} \tag{4.102}$$

Drawing the free body of joint c and summing moments, we get equilibrium as shown in Figure 4.58. Summing moments about joint c gives the equation

$$K_{23} - M_{cb} - M_{cd}^* = 0 \tag{4.103}$$

The required moment equations are

$$M_{cb} = \frac{2EI}{L_{cb}}\left[2r_2 + r_1 + \frac{3(-0.6r_3)}{L_{cb}}\right] = \frac{6EI(-0.6)}{L_{cb}^2} = \frac{6EI(-0.6)}{96^2} \tag{4.104}$$

$$M_{cd}^* = \frac{3EI}{L_{cd}}\left(r_2 + \frac{1.166r_3}{L_{cd}}\right) = \frac{3EI(1.166)}{L_{cd}^2} = \frac{3EI(1.166)}{139.9^2} \tag{4.105}$$

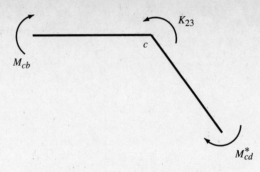

Figure 4.58 Moment equilibrium for joint c.

Substituting equations (4.104) and (4.105) into (4.103), we get

$$K_{23} = 3EI\left[\frac{2(-0.6)}{96^2} + \frac{1.166}{139.9^2}\right] = -1589.6 \text{ kip-in./in.} \tag{4.106}$$

Using the same free body and moments as before, we can draw the free body for finding the horizontal force (K_{33}) (Figure 4.59).

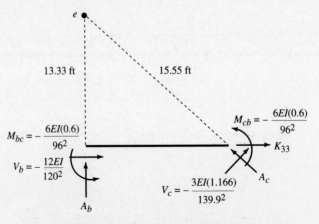

Figure 4.59 Improved moment sum for K_{33} term.

As can be seen in Figure 4.56, we need the perpendicular force (shear force) on the end of a pinned member and a fixed member due to a unit relative displacement. This is similar to the shear force needed for finding the K_{31} term. The difference is that here the cause is a unit relative displacement instead of a rotation. As a result, we need to find two new shear forces. We will use the same method as for the other two forces. First, for the left-hand column we draw a free body with the moments (Figure 4.60). The moments come directly from the slope deflection equation by plugging in a unit δ. Recall that the displaced shape is exaggerated and we are using small displacements, so no axial force is considered. Now by summing moments we can get the end member force.

$$RL + \frac{6EI}{L^2} + \frac{6EI}{L^2} = 0 \tag{4.107}$$

Figure 4.60 Moment summation for shear from unit displacement.

Solving for R, we get

$$R = -\frac{12EI}{L^3} \tag{4.108}$$

For the pinned-end member, we can draw the free body again with the moments shown in Figure 4.61. Summing forces to get the end shear, we have

$$RL + \frac{3EI}{L^2} = 0 \tag{4.109}$$

Solving for the force R, we get

$$R = -\frac{3EI}{L^3} \tag{4.110}$$

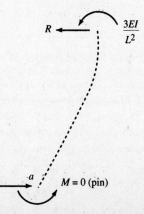

Figure 4.61 Moment summation for shear on pinned member.

Now, using these two shears, we can draw the free body of the beam shown in Figure 4.59. Summing moments about the intersection point, we have

$$K_{33}(13.333)(12) + M_{bc} + M_{cb} - \frac{12EI}{L_{ab}^3}(13.333)(12) - \frac{3EI(1.166)}{L_{cd}^3}(15.55)(12) = 0$$

(4.111)

Substituting the moment equations (4.101) and (4.104), we get

$$K_{33}(160) = -2\left[\frac{6EI(-0.6)}{96^2}\right] + \frac{12EI(160)}{120^3} + \frac{3EI(1.166)(186.6)}{139.9^3} = 15,980.6 \text{ kips/in.}$$

(4.112)

giving

$$K_{33} = 99.9 \text{ kips/in.}$$

(4.113)

Putting the terms into final matrix form, we have

$$\mathbf{K} = \begin{bmatrix} 562,500.0 & 156,250.0 & 195.3 \\ 156,250.0 & 473,329.2 & -1589.0 \\ 195.3 & -1,589.0 & 99.9 \end{bmatrix}$$

(4.114)

This is the final stiffness matrix for the structure. Let's compare it to the one generated by slope deflection equation (4.69). Note that the top two rows are identical. The final row is different, however. The stiffness matrix above is symmetric. This is a requirement for stiffness matrices developed following the definition. While the matrix developed by slope deflection can be considered to be a stiffness matrix, it is not in general symmetric. This is because of the choice of the equilibrium equation for displacement terms. As a result of matrix symmetry, only some of the terms need to be calculated. In most cases, only the moment summation terms would be calculated since they are the easiest (e.g., K_{13}). The symmetric term K_{31}, which requires a more complicated sum, such as that about the intersection point, would just use K_{13}. For the diagonal terms such as K_{33}, the more difficult equation is the only solution.

Condensed Example: Stiffness by Definition—Frame

Given the structure shown at the top of p. 131, find the global stiffness matrix using the stiffness-by-definition method. The first step is to define the DOFs. All possible DOFs will be used as given in the second figure on p. 131. Note that DOF 2 could be eliminated provided that the modified slope deflection equations and appropriate displaced shapes are used. Next, a unit displacement is applied to each DOF, holding the others to zero. A free body of each joint is drawn in order to use the summation of moments as an equilibrium equation to find the rotational stiffness term. The free body of the horizontal member is used for the third equilibrium equation, summation of horizontal forces, to find the translational stiffness term. The next three figures give the displaced shapes and free bodies.

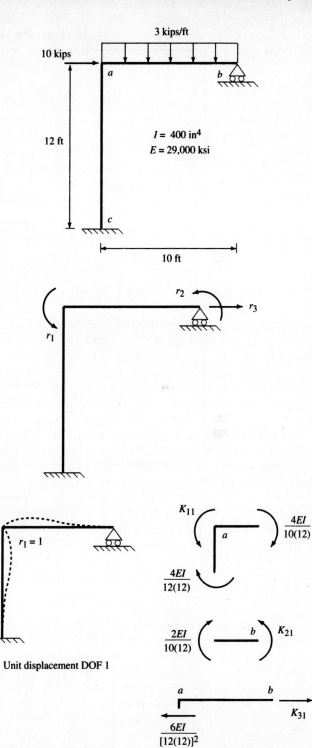

3 kips/ft

10 kips

a

b

12 ft

$I = 400 \text{ in}^4$
$E = 29,000 \text{ ksi}$

c

10 ft

r_2

r_3

r_1

K_{11}

$\dfrac{4EI}{10(12)}$

a

$\dfrac{4EI}{12(12)}$

$r_1 = 1$

Unit displacement DOF 1

$\dfrac{2EI}{10(12)}$

b

K_{21}

a

b

K_{31}

$\dfrac{6EI}{[12(12)]^2}$

Free bodies

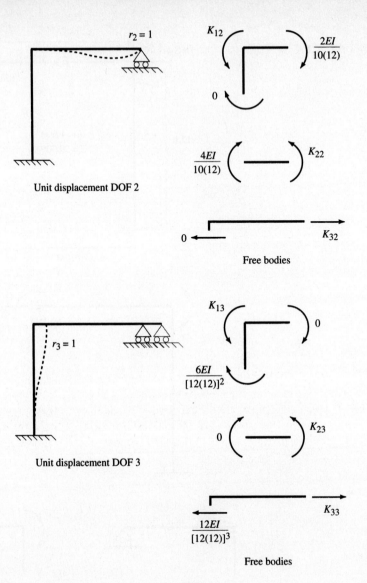

Unit displacement DOF 2

Free bodies

Unit displacement DOF 3

Free bodies

From the free-body diagrams, we can assemble the stiffness matrix directly. It is

$$
\mathbf{K} = \begin{bmatrix}
\dfrac{4EI}{144} + \dfrac{4EI}{120} & \dfrac{2EI}{120} & \dfrac{6EI}{144^2} \\[2ex]
\dfrac{2EI}{120} & \dfrac{4EI}{120} & 0 \\[2ex]
\dfrac{6EI}{144^2} & 0 & \dfrac{12EI}{144^3}
\end{bmatrix}
$$

Condensed Example: Stiffness by Definition—Truss

The stiffness for truss structures can also be formed by using the definition directly. We will find the stiffness for the following truss structure.

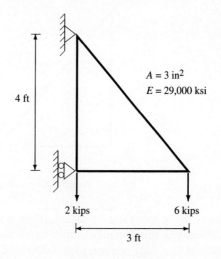

As with any analysis, the first step is to define the DOF for the structure. For truss structures there are only two possible DOFs, translation in the plane. For X–Y coordinate systems, this would be translations in the X and Y directions. Truss members do not have rotational stiffness, and therefore a node *must not* have a rotational DOF. The following DOFs for the structure will be used.

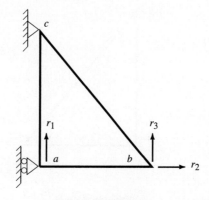

Global DOF definitions

Now, according to the definition, we apply a unit deformation at each DOF and find the forces required to hold that deformed shape. The required displaced shape for DOF

1 is as follows:

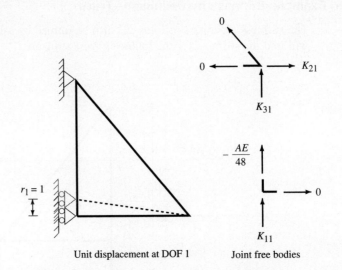

Unit displacement at DOF 1 — Joint free bodies

The stiffness terms are found by taking a free body of the joint and using equilibrium. For truss structures, this becomes simply summation of forces horizontally and vertically. For the DOF above, only one member is deformed and its force is shown. Displacements perpendicular to a member cause no force. The next deformed shape is

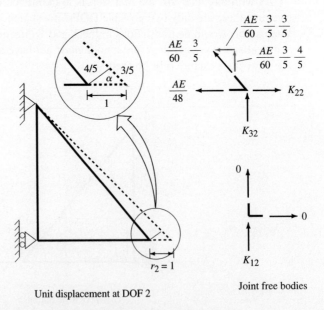

Unit displacement at DOF 2 — Joint free bodies

This displaced shape is slightly more complicated since the displacement is not perpendicular to the slanted member. What is needed is the amount of stretch (or shrinkage) that each member goes through. Again, remember that we are using small-displacement theory and the deformed shapes are greatly exaggerated. Due to small displacements, we assume that the slope of the deformed member is equal to the original member. We draw a perpendicular line from the end of the original member through the new position. The length beyond this line is the amount of extension. This can be calculated by geometry as shown above. Again the stiffness terms are found by equilibrium. Note that the components of the slanted member are used for the equilibrium sums. The final displaced shape and free bodies are

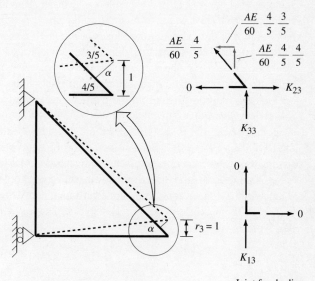

Unit displacement at DOF 3

Joint free bodies

Gathering all the terms into a final matrix, we have

$$
\mathbf{K} = \begin{bmatrix}
\dfrac{AE}{48} & 0 & 0 \\[2ex]
0 & \dfrac{AE}{36} + \dfrac{AE}{60}\left(\dfrac{3}{5}\right)\dfrac{3}{5} & -\dfrac{AE}{60}\left(\dfrac{3}{5}\right)\dfrac{4}{5} \\[2ex]
0 & -\dfrac{AE}{60}\left(\dfrac{3}{5}\right)\dfrac{4}{5} & \dfrac{AE}{60}\left(\dfrac{4}{5}\right)\dfrac{4}{5}
\end{bmatrix}
$$

Substituting in for A and E, we get the final form:

$$
\mathbf{K} = \begin{bmatrix}
1812.5 & 0 & 0 \\
0 & 2938.67 & -696 \\
0 & -696 & 928
\end{bmatrix}
$$

4.8 Structural Loading

In Section 4.7 we discussed how to derive the stiffness matrix by definition. By setting one of the DOFs to a unit value and keeping all the others equal to zero, we found that the terms of the stiffness matrix were equal to the forces required to maintain the deformed shape. Although this generated the stiffness matrix, we still need to look at the load vector. First we must recall that the only types of loads admitted are concentrated loads acting directly on the DOFs. This can be seen by the definition of both the DOF **r** and the load vector **R**. The **r** vector consists of displacements (rotations and translations), and the **R** vector is in a one-to-one correspondence to these. This means that any concentrated load acting at a DOF just needs to be put into the corresponding place in the load vector. The requirement of only concentrated loads at the DOF is too restrictive and must be relaxed. This is true for both distributed loads and concentrated loads not at the DOF. This can easily be done using the same deformed shapes used in deriving the stiffness matrix. Since the shapes were defined as a unit deformation at a single DOF, we can use a work equivalence method for finding the equivalent load value.

Work equivalence says that the work done by the real loads as a result of a displaced shape must be equal to the work by the equivalent loading of only concentrated loads at the DOF. Since the only load considered for the equivalent load is the one corresponding to the DOF where the unit displacement is applied, the load is identically the work done. As an example, for the structure shown in Figure 4.62, only two DOFs are considered, θ_b and δ. If a concentrated moment load is applied, we need to find the equivalent load for the DOF considered. Using the displaced shape for θ_b (Figure 4.63), we see that the real work is

$$W_{\text{real}} = M_c \theta_c \tag{4.115}$$

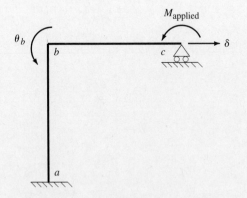

Figure 4.62 Loading example structure.

Notice that for the displaced shape, θ_c has a nonzero value. This is because it is not one of the DOFs and must be left free to rotate. The equivalent work is the desired load for DOF θ_b. This is calculated by

$$W_{\text{equiv}} = M_b \times 1 \tag{4.116}$$

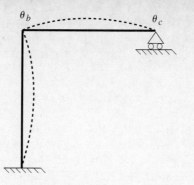

Figure 4.63 Displaced shape for load example.

This is the only unknown force since the other DOF, δ, is not allowed to move, and hence its load generates no work. Now θ_c can be calculated from the slope deflection formula based on a unit rotation of θ_b. Since it is known that the moment M_{cb} is equal to zero, we can solve for θ_c. The required equation is

$$0 = M_{cb} = \frac{2EI}{L_{cb}}\left(2\theta_b + \theta_c + \frac{3\delta}{L_{cb}}\right) \tag{4.117}$$

But $\delta = 0$ and $\theta_b = 1$; therefore, solving for θ_c we get $\theta_c = -\frac{1}{2}$. Setting equation (4.115) equal to (4.116) and substituting for θ_c, we can solve for M_b. This gives

$$R_1 = -\frac{M_c}{2} \tag{4.118}$$

We can do the same for the δ DOF. By applying a unit δ we can see that no rotation occurs at θ_c. Therefore, there is no work generated from the applied moment. The final load vector is, therefore,

$$\mathbf{R} = \begin{bmatrix} -\dfrac{M_c}{2} \\ 0 \end{bmatrix} \tag{4.119}$$

4.9 Distributed Loads

The next step is finding the equivalent loading for a distributed loading on a member. This can be accomplished in two stages. The first stage is to convert all distributed loads into concentrated loads. The second stage is to use the procedure above to convert the concentrated loading to an equivalent loading. Converting distributed loads into equivalent concentrated loads follows the same procedure as above. We are looking for a work equivalent set of loads for the distributed load. This means a set of concentrated loads that causes the same amount of work as the distributed loads. This is a little more complex than before since it now involves integration.

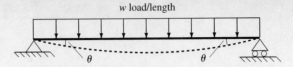

Figure 4.64 Uniformly loaded simple beam.

Take the example of a simply supported beam with a uniform load (Figure 4.64). The only DOFs are the two end rotations. We need to find a set of moments that causes the same work as the distributed loads. We could integrate the displacement times load to get the work caused by the distributed load. However, there is a simpler way to determine these values. We know that for a uniform load on the structure above, the ends will rotate an amount θ. We need to find the concentrated loads that will cause the same end rotations, hence the same work. But we already know these values; they are the fixed-end moments. Clearly, the fixed-end moments are the exact forces required to cause no rotation at the ends. We can show this using the structure shown in Figure 4.65.

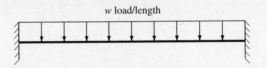

Figure 4.65 Uniformly loaded fixed-end beam.

Using a flexibility approach, we know that we can remove the moments and then solve for the moments that push the displacement back to zero to get the redundant support moments. But since we work with linear elastic structures, the load that causes the final displacement to be zero is also the load that causes the desired rotations. Therefore, the work equivalent loads are exactly the negative of the fixed-end forces. As a result, for the simple beam, the equivalent concentrated loading would be as shown in Figure 4.66. As can be seen, the final displacements will be identical between these loads and the distributed loads.

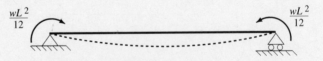

Figure 4.66 Equivalent fixed-end moments.

4.9.1 Distributed Load Example

Let's find the load vector for the structure for which we developed the stiffness matrix in Section 4.8. Since we have translations as well as rotations, we will need to include the fixed-end shears as well as the moments in the load calculations. The structure and load are shown in Figure 4.67. For a triangularly distributed load, we will need to solve the structure shown in Figure 4.68 for the end shears. The fixed-end moments are calculated using Table 4.1. Next, using statics the end reactions are solved for and shown in Figure

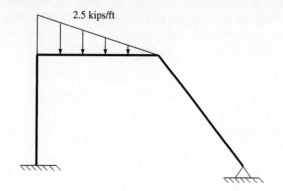

Figure 4.67 Distributed load example.

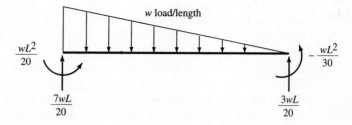

Figure 4.68 Fixed-end forces for triangular load.

4.68. From these reaction forces we can convert the distributed load on the structure to the concentrated loads as shown in Figure 4.69.

Now we have to use the displaced shapes to find the work equivalent loads. From the displaced shape with a unit rotation at DOF 1 (Figure 4.43) we have the work equation

$$R_1 = -\frac{wL^2}{20}(1) = \frac{-(2.5/12)(96)^2}{20} = -96 \text{ kip-in.} \tag{4.120}$$

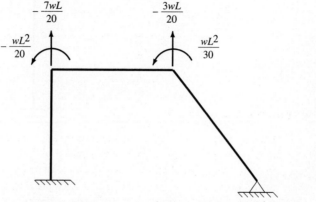

Figure 4.69 Equivalent concentrated loads.

Note that the only movement is the actual unit rotation at DOF 1. Next, from the displaced shape for DOF 2 (Figure 4.51) we have the work equation

$$R_2 = \frac{wL^2}{30}(1) = \frac{(2.5/12)(96)^2}{30} = 64 \text{ kip-in.} \tag{4.121}$$

Again, only the unit rotation contributes to the work. Finally, from the displaced shape for DOF 3 (Figure 4.56) we have the work equation

$$R_3 = -\frac{3wL}{20}[0.6(1)] = \frac{-3(2.5/12)96(0.6)}{20} = -1.8 \text{ kips} \tag{4.122}$$

For DOF 3, notice how the vertical displacement has a work component as a result of the horizontal displacement. This is clearly caused by the slanted member linking the vertical and horizontal DOFs. In addition, since the pinned support is not considered as a DOF, any loads applied to it would have been included in the calculation for R_2 and R_3 since rotations occur there in the displaced shapes. As a result, we can see that at joints, where there are rotational DOFs, any applied moments only contribute load to the DOFs where they are applied. However, at pins, where the DOFs are left out using the modified equation, moments will contribute to other forces. Also notice how, again, the slanted member causes normally vertical forces to have an effect on the horizontal result.

4.10 Problems

4.1. For the following structures, show the required DOFs needed to analyze the structure and draw the deflected shapes. Assume no axial deformations.

(a)

(b)

(c)

(d)

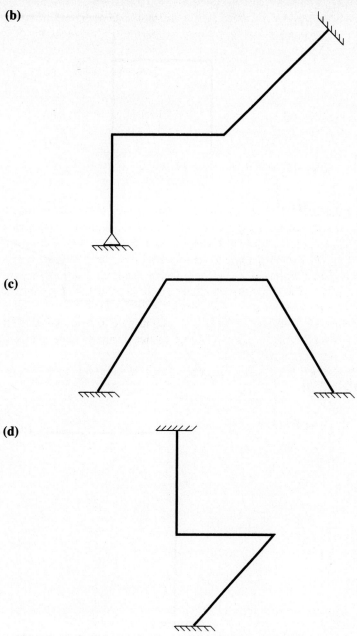

(e)

(f)

(g)

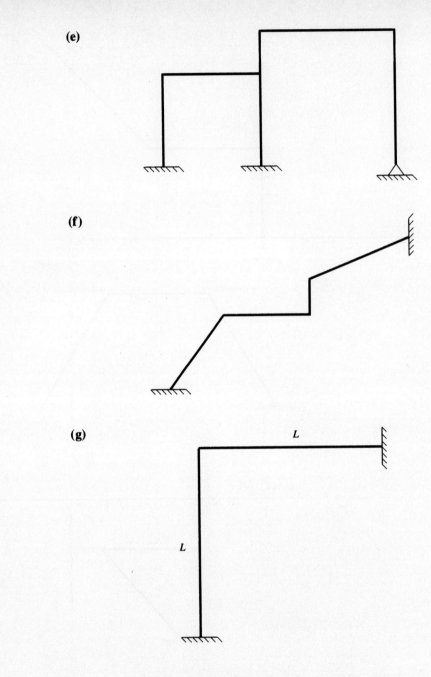

(h)

(i)

(j)

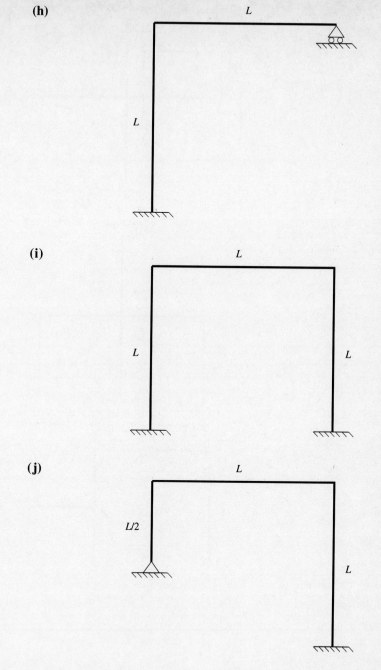

(k)

(l)

(m)

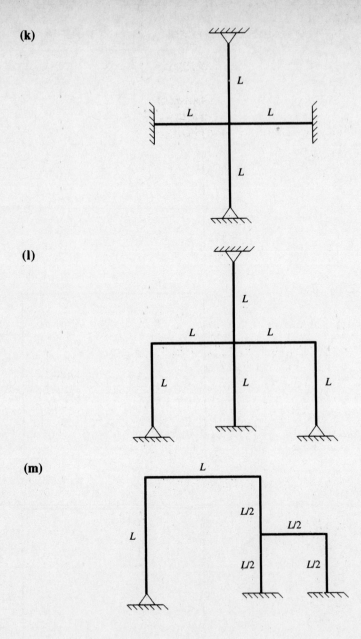

(n)

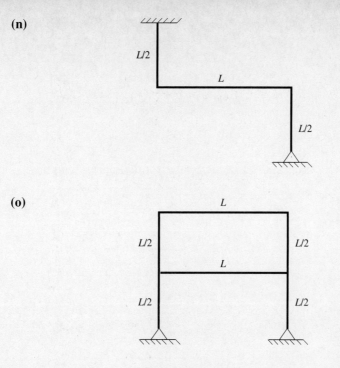

(o)

4.2. Using slope deflection, find the joint rotations and support moments.

(a)

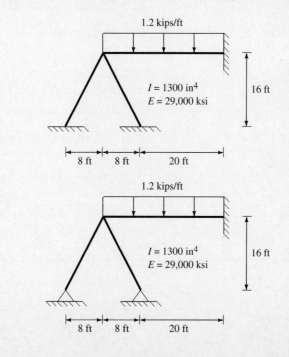

(c)

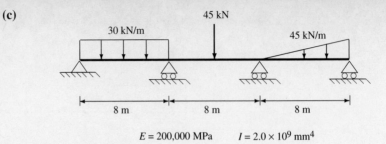

$$E = 200,000 \text{ MPa} \qquad I = 2.0 \times 10^9 \text{ mm}^4$$

4.3. Using slope deflection, solve for the support moments.

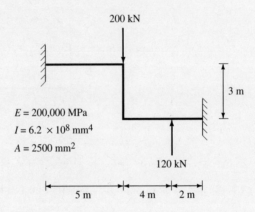

4.4. Using slope deflection, solve for the support reactions and the moment at the corner joint.

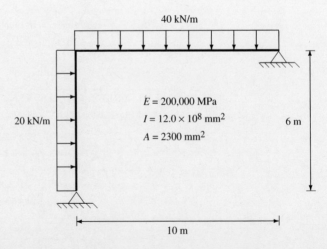

4.5. Using slope deflection, find the moments at the fixed supports and the rotations at the two rollers.

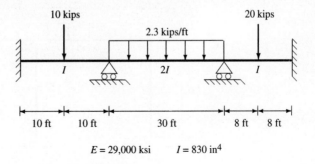

$E = 29,000$ ksi $I = 830$ in^4

4.6. For the following structure, find θ_b and θ_c.

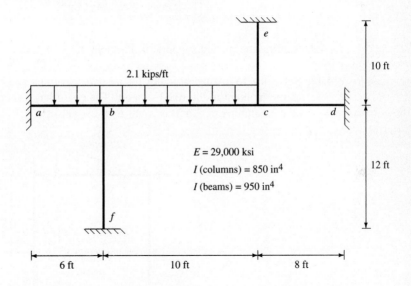

$E = 29,000$ ksi

I (columns) $= 850$ in^4

I (beams) $= 950$ in^4

4.7. For the frame, find the drift (displacement at top of frame). The moment of inertia is 900 in^4 for the columns and 1200 in^4 for the beams; $E = 29,000$ ksi.

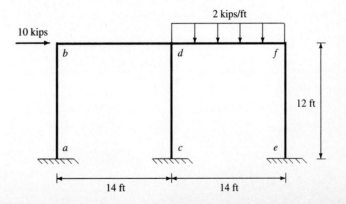

4.8. For the frame with the slanted member, find the rotations θ_b and θ_c. Members *de* and *ef* have $I = 2400$ in⁴, members *ab*, *bd*, and *dc* have $I = 1200$ in⁴; $E = 29,000$ ksi.

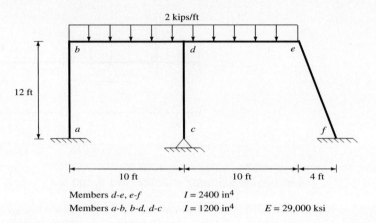

Members *d-e, e-f* $I = 2400$ in⁴
Members *a-b, b-d, d-c* $I = 1200$ in⁴ $E = 29,000$ ksi

4.9. Replace the pinned supports in Problem 4.4 with fixed supports and re-solve.

4.10. Solve the following structure for the two rotations and one displacement as well as the moment at the fixed support. The beam has a moment of inertia twice that of the columns, $I = 450$ in⁴, and $E = 29,000$ ksi.

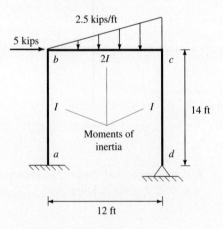

4.11. Solve the following structure for the translation at the top left corner and the support forces.

$E = 200{,}000$ MPa

$I = 7.28 \times 10^8$ mm^4

$A = 1500$ mm^2

30 kN/m

4 m

4 m

12 m

4.12. Using stiffness by definition, find the stiffness for the following structure. Sketch the final deflected shape and find the moment at b for the three load cases.

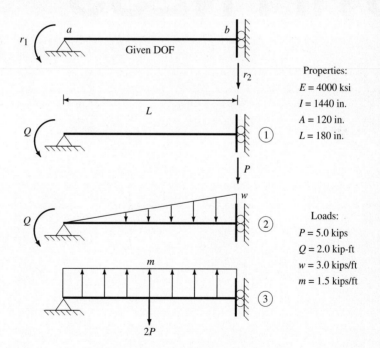

Given DOF

r_1

a b

r_2

L

Q ①

P

w

Q ②

m

$2P$ ③

Properties:

$E = 4000$ ksi

$I = 1440$ in.

$A = 120$ in.

$L = 180$ in.

Loads:

$P = 5.0$ kips

$Q = 2.0$ kip-ft

$w = 3.0$ kips/ft

$m = 1.5$ kips/ft

4.13–4.23. Using the stiffness-by-definition method, find the stiffness for the structures in Problems 4.2 through 4.11.

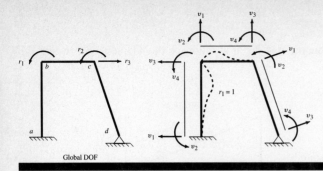

Global DOF

DIRECT STIFFNESS

The stiffness for any structure can be derived by definition. However, this is not the best procedure for hand solutions of general structures. The work required to generate the matrix is not much easier than slope deflection. As a computer method, stiffness by definition is clearly deficient. What is needed is a more systematic method for creating the stiffness and load matrices. This systematic method involves matrix algebra and develops an easier way to create the final set of equilibrium equations.

5.1 Stiffness Transformation Matrix

We start with some definitions and derivations for the required equations. Then we apply the matrix equations to the solution of structural analysis problems. The first definition is the set of element deformations, **v**.

v *is the set of element deformations.*

For the beam element, consistent with the one we have been using from slope deflection formulas, we have the DOFs shown in Figure 5.1 (p. 152). Note that these element deformations are concentrated translations and rotations usually at the ends of an element.

If we look at the same structure as previous examples (Figure 5.2 p. 152), we also have a set of global displacements or degrees of freedom for the structure, **r**. For any element in the structure, there exists a relationship between the element DOFs and the global DOFs. This relationship can be expressed as

$$\mathbf{v} = \mathbf{ar} \tag{5.1}$$

where **a** is defined as the global-to-local transformation matrix. This is what is called the kinematics of the problem, or the displacement relationships. The **a** matrix must be defined using displaced shapes. The displaced shapes are the same ones used for defining the stiffness matrix. From the equation above we can see that the **a** matrix converts a set of global displacements, **r**, into the local or element displacements, **v**, for a particular element.

Similarly, we can define a set of element forces, **S**, that corresponds to the element deformations. There is a one-to-one correspondence between the element deformations and element forces. Therefore, the element forces are defined as shown in Figure 5.3. A similar equation can be written that relates the element forces, **S**, to the global forces, **R**.

$$\mathbf{S} = \mathbf{bR} \tag{5.2}$$

Figure 5.1 Local element DOF definitions.

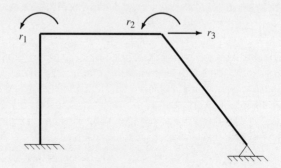

Figure 5.2 Direct stiffness example structure.

Figure 5.3 Local element force definitions.

Therefore, **b** is a transformation matrix that converts global forces, **R**, to the local element forces, **S**. This is an equilibrium relationship. This relationship can be found through the use of statics.

There exists a relationship between the two transformations, **a** and **b**. This can be found using virtual work, the virtual displacements version. The external virtual work and internal virtual work for a structure can be equated and expressed as

$$\delta W_e = \bar{\mathbf{r}}^{\mathsf{T}} \mathbf{R} = \bar{\mathbf{v}}^{\mathsf{T}} \mathbf{S} = \delta W_i \tag{5.3}$$

where the overbar represents "virtual." Substituting for $\mathbf{v}^{\mathsf{T}}$ on the right-hand side using equation (5.1) and canceling $\mathbf{r}^{\mathsf{T}}$ from both sides since it is arbitrary, we get

$$\mathbf{R} = \mathbf{a}^{\mathsf{T}} \mathbf{S} \tag{5.4}$$

Plugging equation (5.2) into (5.4) for **S**, we get

$$\mathbf{R} = \mathbf{a}^{\mathsf{T}} \mathbf{b} \mathbf{R} \tag{5.5}$$

But since the only thing that can be multiplied times a vector and return the vector is the identity matrix, we have

$$\mathbf{a}^\mathsf{T}\mathbf{b} = \mathbf{I} \tag{5.6}$$

We can perform the same process using virtual forces and get the relationship

$$\mathbf{b}^\mathsf{T}\mathbf{a} = \mathbf{I} \tag{5.7}$$

If the **a** and **b** matrices are square, which is true for determinate structures, we see that $\mathbf{a}^{-1}$ equals **b** and $\mathbf{b}^{-1}$ equals **a**. Therefore, the transformation that relates displacements is equal to the transpose of the transformation that relates the forces. This property is called the *contragradient law*.

5.1.1 Definition of the a Matrix

Now let's look at how we define the transformation matrix **a**. By definition, this matrix transforms values in the global coordinate system to values in the local coordinate system. This means that we need to define a local coordinate system. There are only two possibilities for an element. For a horizontal element the two choices are shown in Figure 5.4. Either choice is equally acceptable. The only difference is that the results will have different signs. For the structure used previously there are three element coordinate systems, one for each element, that relate to the three global DOFs. This means that there are three **a** matrices, one for each element. In addition, from equation (5.1) we can see that the size of **a** is the number of element DOFs (EDOFs) by the number of global DOFs (GDOFs). For our problem each **a** matrix is 4×3.

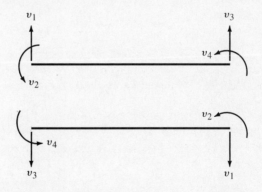

Figure 5.4 Possible horizontal element directions.

Using the same process as deriving the stiffness matrix by definition, we can apply a unit displacement to one DOF, leaving all others zero. Note that setting a displacement equal to 1 gives a column of the **a** matrix just like the stiffness matrix. Before we can do this we need to choose the element definitions. The three displaced shapes for the structure and the chosen element coordinate system are shown in Figure 5.5. Let's find the **a** matrix for the left-hand column of the structure. From the first displaced shape, we can get the terms for the first column. From the figure we can see that the value of the

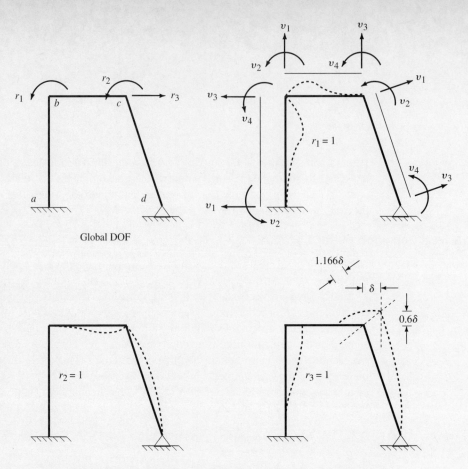

Figure 5.5 Local element definitions and unit displacements.

four element deformations are

$$
\begin{bmatrix} v_1 \\ v_2 \\ v_3 \\ v_4 \end{bmatrix} = \begin{bmatrix} 0 \\ 0 \\ 0 \\ 1 \end{bmatrix}
\tag{5.8}
$$

These values come from just taking the appropriate values from the displaced shapes. The values become the first column of the **a** matrix for this element. From the other displaced shapes we can fill out the other columns. The final matrix for the left column becomes

$$
\mathbf{a}(\text{left column}) = \begin{bmatrix} 0 & 0 & 0 \\ 0 & 0 & 0 \\ 0 & 0 & -1 \\ 1 & 0 & 0 \end{bmatrix}
\tag{5.9}
$$

Using the same procedure, we can get the **a** matrices for the other two elements:

$$\mathbf{a}(\text{beam}) = \begin{bmatrix} 0 & 0 & 0.6 \\ 1 & 0 & 0 \\ 0 & 0 & 0 \\ 0 & 1 & 0 \end{bmatrix} \tag{5.10}$$

$$\mathbf{a}(\text{right column}) = \begin{bmatrix} 0 & 0 & 1.1660 \\ 0 & 1 & 0 \\ 0 & 0 & 0 \\ 0 & -0.5 & -\dfrac{3}{2L}(1.166) \end{bmatrix} \tag{5.11}$$

The terms in the **a** matrix for the slanted column need to be explained. The second column is from the second displaced shape. Note that the pinned end is free to move when a displacement is applied. This is because the pinned-end DOF is not included in the equations, just like the modified equation. As a result, we need to find the rotation at that joint due to a unit rotation at the top. This can be found from the slope deflection formula. Since we know the moment at a pin is equal to zero, we can write

$$0 = M_{dc} = \frac{2EI}{L_{dc}}\left(2\theta_d + \theta_c + \frac{3\delta}{L_{dc}}\right) \tag{5.12}$$

solving for θ_d in terms of θ_c, knowing that $\delta = 0$, we get

$$\theta_d = \frac{\theta_c}{2} \tag{5.13}$$

The third column comes from the third displaced shape. We know from similar triangles that the perpendicular displacement has the value 1.166. We also need to find the rotation at d. Again using the slope deflection equation and the fact that the moment equals zero, we can solve for the rotation θ_d. Note that θ_c is zero by definition since it is a DOF and must be held to zero when a unit displacement for δ is given. Again from the slope deflection formula, equation (5.12), we get

$$2\theta_d = \frac{3\delta}{L_{dc}} \tag{5.14}$$

Solving for θ_c gives

$$\theta_c = \frac{3\delta}{2L_{dc}} \tag{5.15}$$

It is easy to develop the **a** matrices for any structure since they are just geometric values and do not require the formation of any equilibrium equations. The deformed shapes are required, however, for a unit displacement at only a single DOF at a time. From the development above, we can write the definition of the **a** matrix:

a_{ij} *is the displacement at element DOF i due to a unit displacement at global DOF j with all other global DOF equal to zero. (Note: There is one matrix for each element.)*

We now need to continue with some additional derivations to show how to use the **a** matrix and develop the structural stiffness matrix.

5.2   Derivation of Stiffness by Direct Stiffness

The **a** matrix derived in Section 5.1 is a transformation that relates the global displacements of the structure to the local (element) deformations. It can also be thought of as converting the deformations from the global to the local coordinate system. Once we have this relationship, we can derive the stiffness matrix based on this transformation matrix.

   Using the virtual displacements principle, we can equate internal and external virtual work. This is identical to all the derivations done so far. A virtual displacement is applied to the structure. This means that an arbitrary and small set of displacements are applied to the global DOFs. Clearly, these can be transformed to the element DOFs using the **a** matrix. Note that it is only the displacements that are virtual since a transformation is independent of the displacements applied. Therefore, we can write

$$\bar{\mathbf{v}} = \mathbf{a}\bar{\mathbf{r}} \qquad (5.16)$$

The external virtual work is the product of the global loads through the global displacements. For the virtual displacements principle this is written as

$$\delta W_e = \bar{\mathbf{r}}^{\mathrm{T}}\mathbf{R} \qquad (5.17)$$

The internal virtual work is the sum of the work from all the elements in a structure. The work of an individual element is equal to the element displacements times the element forces. The relationship is given as

$$\delta W_i = \sum_{\text{all elements}} \bar{\mathbf{v}}^{\mathrm{T}}\mathbf{S} \qquad (5.18)$$

   We know that we can relate the element displacements to the element forces. This is simply defining a single element as the structure and using stiffness by definition to form the stiffness matrix for the element. This relationship is called the *element stiffness matrix*. We have seen parts of this from the slope deflection equations, where moments are related to the end rotations and displacements. The general form for this is

$$\mathbf{S} = \mathbf{K}_e\mathbf{v} \qquad (5.19)$$

where $\mathbf{K}_e$ is the element stiffness matrix that relates the element DOFs to the element forces. This is the standard form of the stiffness for a structure, where in this case the structure is a single element. Substituting equation (5.19) into (5.18), we get

$$\delta W_i = \sum_{\text{all elements}} \bar{\mathbf{v}}^{\mathrm{T}} \mathbf{K}_e \mathbf{v} \tag{5.20}$$

Substituting equation (5.16) and its transpose into (5.20) and equating internal to external work, we have

$$\bar{\mathbf{r}}^{\mathrm{T}} \mathbf{R} = \sum_{\text{all elements}} \bar{\mathbf{r}}^{\mathrm{T}} \mathbf{a}^{\mathrm{T}} \mathbf{K}_e \mathbf{a} \mathbf{r} \tag{5.21}$$

Since $\bar{\mathbf{r}}^{\mathrm{T}}$ is both arbitrary and small, we can cancel it from both sides of the equation. In addition, we see that $\mathbf{r}$ is a constant in the summation and can be taken outside the summation. This leaves the final form of

$$\mathbf{R} = \left(\sum_{\text{all elements}} \mathbf{a}^{\mathrm{T}} \mathbf{K}_e \mathbf{a} \right) \mathbf{r} \tag{5.22}$$

Since this is the equilibrium relationship for a structure, it is clear that it must conform to the general form for a stiffness solution. Therefore, the global stiffness must be

$$\mathbf{K} = \left(\sum_{\text{all elements}} \mathbf{a}^{\mathrm{T}} \mathbf{K}_e \mathbf{a} \right) \tag{5.23}$$

It is important to recall the sizes for each of the matrices in equation (5.23). The **a** matrix is the number of element DOFs (EDOFs) by the number of global DOFs (GDOFs). The $\mathbf{K}_e$ matrix is the number of element DOFs (EDOFs) by the number of element DOFs (EDOFs). This gives the final global stiffness matrix as GDOFs by GDOFs.

5.3 Two-Dimensional Beam Element Stiffness Derivation

We now need to fill in the terms of the $\mathbf{K}_e$ matrix. Since this is just the stiffness matrix for a single element, we need only apply the definition. According to the definition, we apply a unit displacement at each DOF and find the forces to hold the structure in place. For the given element the displaced shapes shown in Figure 5.6 are needed. Since in all these cases, the displacement is unity, the moments come directly from slope deflection (just as were used in stiffness by definition). The shear forces can then be determined from statics. For each of the shapes of Figure 5.6, the end forces are shown in Figure 5.7. Putting these forces into the correct places in the stiffness matrix, we have the final form for the element stiffness for the beam given in Figure 5.1.

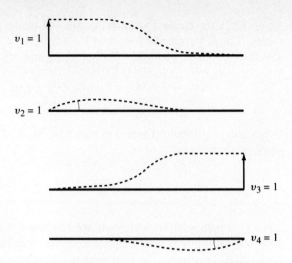

Figure 5.6 Deformed shapes for element stiffness definition.

$$\mathbf{K}_e = \begin{bmatrix} \dfrac{12EI}{L^3} & \dfrac{6EI}{L^2} & -\dfrac{12EI}{L^3} & \dfrac{6EI}{L^2} \\[2mm] \dfrac{6EI}{L^2} & \dfrac{4EI}{L} & -\dfrac{6EI}{L^2} & \dfrac{2EI}{L} \\[2mm] -\dfrac{12EI}{L^3} & -\dfrac{6EI}{L^2} & \dfrac{12EI}{L^3} & -\dfrac{6EI}{L^2} \\[2mm] \dfrac{6EI}{L^2} & \dfrac{2EI}{L} & -\dfrac{6EI}{L^2} & \dfrac{4EI}{L} \end{bmatrix} \qquad (5.24)$$

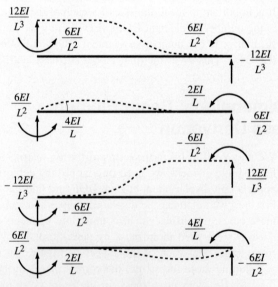
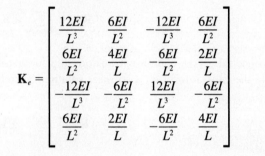

Figure 5.7 Forces required for unit deformations.

5.4   Direct Stiffness Solution Procedure

The solution process becomes more systematic and mechanical when using this formulation. The only hand work comes from defining the geometric relationship of the transformation matrix **a**. The process can now be reduced to the following steps:

1. Define the structural (global) DOFs (**r**).

2. Find the **a** matrix for each element. This is done by applying a unit displacement to each DOF, holding all others still, and finding the corresponding element displacements. This fills a column of the **a** matrix for each displacement applied. Repeat until all columns of all **a** matrices are filled.

3. Sum $\mathbf{a}^T\mathbf{K}_e\mathbf{a}$ for all elements in the structure.

4. Form the load vector **R**.

5. Solve the set of equations $\mathbf{Kr} = \mathbf{R}$.

6. Recover the element forces.

There are still two steps that need to be explained to complete the solution process. This first is formation of the load vector **R**. The load vector can be considered to be composed of two parts, the concentrated loads applied directly to global DOFs and the equivalent loads that were calculated previously by equivalent work. The concentrated loads again are just placed into the corresponding location for the DOF at which they are applied.

5.5   Beam Element Loads

In Section 4.9, distributed loads were first converted into concentrated loads equal to the negative of the fixed-end forces. Then an equivalent work procedure was applied to convert these into loads applied to the global DOF. Distributed loads can also be handled by matrix operations. Again the loads are first converted into the concentrated equivalent fixed-end forces. But now we can use the transformation matrix to convert these directly into the equivalent loads using the relationship

$$\mathbf{R} = \mathbf{a}^T\mathbf{S} \tag{5.25}$$

This equation transforms the element end forces into the global loads. We define a new set of element loads, $\mathbf{S}^f$, as the equivalent loads for a load applied within the span of an element. Clearly, these forces are just the negative of the fixed-end forces:

$$\mathbf{S}^f = -(\text{fixed-end forces}) \tag{5.26}$$

Since the equation is general, it will clearly convert these end forces into global loads. It can be shown that this equation is identical to finding the work equivalent loads for a deflected shape. Note that the global loads in equation (5.25) are only the contribution of the element considered. Therefore, all the elements with internal (distributed) loads need to be included and hence their individual contributions must be summed. Therefore, the equation to convert all element loads to global is

$$\mathbf{R} = \sum_{\text{all elements}} \mathbf{a}^{\mathrm{T}} \mathbf{S}^f \tag{5.27}$$

5.6   Beam Element Force Recovery

The only step left to be explained is the final step of the procedure, force recovery. This step takes the final structural displacements, $\mathbf{r}$, and converts them into element end forces. The equation to use is

$$\mathbf{S} = \mathbf{K}_e \mathbf{v} \tag{5.28}$$

If we substitute equation (5.1) for $\mathbf{v}$ into (5.28), we get

$$\mathbf{S} = \mathbf{K}_e \mathbf{a} \mathbf{r} \tag{5.29}$$

This equation directly converts the global DOF displacements into element forces in the local element directions. This equation must be performed for each element in the structure for which forces are desired. The combination of terms $\mathbf{K}_e \mathbf{a}$ is usually referred to as the *element force transformation matrix*. Notice that this is also part of the multiplication required in forming equation (5.23). If we give the force transformation matrix a new symbol, $\mathbf{FT}$, we can rewrite (5.23) as

$$\mathbf{K} = \sum_{\text{all elements}} \mathbf{a}^{\mathrm{T}}(\mathbf{FT}) \tag{5.30}$$

where

$$\mathbf{FT} = \mathbf{K}_e \mathbf{a} \tag{5.31}$$

This is the usual method followed by linear structural analysis programs. This is because it is more efficient to multiply the matrices only once to save computer effort.

There is one additional correction that needs to be applied to equation (5.29) when the applied load includes the fixed-end forces generated by distributed loads. This is best explained by an example. Given a simple beam with a uniform distributed load, we want to find the moments at the end of the beam. The structure is shown in Figure 5.8. To solve this structure we follow the standard procedure. We first convert the distributed load to the equivalent loads (opposite the fixed-end forces), which are given in Figure 5.9. Next, we form the stiffness matrix for the structure and generate the final equations:

$$\frac{EI}{L} \begin{bmatrix} 4 & 2 \\ 2 & 4 \end{bmatrix} \begin{bmatrix} r_1 \\ r_2 \end{bmatrix} = \begin{bmatrix} -\dfrac{wL^2}{12} \\ \dfrac{wL^2}{12} \end{bmatrix} \tag{5.32}$$

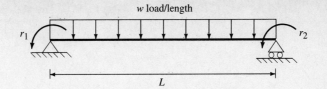

Figure 5.8 Uniform load example.

Figure 5.9 Equivalent concentrated loads.

Solving this set of simultaneous equations for the displacements, we get

$$r_1 = -r_2 = -\frac{wL^3}{24EI} \tag{5.33}$$

Performing the final step of force recovery using only equation (5.29), we get

$$\mathbf{S} = \frac{EI}{L}\begin{bmatrix} 4 & 2 \\ 2 & 4 \end{bmatrix}\begin{bmatrix} 1 & 0 \\ 0 & 1 \end{bmatrix}\begin{bmatrix} r_1 \\ r_2 \end{bmatrix} \tag{5.34}$$

Note that the **a** matrix is the identity matrix since the local element DOFs are identical to the global DOF. If we plug in the values for r_1 and r_2, we get

$$\mathbf{S} = \begin{bmatrix} -\dfrac{wL^2}{12} \\ \dfrac{wL^2}{12} \end{bmatrix} \tag{5.35}$$

But we know that the moments on the end of a simply supported beam are equal to zero. The error lies in the fact that the answers are correct for the structure solved. The actual structure solved was shown with the concentrated loads, Figure 5.9.

For this loading there are end moments that are exactly equal to those recovered above. However, this is not the original problem. This is the converted problem since distributed loads have to be accounted for by an equivalent load. To correct these recovered forces, we just need to subtract off the equivalent loads applied to the structure and add back the original distributed load. Therefore, the final form of the equation for recovering element loads is

$$\mathbf{S} = \mathbf{K}_e \mathbf{a r} - \mathbf{S}^f = (\mathbf{F T})\mathbf{r} - \mathbf{S}^f \tag{5.36}$$

5.7 Final Direct Stiffness Solution Procedure

The final solution procedure can be summarized as follows:

1. Define the global DOF, $\mathbf{r}$, for the structure.

2. Find the $\mathbf{a}$ matrix for each element in the structure.

3. Sum $\mathbf{a}^T\mathbf{K}_e\mathbf{a}$ for all elements.

4. Form the load, $\mathbf{R} = \mathbf{R}_{\text{concentrated}} + \Sigma\, \mathbf{a}^T\mathbf{S}^f$.

5. Solve $\mathbf{Kr} = \mathbf{R}$.

6. Recover the element force, $\mathbf{S} = \mathbf{K}_e\mathbf{ar} - \mathbf{S}^f$.

Using this procedure, general structures can be analyzed using the direct stiffness method. Now only the geometric relationship for developing the transformation matrix, $\mathbf{a}$, needs to be done by hand. All the other work is accomplished by linear algebra. Also notice that the element stiffness matrix never has to be rederived—only different properties used.

5.8 Direct Stiffness Example

As an example of use of the direct stiffness method, the structure shown in Figure 5.10 will be solved using the direct stiffness method. The first step is to choose the global DOFs. If we include both rotations and the single displacement, we have three DOFs. They are shown in Figure 5.11. The second step is to form the $\mathbf{a}$ matrices for each element. A subscript will be used to denote the element to which the $\mathbf{a}$ matrix belongs. The three displaced shapes and the element coordinate systems we need for filling out the $\mathbf{a}$ matrices are given in Figures 5.12, 5.13, and 5.14.

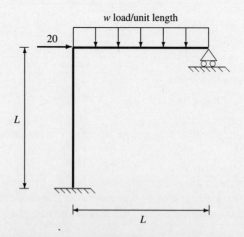

Figure 5.10 Direct stiffness example.

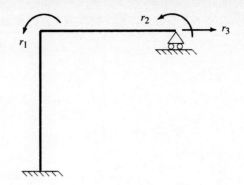

Figure 5.11 DOF for direct stiffness example.

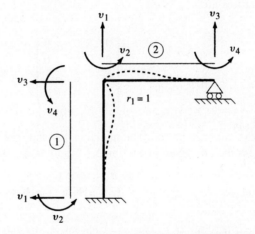

Figure 5.12 Unit displacement at DOF 1.

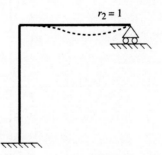

Figure 5.13 Unit displacement at DOF 2.

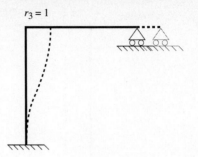

Figure 5.14 Unit displacement at DOF 3.

The element coordinate systems are chosen and shown on the first displaced shape. Looking at the displaced shapes, we can fill in the terms of the **a** matrices. The required matrices are

$$\mathbf{a}_1 = \begin{bmatrix} 0 & 0 & 0 \\ 0 & 0 & 0 \\ 0 & 0 & -1 \\ 1 & 0 & 0 \end{bmatrix} \qquad \mathbf{a}_2 = \begin{bmatrix} 0 & 0 & 0 \\ 1 & 0 & 0 \\ 0 & 0 & 0 \\ 0 & 1 & 0 \end{bmatrix} \tag{5.37}$$

Next, we must perform the muliplication and then sum the results for each element. The $\mathbf{K}_e$ is the one given in Section 5.7. The final stiffness is then found using the equation

$$\mathbf{K} = \mathbf{a}_1^T \mathbf{K}_e \mathbf{a}_1 + \mathbf{a}_2^T \mathbf{K}_e \mathbf{a}_2 = \mathbf{K}_1 + \mathbf{K}_2 \tag{5.38}$$

where the subscript again represents which element produced the contribution to the global stiffness. If we perform the multiplication, we get the contribution of each element to the global stiffness:

$$\mathbf{K}_1 = \begin{bmatrix} \dfrac{4EI}{L} & 0 & \dfrac{6EI}{L^2} \\ 0 & 0 & 0 \\ \dfrac{6EI}{L^2} & 0 & \dfrac{12EI}{L^3} \end{bmatrix} \tag{5.39}$$

$$\mathbf{K}_2 = \begin{bmatrix} \dfrac{4EI}{L} & \dfrac{2EI}{L} & 0 \\ \dfrac{2EI}{L} & \dfrac{4EI}{L} & 0 \\ 0 & 0 & 0 \end{bmatrix} \tag{5.40}$$

The final stiffness matrix is the sum of these two element contributions:

$$
\mathbf{K} = \begin{bmatrix}
\dfrac{8EI}{L} & \dfrac{2EI}{L} & \dfrac{6EI}{L^2} \\[3mm]
\dfrac{2EI}{L} & \dfrac{4EI}{L} & 0 \\[3mm]
\dfrac{6EI}{L^2} & 0 & \dfrac{12EI}{L^3}
\end{bmatrix}
\tag{5.41}
$$

If we look carefully at the results, we can learn some things about the structure from the matrices. A diagonal term in a stiffness matrix says there is an element connected to that DOF. For a structure to be stable, there should always be a nonzero diagonal in the global stiffness. The element stiffness may, however, have zero-diagonal stiffness terms. This is because an element does not connect to all the DOFs of a structure.

An off-diagonal term in a stiffness matrix means that an element is linking the two DOFs where the term is located. For example, the K_{12} has a nonzero value. This means that some element (or elements) connects DOF 1 and DOF 2. We can look at the element contributions to see which one added to this term. Clearly, it is element 2 that links the two rotations. Linking a DOF says that if there is a force at one DOF, there will be one at the other. We also see that the K_{23} term is zero. This means that DOF 2 and DOF 3 are not connected and can occur independently of one another.

Next we need to find the load vector $\mathbf{R}$. First to consider are the concentrated loads acting directly at the DOF. For our problem there is only one 20-kip load that corresponds to DOF 3. Therefore, the contribution to the load vector due to concentrated load is

$$
\mathbf{R}_{\text{concentrated}} = \begin{bmatrix} 0 \\ 0 \\ 20 \end{bmatrix}
\tag{5.42}
$$

Next we need to include the contribution to the load vector from loads within the span of an element. Since only element 2 has a load within the span, it is the only one we need to work with. The procedure says first to create the element concentrated load vector $\mathbf{S}^f$. Recall that these are the negative of the fixed-end forces. Since our load is a uniform load, the vector is

$$
\mathbf{S}_2^f = \begin{bmatrix}
-\dfrac{wL}{2} \\[3mm]
-\dfrac{wL^2}{12} \\[3mm]
-\dfrac{wL}{2} \\[3mm]
\dfrac{wL^2}{12}
\end{bmatrix}
\tag{5.43}
$$

The equation to convert this into the element's contribution to the global load vector is

$$\mathbf{R}_2 = \mathbf{a}_2^{\mathsf{T}} \mathbf{S}_2^f \tag{5.44}$$

If we multiply out the matrices, we get

$$\mathbf{R}_2 = \begin{bmatrix} -\dfrac{wL^2}{12} \\ \dfrac{wL^2}{12} \\ 0 \end{bmatrix} \tag{5.45}$$

This result shows that the element contributes directly to the first two DOFs. The final form of the load vector is

$$\mathbf{R} = \mathbf{R}_{\text{concentrated}} + \mathbf{R}_1 + \mathbf{R}_2 \tag{5.46}$$

which gives a final value of

$$\mathbf{R} = \begin{bmatrix} -\dfrac{wL^2}{12} \\ \dfrac{wL^2}{12} \\ 20 \end{bmatrix} \tag{5.47}$$

If we solve the set of equations, we can then recover the element end forces (shears and moments) using the force transformation matrix. Remember to subtract the $\mathbf{S}^f$ for element 2 since it has a distributed load.

This type of procedure is very practical when using the CAL-90 program. The **a** matrix is fairly easy to form. The element stiffness matrix, $\mathbf{K}_e$, is the same form for all elements, just needing values for E, I, and L. The loads are also formed very easily. The rest of the procedure consists of matrix multiplication, transpose, and solution, which are all handled by CAL-90. As a result, all of the homework assignments should be done using CAL-90. Assuming that $E = 29,000$ ksi, $I = 850$ in⁴, $L = 120$ in., and $w = 2.3$ kips/ft, the CAL-90 commands to perform the example analysis are as follows:

```
EXAM1
ZERO K R=3 C=3    : Create a zeroed global stiffness matrix
LOAD A1 R=4 C=3   : Load the a matrix for element 1
0 0 0
0 0 0
0 0 -1
1 0 0
PRINT A1
LOAD A2 R=4 C=3   : Load the a matrix for element 2
0 0 0
```

```
1 0 0
0 0 0
0 1 0
PRINT A2
LOAD KE R=4 C=4  : Load the element stiffness matrix
171.18       10270.83    -171.18      10270.83
10270.83   821666.66 -10270.83   410833.33
-171.18   -10270.83      171.18    -10270.83
10270.83   410833.33 -10270.83   821666.66
PRINT KE
TRAN A1 A1T
MULT KE A1 KEA1
MULT A1T KEA1 K1
PRINT K1
TRAN A2 A2T
MULT KE A2 KEA2
MULT A2T KEA2 K2
PRINT K2
ADD K K1
ADD K K2
LOAD R R=3 C=1    : Load the concentrated loads
0
0
20
LOAD SF2 R=4 C=1 : Load element fixed-end forces
-13.8
-276
-13.8
276
PRINT SF2
MULT A2T S2F S2 : Transform elem. load to global directions
ADD R S2        : Add concentrated + elem. contributions
SOLVE K R       : Solve for displacements
PRINT R
MULT KEA1 R S1  : Recover element 1 forces
PRINT S1
MULT KEA2 R S2
SUB S2 SF2      : Recover element 2 forces
PRINT S2
RETURN
```

If these commands are entered into a file named INPUT and CAL-90 started with the command CAL-90> "SUBMIT EXAM1" issued, the OUTPUT file will contain the results of the analysis.

Condensed Example: Frame Using the Full a Matrix

Given the following frame structure, find the moment in the top beam over the center support.

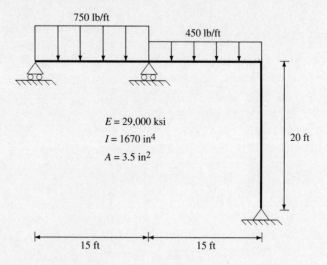

To analyze the structure, we need to define the DOFs and the element directions for each element. The following figure gives the DOF and element numbering.

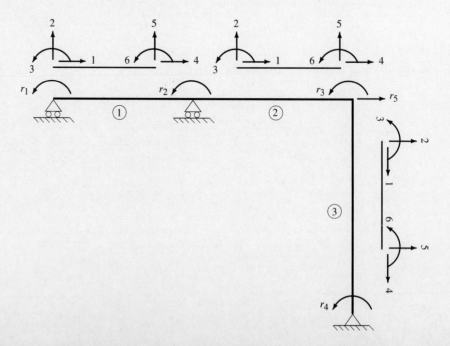

Defined DOF, element numbers, and local element definitions.

The definition of the **a** matrix says to apply a unit displacement to each DOF and find the displacement at the corresponding element DOF. The following two figures give the unit displaced shapes.

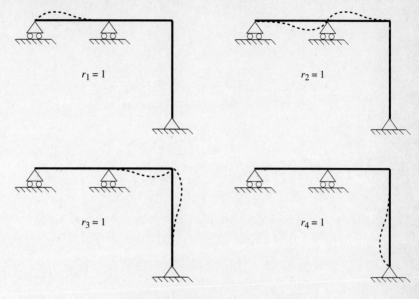

Unit displacements for each DOF.

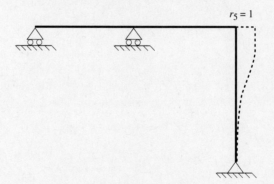

Unit displacement for DOF 5.

Using these figures, we can form the required **a** matrices:

$$\mathbf{a}_1 = \begin{bmatrix} 0 & 0 & 0 & 0 & 1 \\ 0 & 0 & 0 & 0 & 0 \\ 1 & 0 & 0 & 0 & 0 \\ 0 & 0 & 0 & 0 & 1 \\ 0 & 0 & 0 & 0 & 0 \\ 0 & 1 & 0 & 0 & 0 \end{bmatrix}$$

$$\mathbf{a}_2 = \begin{bmatrix} 0 & 0 & 0 & 0 & 1 \\ 0 & 0 & 0 & 0 & 0 \\ 0 & 1 & 0 & 0 & 0 \\ 0 & 0 & 0 & 0 & 1 \\ 0 & 0 & 0 & 0 & 0 \\ 0 & 0 & 1 & 0 & 0 \end{bmatrix}$$

$$\mathbf{a}_3 = \begin{bmatrix} 0 & 0 & 0 & 0 & 0 \\ 0 & 0 & 0 & 0 & 1 \\ 0 & 0 & 1 & 0 & 0 \\ 0 & 0 & 0 & 0 & 0 \\ 0 & 0 & 0 & 0 & 0 \\ 0 & 0 & 0 & 1 & 0 \end{bmatrix}$$

The final data needed are the equivalent concentrated loads for the distributed loading, $\mathbf{S}^f$. These loads are given in the following figure:

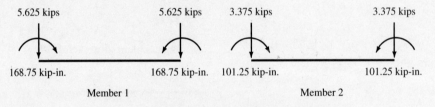

Equivalent loads for beams.

Using these **a** matrices, the loads, the definition of the stiffness [equation (5.23)], and the analysis procedure given in Section 5.7, we can create the following CAL-90 input file:

```
Example: Stiffness Using Full a Matrix
qe6
load a1 r=6 c=5      : Transformation for beam 1
0 0 0 0 1
0 0 0 0 0
1 0 0 0 0
0 0 0 0 1
0 0 0 0 0
0 1 0 0 0
load a2 r=6 c=5      : Transformation for beam 2
0 0 0 0 1
0 0 0 0 0
0 1 0 0 0
0 0 0 0 1
0 0 0 0 0
0 0 1 0 0
```

```
load a3 r=6 c=5      : Transformation for column
0 0 0 0 0
0 0 0 0 1
0 0 1 0 0
0 0 0 0 0
0 0 0 0 0
0 0 0 1 0
load k12e r=6 c=6   : Stiffness for beams
563.88      0          0       -563.88      0          0
   0       99.65     8968.52      0       -99.65     8968.52
   0      8968.52   1076222.22    0      -8968.52   538111.11
-563.88     0          0        563.88      0          0
   0      -99.65    -8968.52      0        99.65    -8968.52
   0      8968.52    538111.11    0      -8968.52  1076222.22
load k3e   r=6 c=6    : Stiffness for column
422.917     0          0       -422.917     0          0
   0       42.04     5044.79      0        -42.04     5044.79
   0      5044.79   807166.66     0       -5044.79   403583.33
-422.917    0          0        422.917     0          0
   0       -42.04    -5044.79      0        42.04     -5044.79
   0      5044.79   403583.33     0       -5044.79   807166.66
C----- Form contribution for beams   (AtKA)
C-- Beam 1
tran a1 a1t
mult a1t k12e ak
mult ak a1 k1
C-- Beam 2
tran a2 a2t
mult a2t k12e ak
mult ak a2 k2
C-- Column
tran a3 a3t
mult a3t k3e ak
mult ak a3 k3
C-- Add individual stiffness -----
add k1 k2
add k1 k3
p k1   : Final total stiffness
C-- Put in load vector -----
load sf1 r=6 c=1    : Load fixed-end loads beam 1
0
-5.625
-168.75
0
-5.625
168.75
load sf2 r=6 c=1
```

```
0
-3.375
-101.25
0
-3.375
101.25
C-- Form global load using a matrices -----
mult a1t sf1 r1
mult a2t sf2 r2
add r1 r2
p r1
C--- Solve for displacements -----
solve k1 r1
p r1
C--- Recover element forces -----
mult a1 r1 ar
mult k12e ar f1
sub f1 sf1
p f1
C--- Beam 2 ----
mult a2 r1 ar
mult k12e ar f2
sub f2 sf2
p f2
C---- Column  (no fixed-end loads)
mult a3 r1 ar
mult k3e ar f3
p f3
return
```

The data above are contained in a file named ''INPUT.'' To execute the procedure you need to use the CAL-90 SUBMIT command.

CAL-90> SUBMIT QE6

The complete analysis will be run and the results placed in a file named ''OUTPUT.'' The moment desired is in either member 1 or member 2. The output for these two members follows.

```
MULT A1 R1 AR
MULT K12E AR F1
SUB F1 SF1
P F1
PRINT OF ARRAY NAMED ''F1   ''
COL# =            1
ROW   1      .00000
ROW   2     4.50000
```

```
ROW    3       .00000
ROW    4       .00000
ROW    5      6.75000
ROW    6   -202.49991
```

```
MULT A2 R1 AR
MULT K12E AR F2
SUB F2 SF2
P F2
PRINT OF ARRAY NAMED ''F2    .''
```

```
COL# =              1
ROW    1       .00000
ROW    2      4.50000
ROW    3    202.49991
ROW    4       .00000
ROW    5      2.25000
ROW    6      -.00035
```

From both of the element force prints we can see that the moment over the support is −202.5.

Condensed Example: Truss by Direct Stiffness Using the Full a Matrix

Using the full **a** matrix technique, find the stiffness for the following structure. This is the same structure as that used in the example of stiffness by definition. The process is identical to frame structures accounting for the correct global DOFs and element DOFs.

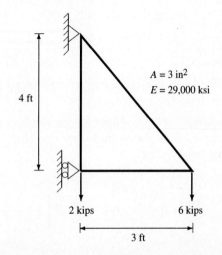

$A = 3 \ \text{in}^2$
$E = 29{,}000 \ \text{ksi}$

4 ft

2 kips   6 kips

3 ft

Again we must define the DOFs for the structure. They are:

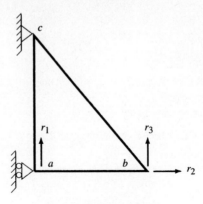

Global DOF definitions.

Like frames, we need to apply a unit displacement to each global DOF and find the corresponding element DOF for the terms in the **a** matrix. To do this we must define the element DOFs. For a truss element, there are only two axial DOFs. We must also number the elements. The chosen element orientations, local DOFs, and element numbers are given below.

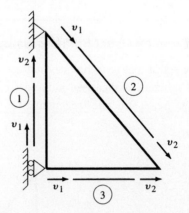

Local element DOF and element numbers.

The required displaced shapes for the structure are shown on p. 175.

From each displaced shape, we can fill a column of each element's **a** matrix. The resulting matrices are

$$\mathbf{a}_1 = \begin{bmatrix} 0 & 0 & 0 \\ 1 & 0 & 0 \end{bmatrix} \qquad \mathbf{a}_2 = \begin{bmatrix} 0 & 0 & 0 \\ 0 & \dfrac{3}{5} & -\dfrac{4}{5} \end{bmatrix} \qquad \mathbf{a}_3 = \begin{bmatrix} 0 & 0 & 0 \\ 0 & 1 & 0 \end{bmatrix}$$

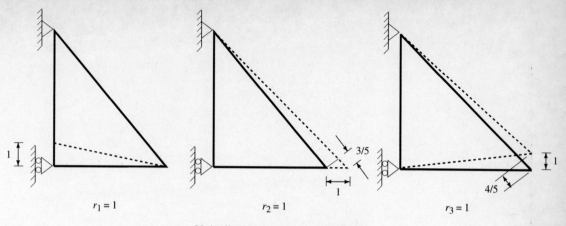

$r_1 = 1$ $r_2 = 1$ $r_3 = 1$

Unit displacements at each DOF.

We also need the stiffness matrix for an axial member with the DOFs shown above. The required stiffness is

$$\mathbf{K}_{\text{truss}} = \begin{bmatrix} \dfrac{AE}{L} & -\dfrac{AE}{L} \\ -\dfrac{AE}{L} & \dfrac{AE}{L} \end{bmatrix}$$

Using the **a** matrices and the stiffness matrix, we can create the required CAL-90 input file to create the stiffness matrix. The required input follows.

```
Example: Truss Using Full a Matrix
qe12
load a1 r=2 c=3
0 0 0
1 0 0
p a1
load a2 r=2 c=3
0 0 0
0 3/5 -4/5
p a2
load a3 r=2 c=3
0 0 0
0 1 0
p a3
load k1 r=2 c=2
1812.5 -1812.5
-1812.5 1812.5
load k2 r=2 c=2
```

```
1450 -1450
-1450 1450
load k3 r=2 c=2
2416.67 -2416.67
-2416.67 2416.67
C----- Form atka for each matrix and add them -----
tran a1 a1t
mult a1t k1 ak1
mult ak1 a1 k11
tran a2 a2t
mult a2t k2 ak2
mult ak2 a2 k22
tran a3 a3t
mult a3t k3 ak3
mult ak3 a3 k33
add k11 k22
add k11 k33
p k11
return
```

To use the file, we start CAL-90 and use the command

CAL-90> SUBMIT QE12

This will run the complete procedure and print the final stiffness to the OUTPUT file. The final result from this file is

```
PRINT OF ARRAY NAMED ``K11   ''
COL# =             1            2            3
ROW   1  1812.50000       .00000       .00000
ROW   2       .00000  2938.67000  -696.00000
ROW   3       .00000  -696.00000   928.00000
RETURN
```

5.9 The a Matrix Revisited

To make the process of direct stiffness even simpler, we need to look at the transformation matrix more closely. To do this, let's see what the matrix would look like for the structure shown in Figure 5.15. According to the definition, we need to apply a unit displacement to DOF 1 and fill out column 1 for each element. The displaced shape required is shown in Figure 5.16. We can see that only two elements have terms that are nonzero. It should be obvious that a similar result will occur for all of the DOFs. This is because an element is directly affected only by a small number of DOFs. As a matter of fact, it is the **a** matrix that tells us which global DOFs cause an element to deform. Also notice that the matrix is 4×12 for this example structure.

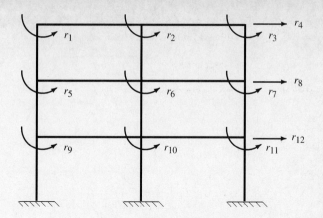

Figure 5.15 Example structure for **a** matrix.

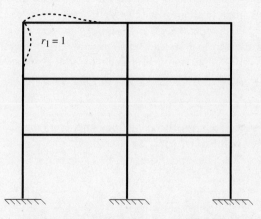

Figure 5.16 Unit displacement for DOF 1.

Now let's look at the slanted member structure used to derive the stiffness matrix by definition. If we apply a unit displacement to the translational DOF, we get the shape shown in Figure 5.17. To fill out the column of the **a** matrix for this shape, we need to use geometry to find the value of the perpendicular displacement at the end of each member. This geometric calculation also becomes a part of the **a** matrix. Therefore, the **a** matrix can be considered to be composed of two parts. The first part tells us to which global DOF the element DOFs are connected. The second part accounts for the geometry, or coordinate rotation change between the global DOF and the element DOF. As a result, let's break the **a** matrix up into these two component parts:

$$\mathbf{a} = \mathbf{a}_\alpha \mathbf{a}_L \tag{5.48}$$

where $\mathbf{a}_\alpha$ is the part of the transformation that accounts for the rotation between the element coordinate system and the global coordinate system and $\mathbf{a}_L$ is the part of the

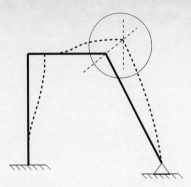

Figure 5.17 Geometric contribution of **a** matrix.

matrix that connects the element DOF to the global DOF. Note that the full **a** matrix is the multiplication of the two parts.

Let's look at each of these parts and see how to form them. The rotation portion can be used to express the transformation from an intermediate set of element DOFs that has the same coordinate directions as the global DOFs. Let's define this intermediate coordinate set:

v* *is a vector of element displacements in the global coordinate directions.*

For example, for the slanted member in Figure 5.17, the two coordinate systems are as shown in Figure 5.18. The $\mathbf{a}_\alpha$ matrix is the transformation matrix that converts from one system to the other. This can be expressed by the equation

$$\mathbf{v} = \mathbf{a}_\alpha \mathbf{v}^*$$

(5.49)

Note that this is in the same form as the original in terms of global to local or **v*** to **v**. The same result can be used to convert in the opposite direction, as was done with forces.

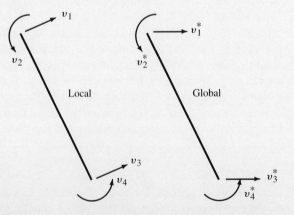

Figure 5.18 Slanted member's local and global DOFs.

This was done using the transpose and the equation

$$\mathbf{v}^* = \mathbf{a}_\alpha^\mathsf{T} \mathbf{v} \tag{5.50}$$

5.9.1 Inclusion of Axial Deformations

It is now advantageous to expand our element deformations to include the axial deformation. This makes the transformation more general. In addition, it gets rid of the problem of finding the independent and dependent global displacements as was done for the slanted member example. However, it adds to the total number of DOFs since we now need to include the axial deformation in our formulation. The advantages outweigh the disadvantages because of the generalized methods that can be developed and employed. The new form that we will use for the rest of the book for the element DOFs is shown in Figure 5.19.

Figure 5.19 Definition of DOFs for planar beam element.

The element forces, **S**, have the corresponding numbers. The stiffness matrix can be formed for this element in the same manner as the previous one. The displaced shapes for the previous element (Figure 5.6) are identical. All the stiffness terms from the previous matrix are also identical. The only difference is the addition of the axial terms. For these added DOFs the deformed shapes and corresponding forces for DOFs 1 and 4 are shown in Figure 5.20. The resulting stiffness matrix for this element is

$$\mathbf{K}_e = \begin{bmatrix} \dfrac{AE}{L} & 0 & 0 & -\dfrac{AE}{L} & 0 & 0 \\[2ex] 0 & \dfrac{12EI}{L^3} & \dfrac{6EI}{L^2} & 0 & -\dfrac{12EI}{L^3} & \dfrac{6EI}{L^2} \\[2ex] 0 & \dfrac{6EI}{L^2} & \dfrac{4EI}{L} & 0 & -\dfrac{6EI}{L^2} & \dfrac{2EI}{L} \\[2ex] -\dfrac{AE}{L} & 0 & 0 & \dfrac{AE}{L} & 0 & 0 \\[2ex] 0 & -\dfrac{12EI}{L^3} & -\dfrac{6EI}{L^2} & 0 & \dfrac{12EI}{L^3} & -\dfrac{6EI}{L^2} \\[2ex] 0 & \dfrac{6EI}{L^2} & \dfrac{2EI}{L} & 0 & -\dfrac{6EI}{L^2} & \dfrac{4EI}{L} \end{bmatrix} \tag{5.51}$$

Notice that as a result of the addition of the axial DOFs, we can also develop a stiffness matrix just for truss elements. The element would consist only of the axial DOFs.

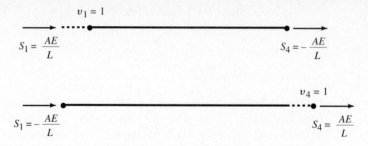

$$v_1 = 1$$

$$S_1 = \frac{AE}{L} \qquad\qquad S_4 = -\frac{AE}{L}$$

$$v_4 = 1$$

$$S_1 = -\frac{AE}{L} \qquad\qquad S_4 = \frac{AE}{L}$$

Figure 5.20 Additional axial DOFs for beam element.

Its DOFs and the element stiffness are depicted in Figure 5.21. The resulting stiffness matrix, which is just the axial portion of the beam matrix, is

$$\mathbf{K}_e = \begin{bmatrix} \dfrac{AE}{L} & -\dfrac{AE}{L} \\ -\dfrac{AE}{L} & \dfrac{AE}{L} \end{bmatrix} \tag{5.52}$$

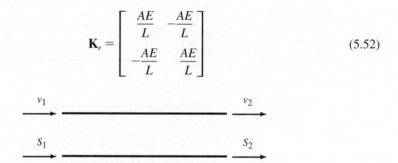

$$v_1 \qquad\qquad\qquad\qquad v_2$$

$$S_1 \qquad\qquad\qquad\qquad S_2$$

Figure 5.21 Truss (axial) element DOFs and forces.

5.9.2   Derivation of Rotational Transformation

Now we can define the $\mathbf{a}_\alpha$ matrix. As given in equation (5.49), we can relate the element DOFs in the global coordinate system, $\mathbf{v}^*$, to the element DOFs in the local coordinate system, $\mathbf{v}$. If both translations are included, the required element DOFs appear as shown in Figure 5.22. The $\mathbf{v}^*$ displacements are always in the global directions. As can be seen, this is nothing but a coordinate rotation in two dimensions. Therefore, the $\mathbf{a}_\alpha$ is a coordinate rotation matrix. It can be developed from geometry knowing the angle between the two coordinate systems. The result of redrawing one of the ends is shown in Figure 5.23.

The matrix form for the rotation of coordinates has the form

$$\begin{bmatrix} v_1 \\ v_2 \end{bmatrix} = \begin{bmatrix} a & b \\ c & d \end{bmatrix} \begin{bmatrix} v_1^* \\ v_2^* \end{bmatrix} \tag{5.53}$$

where we want to transform from the global to the local system, just as in the original $\mathbf{a}$ transformation definition. In this case we need to find the vector components of the $\mathbf{v}^*$ unit vectors in the $\mathbf{v}$ coordinate. The geometry for finding the vectors is shown in Figure 5.24. From Figure 5.24 we can place the components into the matrix to form the transforma-

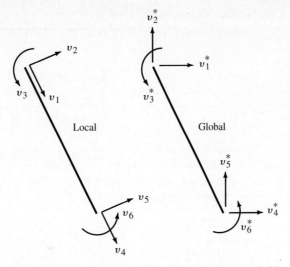

Figure 5.22 Complete element local and global DOFs.

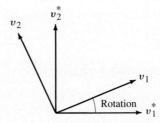

Figure 5.23 Rotation of local to global directions.

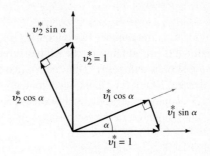

Figure 5.24 Components of global unit vectors in local directions.

tion from global to local coordinates, which is

$$\begin{bmatrix} v_1 \\ v_2 \end{bmatrix} = \begin{bmatrix} \cos \alpha & \sin \alpha \\ -\sin \alpha & \cos \alpha \end{bmatrix} \begin{bmatrix} v_1^* \\ v_2^* \end{bmatrix} \qquad (5.54)$$

Clearly, the rotational DOF term does not change with a rotation in coordinates. As a result, the rotation of the DOF at each end of the beam can be formed as

$$
\begin{bmatrix} v_1 \\ v_2 \\ v_3 \end{bmatrix} = \begin{bmatrix} \cos\alpha & \sin\alpha & 0 \\ -\sin\alpha & \cos\alpha & 0 \\ 0 & 0 & 1 \end{bmatrix} \begin{bmatrix} v_1^* \\ v_2^* \\ v_3^* \end{bmatrix}
\tag{5.55}
$$

The final transformation for the entire element can be developed by realizing that the DOFs at each end of the beam are independent from those at the other end. A rotation at one end does not affect the other directly. We can put the rotation matrix together to form the element rotation matrix, $\mathbf{a}_\alpha$:

$$
\begin{bmatrix} v_1 \\ v_2 \\ v_3 \\ v_4 \\ v_5 \\ v_6 \end{bmatrix} = \begin{bmatrix} \cos\alpha & \sin\alpha & 0 & 0 & 0 & 0 \\ -\sin\alpha & \cos\alpha & 0 & 0 & 0 & 0 \\ 0 & 0 & 1 & 0 & 0 & 0 \\ 0 & 0 & 0 & \cos\alpha & \sin\alpha & 0 \\ 0 & 0 & 0 & -\sin\alpha & \cos\alpha & 0 \\ 0 & 0 & 0 & 0 & 0 & 1 \end{bmatrix} \begin{bmatrix} v_1^* \\ v_2^* \\ v_3^* \\ v_4^* \\ v_5^* \\ v_6^* \end{bmatrix}
\tag{5.56}
$$

5.9.3   Derivation of the Location Transformation

The second part of the transformation matrix is the $\mathbf{a}_L$ portion. This part converts the global DOFs, $\mathbf{r}$, to the element DOFs in global coordinates, $\mathbf{v}^*$. Again we have the form global to local. In equation form we have

$$
\mathbf{v}^* = \mathbf{a}_L \mathbf{r}
\tag{5.57}
$$

The size of $\mathbf{a}_L$ is the number of element DOFs by the number of global DOFs. Since the $\mathbf{v}^*$ components are already in the global directions, this matrix is a Boolean matrix or contains just ones and zeros. This matrix just pulls out the global DOFs that correspond to the element DOFs (in global coordinates).

The $\mathbf{v}^*$ components at the end of an element are always in the direction of the global system and numbered in the form shown in Figure 5.25. If we use the same structure as that shown in Figure 5.26, we can form the $\mathbf{a}_L$ for the left-hand column. For this structure the global DOFs are numbered as shown. Here we will number *all* DOFs. We will still be able to take advantage of the pinned end later. Note that if we wish to ignore axial deformations, as by using the DOFs of Figure 5.5, we would have to include an additional (third) transformation that would enforce the no-axial-deformation constraint. Choosing the DOFs shown in Figure 5.26 will require using a large area to simulate no axial deformations.

Also shown are the $\mathbf{v}^*$ DOFs on the element next to the structure. By inspection we can see that only r_1, r_2, and r_3 are connected to the left column. In forming the $\mathbf{a}_L$ matrix

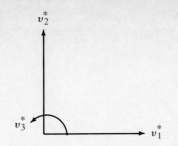

Figure 5.25 Global direction and numbering of DOFs.

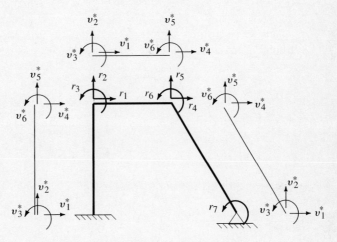

Figure 5.26 Global element DOF definitions for structure.

for an element it is only the directly connected DOFs that should be included. In matrix form these DOFs are included as

$$
\begin{bmatrix} v_1^* \\ v_2^* \\ v_3^* \\ v_4^* \\ v_5^* \\ v_6^* \end{bmatrix} = \begin{bmatrix} 0 & 0 & 0 & 0 & 0 & 0 & 0 \\ 0 & 0 & 0 & 0 & 0 & 0 & 0 \\ 0 & 0 & 0 & 0 & 0 & 0 & 0 \\ 1 & 0 & 0 & 0 & 0 & 0 & 0 \\ 0 & 1 & 0 & 0 & 0 & 0 & 0 \\ 0 & 0 & 1 & 0 & 0 & 0 & 0 \end{bmatrix} \begin{bmatrix} r_1 \\ r_2 \\ r_3 \\ r_4 \\ r_5 \\ r_6 \\ r_7 \end{bmatrix} \tag{5.58}
$$

The 1s in the matrix are located wherever an element DOF directly connects or touches a global DOF. The row corresponds to the element DOF number and the column corresponds to the global DOF number. From the figure we can see that element DOF 4 connects directly to global DOF 1. Therefore, there is a 1 in row 4, column 1 of the matrix. This matrix is very simple to form since it only requires looking at the connection of an element to the structure. The matrices can easily be developed for the other elements:

$$\begin{bmatrix} 1 & 0 & 0 & 0 & 0 & 0 & 0 \\ 0 & 1 & 0 & 0 & 0 & 0 & 0 \\ 0 & 0 & 1 & 0 & 0 & 0 & 0 \\ 0 & 0 & 0 & 1 & 0 & 0 & 0 \\ 0 & 0 & 0 & 0 & 1 & 0 & 0 \\ 0 & 0 & 0 & 0 & 0 & 1 & 0 \end{bmatrix} \qquad \begin{bmatrix} 0 & 0 & 0 & 0 & 0 & 0 & 0 \\ 0 & 0 & 0 & 0 & 0 & 0 & 0 \\ 0 & 0 & 0 & 0 & 0 & 0 & 1 \\ 0 & 0 & 0 & 1 & 0 & 0 & 0 \\ 0 & 0 & 0 & 0 & 1 & 0 & 0 \\ 0 & 0 & 0 & 0 & 0 & 1 & 0 \end{bmatrix} \qquad (5.59)$$

$$\text{(top beam)} \qquad\qquad\qquad \text{(right column)}$$

Now let's substitute the new form of the **a** matrix into the stiffness matrix equation. Substituting equation (5.48) into (5.23) for **a** gives us the form

$$\mathbf{K} = \sum_{\text{all elements}} \mathbf{a}_L^T \mathbf{a}_\alpha^T \mathbf{K}_e \mathbf{a}_\alpha \mathbf{a}_L \qquad (5.60)$$

Now let's give the middle portion

$$\mathbf{K}_{xy} = \mathbf{a}_\alpha^T \mathbf{K}_e \mathbf{a}_\alpha \qquad (5.61)$$

a separate name: the *element stiffness in global coordinates* (K_{xy}). It represents the element stiffness matrix in global (**v***) coordinates. This is shown in Figure 5.27.

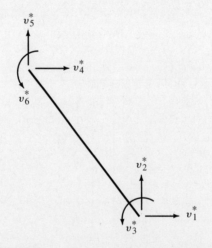

Figure 5.27 DOF for element stiffness in global directions.

This stiffness is identical to every stiffness matrix in the past. It can be defined by applying a unit displacement to a single DOF, all others remaining zero, and calculating the forces to cause the displaced shape. As an example, let's apply a unit displacement to DOF 2. The displaced shape is shown in Figure 5.28. Notice that we now have both

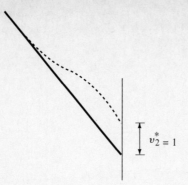

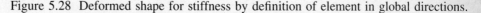

$v_2^* = 1$

Figure 5.28 Deformed shape for stiffness by definition of element in global directions.

an axial and a shear deformation in the element. This is a little more difficult to calculate than the terms in the previous element. An important point to make is that the rotation of an element, slanting, causes coupling between the axial and shear DOFs of the element. This coupled effect is what makes the hand solutions of problems with slanted members so difficult.

If we substitute our definition of the element stiffness in global directions (5.61) into equation (5.60), we get

$$\mathbf{K} = \sum_{\text{all elements}} \mathbf{a}_L^T \mathbf{K}_{xy} \mathbf{a}_L \tag{5.62}$$

This is the form most often used in finite-element programs. The element stiffness is assumed to be transformed into global directions. It is then assembled into the global stiffness through use of the location or connectivity transformation.

5.9.4   Stiffness Assembly

Let's look more closely at what the $\mathbf{a}_L$ matrix does. To do this, let's look at a new structure with more DOFs. Let's also consider only the upper left-hand column. The structure and the element to be considered are shown in Figure 5.29. The global DOFs are numbered in an arbitrary manner. Note that the order in which the global DOFs are numbered is not important to the solution process. It does have an effect if sparse matrix methods are used.

The $\mathbf{a}_L$ matrix for the element and structure shown is

$$
\begin{bmatrix}
0 & 0 & 0 & 1 & 0 & 0 & 0 & 0 & 0 & 0 & 0 & 0 \\
0 & 0 & 0 & 0 & 1 & 0 & 0 & 0 & 0 & 0 & 0 & 0 \\
0 & 0 & 0 & 0 & 0 & 1 & 0 & 0 & 0 & 0 & 0 & 0 \\
0 & 0 & 0 & 0 & 0 & 0 & 0 & 0 & 0 & 1 & 0 & 0 \\
0 & 0 & 0 & 0 & 0 & 0 & 0 & 0 & 0 & 0 & 1 & 0 \\
0 & 0 & 0 & 0 & 0 & 0 & 0 & 0 & 0 & 0 & 0 & 1
\end{bmatrix}
\tag{5.63}
$$

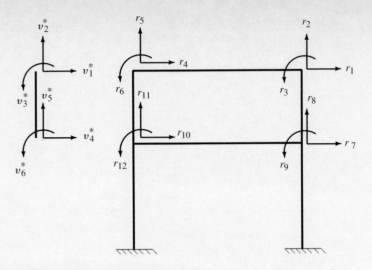

Figure 5.29 Structure and element for forming the $\mathbf{a}_L$ matrix.

If we carry out the multiplication in equation (5.62), we can get this element's contribution to the global stiffness. Doing so, we get

$$
\begin{bmatrix}
0 & 0 & 0 & 0 & 0 & 0 & 0 & 0 & 0 & 0 & 0 & 0 \\
0 & 0 & 0 & 0 & 0 & 0 & 0 & 0 & 0 & 0 & 0 & 0 \\
0 & 0 & 0 & 0 & 0 & 0 & 0 & 0 & 0 & 0 & 0 & 0 \\
0 & 0 & 0 & K_{11} & K_{12} & K_{13} & 0 & 0 & 0 & K_{14} & K_{15} & K_{16} \\
0 & 0 & 0 & K_{21} & K_{22} & K_{23} & 0 & 0 & 0 & K_{24} & K_{25} & K_{26} \\
0 & 0 & 0 & K_{31} & K_{32} & K_{33} & 0 & 0 & 0 & K_{34} & K_{35} & K_{36} \\
0 & 0 & 0 & 0 & 0 & 0 & 0 & 0 & 0 & 0 & 0 & 0 \\
0 & 0 & 0 & 0 & 0 & 0 & 0 & 0 & 0 & 0 & 0 & 0 \\
0 & 0 & 0 & 0 & 0 & 0 & 0 & 0 & 0 & 0 & 0 & 0 \\
0 & 0 & 0 & K_{41} & K_{42} & K_{43} & 0 & 0 & 0 & K_{44} & K_{45} & K_{46} \\
0 & 0 & 0 & K_{51} & K_{52} & K_{53} & 0 & 0 & 0 & K_{54} & K_{55} & K_{56} \\
0 & 0 & 0 & K_{61} & K_{62} & K_{63} & 0 & 0 & 0 & K_{64} & K_{65} & K_{66}
\end{bmatrix}
\tag{5.64}
$$

where the terms K_{ij} are the element stiffness terms from the $\mathbf{K}_{xy}$ in the i,j location (i.e., K_{23} is the term in the second row, third column of $\mathbf{K}_{xy}$). Notice that all the $\mathbf{a}_L$ matrix does is to split up the stiffness and put it into the appropriate location in the gloal stiffness matrix. Therefore, $\mathbf{a}_L$ just rearranges and expands the element stiffness (in global directions). Again this final stiffness is the element's contribution to the global stiffness matrix. When all the elements are added, we have the final global stiffness.

The first thing to notice is that the $\mathbf{a}_L$ matrix contains a large number of zeros. Next is that it has a maximum of only six ones. This is all that the matrix can ever contain.

Therefore, it is very inefficient to multiply using this $\mathbf{a}_L$ matrix. If this is all the $\mathbf{a}_L$ matrix does, perhaps there is an easier way. The answer to this type of question (in a textbook) is always "yes." If we look carefully at the $\mathbf{a}_L$ matrix above, we can see that the columns represent the global DOFs and the rows represent the element DOFs. If a global DOF connects to an element DOF, then in the column corresponding to that global DOF there is a 1 in the row that corresponds to the element DOF to which it connects. For example, global DOF 10 connects to element DOF 4. Therefore, there is a 1 in column 10, row 4. This information can be contained in a single vector that is the number of element DOFs long. This vector is called the *location vector* (**LV**). In each row of the location vector we keep the global DOFs to which it connects. The entire $\mathbf{a}_L$ matrix for the previous structure can be represented by

$$\mathbf{LV} = \begin{bmatrix} 4 \\ 5 \\ 6 \\ 10 \\ 11 \\ 12 \end{bmatrix} \tag{5.65}$$

where again, when we look at row 4 (element DOF) we see that it connects directly to global DOF 10. From this vector we could always construct the true $\mathbf{a}_L$ matrix.

We can use this structure to perform the same function as equation (5.62). If we use this vector as a location pointer to where the stiffness terms should be *added*, we can perform the process. To do this we put the location vector across the top row and down the left-hand column of the element stiffness ($\mathbf{K}_{xy}$). We can now use these vectors as locators as to where to add the term in the global stiffness matrix.

$$
\begin{array}{c|cccccc}
\mathbf{LV} & 4 & 5 & 6 & 10 & 11 & 12 \\[4pt]
4 & K_{11} & K_{12} & K_{13} & K_{14} & K_{15} & K_{16} \\
5 & K_{21} & K_{22} & K_{23} & K_{24} & K_{25} & K_{26} \\
6 & K_{31} & K_{32} & K_{33} & K_{34} & K_{35} & K_{36} \\
10 & K_{41} & K_{42} & K_{43} & K_{44} & K_{45} & K_{46} \\
11 & K_{51} & K_{52} & K_{53} & K_{54} & K_{55} & K_{56} \\
12 & K_{61} & K_{62} & K_{63} & K_{64} & K_{65} & K_{66}
\end{array}
\tag{5.66}
$$

Now, to see where any term of the element stiffness should be added, look at the **LV** terms in the same row and column. For example, let's find where the element K_{32} goes. Looking at the **LV** numbers in row 3 and column 2, we get 6 and 5. Therefore, it should be added into location (6, 5) of the global stiffness. If you look at the expanded matrix, equation (5.64), for this element, you can see that this is correct. This process is called *assembly*. It is this method that all finite-element and structural analysis programs use. They *never* multiply out equation (5.62). The elements are assembled into the global stiffness using this pointer system.

5.9.5 Force Recovery Using Transformations

Force recovery also changes as a result of splitting up the **a** matrix. If we substitute the new definition for **a** into the force recovery equation, we get

$$\mathbf{S} = \mathbf{K}_e \mathbf{a}_\alpha \mathbf{a}_L \mathbf{r} - \mathbf{S}^f \tag{5.67}$$

Let's reorganize some of the terms by redefining the force transformation matrix:

$$\mathbf{FT}_\alpha = \mathbf{K}_e \mathbf{a}_\alpha \tag{5.68}$$

This is the force transformation matrix that converts the $\mathbf{v}^*$ displacements into element forces. This is a direct transformation from displacements in global coordinates into local element forces. The subscript α is used to signify that only the $\mathbf{a}_\alpha$ portion of the **a** matrix is included. As can be seen, the other portion of the equation, $\mathbf{a}_L \mathbf{r}$, converts the global displacements into $\mathbf{v}^*$.

Let's look more closely at what the $\mathbf{a}_L \mathbf{r}$ portion does. Earlier we saw that the $\mathbf{a}_L$ matrix chose the global displacements that connected directly to the element displacements. If we use the two-story single structure shown in Figure 5.29, we can see what happens. The $\mathbf{a}_L$ matrix for the element shown is given in equation (5.63). There are 12 DOFs for the total structure; therefore, the **r** vector is 12 long. If we carry out the multiplication of $\mathbf{a}_L \mathbf{r}$, we get

$$\mathbf{v}^* = \mathbf{a}_L \mathbf{r} = \begin{bmatrix} r_4 \\ r_5 \\ r_6 \\ r_{10} \\ r_{11} \\ r_{12} \end{bmatrix} \tag{5.69}$$

This says that a few of the global DOFs are identical to the $\mathbf{v}^*$ of the element. This is exactly true since an element is connected to a structure and the global displacements are common to the elements. Therefore, all the $\mathbf{a}_L$ matrix does is to pull out the correct global displacements for the particular element.

This can be handled much more efficiently using the same pointer system as that used for assembly. For assembly of the element stiffness matrix into the global stiffness we used the **LV** vector for the element. By putting the vector down and across we had a two-dimensional pointer. Here, we only need to use a single LV vector to tell which global displacements to place into the $\mathbf{v}^*$ vector. Recall that the **LV** vector for this element is given in equation (5.65). This easily shows that we should put the global displacements in the order shown above. Using this method of pulling the correct displacements, the $\mathbf{a}_L$ does not need to be multiplied.

5.9.6 Split Transformation Analysis Procedure

The analysis procedure that will be followed is similar to most linear static analysis programs. It is based on the methods described in Section 5.9 for splitting up the **a** matrix.

The procedure is as follows:

1. Form all element-related matrices for each element in the structure. [Form the element stiffness for all elements in global coordinates ($\mathbf{K}_{xy}$), and the location vector (**LV**) and element contribution to the global load vector ($\mathbf{a}_\alpha^\mathrm{T}\mathbf{S}^f$).]
   (a) Form the element stiffness in local coordinates ($\mathbf{K}_e$).
   (b) Rotate the stiffness to global coordinates ($\mathbf{a}_\alpha^\mathrm{T}\mathbf{K}_e\mathbf{a}_\alpha$). [Usually, save $\mathbf{K}_e\mathbf{a}_\alpha$ (force transformation).]
   (c) Form the location vector (**LV**) for each element. (This is the compact form of the $\mathbf{a}_L$ matrix.)
   (d) Form the element load contribution to the global loads ($\mathbf{a}_\alpha^\mathrm{T}\mathbf{S}^f$). (Note that this will also have to be placed into the proper locations in the global load vector, the opposite of pulling out the proper displacements using the **LV** vector.)

2. Assemble all element stiffnesses ($\mathbf{K}_{xy}$) into the global stiffness as well as the element load contributions into the global load vector. Use the **LV** vector pointer method.

3. Add all concentrated loads to the global load vector.

4. Solve for the global displacements (**r**).

5. Recover the local element forces ($\mathbf{S} = \mathbf{FT}_\alpha\mathbf{v}^* - \mathbf{S}^f$). (Again, use the **LV** pointer method.)

Note that there are many variations on this basic procedure that will affect the performance, but not the outcome, of the analysis. It is possible to combine steps 1 and 2 into a single step so that an element is formed and assembled immediately. This would then be done repetitively for all elements. This will have some advantages in use of memory and computer input/output but will also limit the flexibility of the program. Other portions can also be combined and their order changed to take advantage of specific options available to a particular computer.

5.10 Stiffness Analysis Using CAL-90

The procedure described in Section 5.9.6 will be implemented using the direct stiffness commands in CAL-90. The splitting of the **a** matrix and the use of the **LV** pointer has reduced the amount of hand work to a minimum. The work that needs to be done by hand consists of numbering the DOFs, forming the **LV** pointer vector for each element, and forming the member load ($\mathbf{S}^f$) vector and its required rotation matrix. The rest is taken care of by the CAL-90 commands.

There are five new commands that we will be using to perform an analysis: FRAME, ADDK, ADDL, MEMFRC, and LOADI. The FRAME command performs step 1 of the analysis procedure, ADDK performs step 2, ADDL performs step 3, and MEMFRC performs a portion of step 5. The SOLVE command will again be used to find displacements. The LOADI command is equivalent to the LOAD command except that it loads an integer matrix with R1 rows and C1 columns. LOADI is needed to load the **LV** pointer vector used to assemble the element stiffness and load matrices. The manual pages for the new commands are discussed below. Note the convention of symbolic names for the required matrices just as in previous commands.

5.10.1   Direct Stiffness Operations

FRAME M1+ M2+ X=XI,XJ Y=YI,YJ E=E1 I=I1 A=A1

where   M1   is the element stiffness matrix in global directions ($\mathbf{K}_{xy} = \mathbf{a}_\alpha \mathbf{K}_e \mathbf{a}_\alpha$).

M2   is the force transformation matrix ($\mathbf{FT}_\alpha = \mathbf{K}_e \mathbf{a}_\alpha$).

XI   is the X coordinate of the ith node.

XJ   is the X coordinate of the jth node.

YI   is the Y coordinate of the ith node.

YJ   is the Y coordinate of the jth node.

E1   is the modulus of elasticity, E.

I1   is the moment of inertia, I.

A1   is the axial area, A.

This operation forms the 6×6 stiffness matrix M1 and the 6×6 force transformation matrix M2. The stiffness formed is for a two-dimensional beam element. The degrees of freedom are defined as shown in Figure 5.30. The force transformation matrix may be used to recover the element forces once the final displacements are found. The element forces are defined by Figure 5.31. The directions for the axial forces (F_1 and F_4) are always positive-going from nodes I to J. The shears are positive at 90° counterclockwise from the axial forces. The moments are positive if counterclockwise.

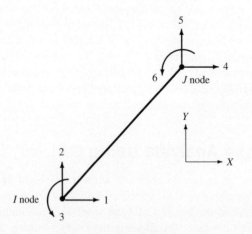

Figure 5.30 Global DOF definition for beam element.

ADDK M1+ M2 M3 N=N1

where   M1   is the global stiffness matrix (defined previously by the ZERO command).

M2   is the element stiffness matrix to be added ($\mathbf{K}_{xy}$).

M3   is an integer array in which column N1 contains the global degree of

I to J is positive for axial force
Shear is positive counterclockwise 90°

Figure 5.31 Definition of force results for beam.

freedom numbers to which the element degrees of freedom correspond (matrix containing the **LV** vectors).

N1   is the column of M3 to be used as the **LV** vector.

This command adds an element stiffness matrix to the total global stiffness matrix. This is a general command that works with a stiffness matrix of any size. If the matrix M2 is $N \times N$, the matrix M3 must have N rows.

ADDL M1+ M2 M3 N=N1

where   M1   is the global load matrix. (The matrix *must* have been defined previously by the ZERO or LOAD command.)

M2   is the element load matrix to be added ($\mathbf{a}_\alpha^T \mathbf{S}^f$).

M3   is an integer array in which column N1 contains the global DOF numbers to which the element degrees of freedom correspond (matrix containing the **LV** vectors).

N1   is the column of M3 to be used as the **LV** vector.

This command adds an element load matrix to the total global load matrix. This is a general command that works with a load matrix of any size. If the matrix M2 is $N \times N$, the matrix M3 must have N rows.

MEMFRC M1 M2 M3 M4+ N=N1

where   M1   is the element force transformation array ($\mathbf{FT}_\alpha = \mathbf{K}_e \mathbf{a}_\alpha$), usually from FRAME.

M2   is the matrix containing the global joint displacements.

M3   is the integer array in which column N1 contains the global DOF numbers that correspond to the element DOFs. (This is the same as that used in ADDK and ADDL.)

M4   is the element force array created. The force directions are defined under the FRAME command.

N1   is the column number of M3 to be used as the **LV** vector.

This command forms the array M4, which contains the local element forces.

5.10.2 Example of CAL-90 Direct Stiffness Commands

The structure shown in Figure 5.32 will be analyzed to demonstrate the CAL-90 direct stiffness commands. The first work to be done when analyzing a structure using the CAL-

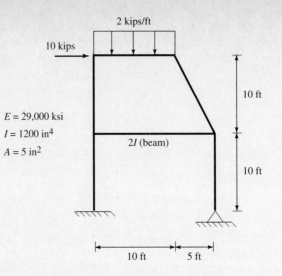

Figure 5.32 CAL-90 direct stiffness commands example.

90 commands is to number the global DOFs. The choice of numbering is arbitrary; however, all DOFs must be numbered when using the FRAME command. This is because FRAME assumes that all DOFs exist for an element (i.e., no pins). Step 1 is to form the element stiffness ($\mathbf{K}_{xy}$). This is handled by the FRAME command. The frame command takes care of rotating the element to global directions. The rotation of a stiffness matrix in two dimensions requires an angle ($\mathbf{a}_\alpha$). FRAME uses the coordinates of the two ends of the beam to calculate the angle. A beam has an *I* end and a *J* end. The *I* end always has DOF 1-2-3, and the *J* end always has 4-5-6 (as shown in Figure 5.30). This means that we will need the coordinates at the ends of each member to define the *I* and *J* ends for each member. Figure 5.33 shows the coordinates and the *I–J* end choices for our example.

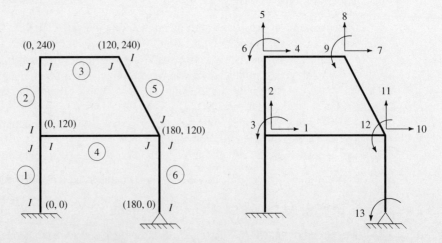

Figure 5.33 DOFs, elements, *I–J* ends, and coordinate definitions.

The I–J end is an arbitrary choice; however, it is beneficial to be consistent in this choice (such as I left and J right or I bottom and J top). This will help avoid errors. In addition, it will allow reuse of some of the work. Recall what properties make one element different from another. The only things that can change are A, E, I, and the angle. If A, E, I, and the angle are the same, the element stiffnesses ($\mathbf{K}_{xy}$) are the same. Choosing the I and J locations on an element to be the same makes the angle the same. Looking at Figure 5.33, we can see that elements 1, 2, and 6 are identical. Since we chose I and J the same, the result from the FRAME command is also the same. Therefore, we only need to call FRAME once for these elements. Procedures like this can save an enormous amount of computer time.

First, we ZERO out the matrix that will contain the final stiffness matrix. Since there are 13 DOFs, the matrix needs to be 13×13. Then we will LOADI, the pointer matrix **LM**. This matrix will contain the pointer vectors **LV**, one column for each element. There are six element DOFs, therefore six rows in the **LM** matrix. There are six elements in the structure, therefore six columns in the **LM** matrix. Placing all the **LV** vectors in one matrix makes it easy to load and refer to. This is not required; each element could have a different pointer matrix. The commands to perform these steps are

```
ZERO  K  R=13  C=13
LOADI LM R=6  C=6
   0   1   4   1   7   0
   0   2   5   2   8   0
   0   3   6   3   9  13
   1   4   7  10  10  10
   2   5   8  11  11  11
   3   6   9  12  12  12
```

Next we need to form the element stiffnesses and assemble them, steps 1 and 2 of the procedure. For our structure there are only four unique elements. Since the global stiffness matrix ($\mathbf{K}$) must remain in memory at the same time as at least one element stiffness, we can combine steps 1 and 2 to same some memory and steps. This may be critical in larger applications. Therefore, we will use the FRAME command to form an element stiffness (K1) and then assemble it immediately. Then in the subsequent call to FRAME we will reuse the same matrix for the element stiffness. This will avoid having to store all element stiffnesses in memory. The required steps are

```
FRAME K1 T1 X=0,0 Y=0,10*12 E=29000 I=1200 A=5
ADDK K K1 LM N=1
ADDK K K1 LM N=2
ADDK K K1 LM N=6
FRAME K1 T3 X=0,10*12 Y=20*12,20*12 E=29000 I=2400 A=5
ADDK K K1 LM N=3
FRAME K1 T4 X=0,15*12 Y=0,0 E=29000 I=1200 A=5
ADDK K K1 LM N=4
FRAME K1 T5 X=10*12,15*12 Y=20*12,10*12 E=29000 I=1200 A=5
ADDK K K1 LM N=5
```

Note that the stiffness for the three identical columns is formed with the same FRAME command. Then it is assembled using a different column of the **LM** matrix. Also note that all the FRAME commands use K1 for the element stiffness matrix. This is because once the element has been assembled, it is no longer needed and can be reused to save space. In addition, note that for element 4 we do not use the correct Y coordinates. This is because the coordinates are used only for the angle and the element length. The fact that both Y coordinates are the same (horizontal element) is all that matters.

Next we need to form the global load vector, step 3. This consists of the concentrated and element load portions. First we LOAD the concentrated portion directly. Then we use ADDL to assemble the distributed load portion for element 3. The required steps are

```
LOAD RHS R=13 C=1
0
0
0
10
0
0
0
0
0
0
0
0
0
load SF3 R=6 C=1
0
-10
-200
0
-10
 200
ADDL RHS SF3 LM N=3
```

Notice that SF3 is the element load vector ($\mathbf{S}^f$) for element 3. The ADDL command assumes that the element load vector is already given in global coordinates. Normally, this is not the case; it is given in element coordinates. Therefore, it needs to be rotated to the global directions. The equation for this rotation is

$$\mathbf{S}^f(\text{global}) = \mathbf{a}_\alpha^\mathrm{T} \mathbf{S}^f$$

where this is the same $\mathbf{a}_\alpha$ matrix as was used by the FRAME command. However, the FRAME command does not save the $\mathbf{a}_\alpha$ matrix. Therefore, normally you would have to LOAD and then MULT the matrix to get the global vector. In this particular case, the rotation angle is zero. As a result, the $\mathbf{a}_\alpha$ matrix is the identity matrix.

We are now ready to solve for the global displacements, step 4. This is done using the SOLVE command. After that we want to recover the element forces for all members.

To recover the displacement portion of the forces, we use the MEMFRC command, step 5. This pulls out the correct displacements and multiplies the force transformation matrix. The final steps are

```
SOLVE K RHS
MEMFRC T1 RHS LM F1 N=1
MEMFRC T1 RHS LM F2 N=2
MEMFRC T3 RHS LM F3 N=3
SUB F3 SF3
MEMFRC T4 RHS LM F4 N=4
MEMFRC T5 RHS LM F5 N=5
MEMFRC T1 RHS LM F6 N=6
```

Note that for the elements that are the same, we use the same force transformation matrix with a different column of the **LM** matrix. We still need to subtract back off the element forces from any element with element loading. In our case, this is only element 3. Note that the forces being subtracted are the local version, not the rotated loads added using the ADDL command. Again, in this particular example they are the same.

At any time the result of a command can be printed. This is helpful in seeing the results of a process. In addition, the ADDK, ADDL, SOLVE, and MEMFRC commands are general. They will work with any size stiffness, load, and force transformation matrix, provided that the sizes are compatible. This allows for the development of other elements within the CAL-90 system.

Condensed Example: Frame Analysis Using CAL-90

Given the following frame structure, find the moments in the beam member at the joint where the 6-kip load is applied. Use the method with the two-part **a** matrix ($\mathbf{a}_\alpha \mathbf{a}_L$).

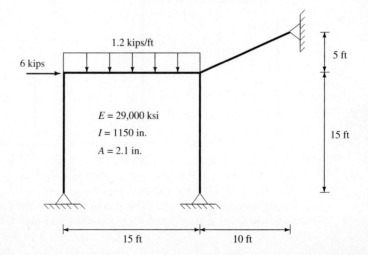

Recall that when using the two-part **a** matrix, the $\mathbf{a}_\alpha$ portion is a rotation and is taken care of by the FRAME command. The $\mathbf{a}_L$ portion is the assembly portion and is taken care of by the ADDK command. The first step is to number the global DOF for the structure shown below. Notice that all possible DOFs are numbered, including axial. This simplifies the process by not requiring additional constraint transformations.

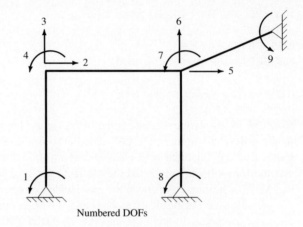

Numbered DOFs

Next we need to number the elements and the i–j ends of each member. We also need to convert any distributed load into its equivalent concentrated loads, $\mathbf{S}^f$. The coordinates of the nodes are also given. The following figure gives all these data.

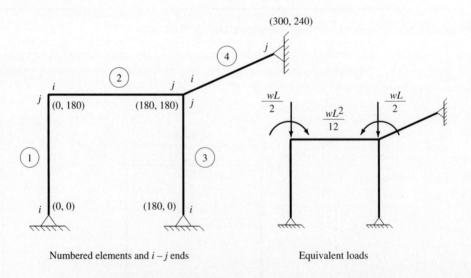

Numbered elements and $i - j$ ends   Equivalent loads

The next step is to form the **LM** matrix (connectivity) for each element. Recall that this is just the X, Y, and θ global DOFs at the i and then the j ends. The required **LM** matrix, one column for each element, is

$$
\mathbf{LM} = \begin{bmatrix} 0 & 2 & 0 & 5 \\ 0 & 3 & 0 & 6 \\ 1 & 4 & 8 & 7 \\ 2 & 5 & 5 & 0 \\ 3 & 6 & 6 & 0 \\ 4 & 7 & 7 & 9 \end{bmatrix}
$$

Using the procedure given in Section 5.9.6, we can create a CAL-90 batch file that can be used to analyze the structure. One possible sequence of CAL-90 commands is the following:

```
Example: Frame Using CAL-90 and Direct Stiffness Commands
qe2
zero kg r=9 c=9   : Zero global stiffness
loadi lm r=6 c=4  : Load connectivity matrix
0 2 0 5
0 3 0 6
1 4 8 7
2 5 5 0
3 6 6 0
4 7 7 9
C-- Form stiffness columns
frame k13 t13 x=0,0 y=0,180 e=29000 i=1150 a=2.1
addk kg k13 lm n=1        : Assemble column 1
addk kg k13 lm n=3        : Assemble column 3
C-- Form stiffness beam
frame k2 t2 x=0,180 y=180,180 e=29000 i=1150 a=2.1
addk kg k2 lm n=2         : Assemble beam
C-- Form slanted member
frame k4 t4 x=180,300 y=180,240 e=29000 i=1150 a=2.1
addk kg k4 lm n=4         : Assemble slanted member
print kg
C-- Form load vector
load rhs r=9 c=1          : Concentrated load portion
0
6
0
0
0
0
0
0
```

```
0
load sf1 r=6 c=1              : Distributed member load
0
-0.1*180/2
-0.1*180*180/12
0
-0.1*180/2
0.1*180*180/12
addl rhs sf1 lm n=2          : Assemble member load to global
p rhs
C-- Solve for displacements
solve kg rhs
p rhs                        : Print displacements
C-- Recover forces
memfrc t13 rhs lm f1 n=1     : Recover member 1
p f1
memfrc t2 rhs lm f2 n=2      : Recover member 2
sub f2 sf1                   : Subtract fixed-end loads
p f2
memfrc t13 rhs lm f3 n=3     : Recover member 3
p f3
memfrc t4 rhs lm f4 n=4      : Recover member 4
p f4
return
```

The file above is named "INPUT" and can be processed using the SUBMIT command. From within CAL-90, execute the command

CAL-90 > SUBMIT QE2

The results will be placed in the file named "OUTPUT." From this file we can find where the forces for the beam member are printed (P F2). This portion of the output is

```
MEMFRC T2 RHS LM F2 N=2
SUB F2 SF1
P F2

PRINT OF ARRAY NAMED ''F2   ''

Col# =              1
ROW   1      6.51672
ROW   2      8.48702
ROW   3     93.01022
ROW   4     -6.51672
ROW   5      9.51298
ROW   6   -185.34611
```

Since we want the moment at the i end, we need the third force in the vector. The value is: 93.01 kip-in. Notice that the S^f vector had to be subtracted from the forces recovered directly from the displacements.

Condensed Example: Truss by Direct Stiffness Using the Split a Matrix

Using the two-part **a** matrix ($\mathbf{a}_\alpha \mathbf{a}_L$) method and CAL-90, find the stiffness for the following truss.

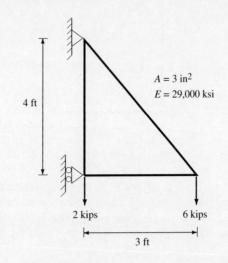

$$A = 3 \text{ in}^2$$
$$E = 29{,}000 \text{ ksi}$$

4 ft

2 kips   6 kips

3 ft

As always, the first step is to identify the global DOFs. The following DOFs are used:

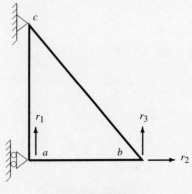

Global DOF definitions

To solve this structure, we would like to have a TRUSS command that creates the rotated element stiffness ($\mathbf{K}_{xy}$) and the force transformation matrix, just like the FRAME command. However, there is no such command. As a result, we will use the FRAME command but set the moment of inertia equal to zero. Remember, a frame element stiffness consists of the bending portion and the uncoupled axial portion. The area controls the axial portion and the moment of inertia controls the bending portion. To use the FRAME command, we need to create the **LM** (connectivity) matrix. To do this we need to number the members and the $i—j$ ends of the members. The following convention is used:

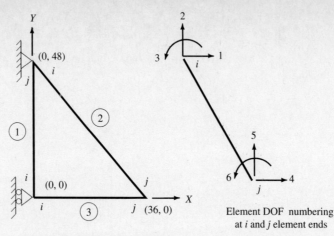

Element numbers and $i-j$ ends

Element DOF numbering
at i and j element ends

Note that the element DOF numbering (in global directions) used is that for the frame element. However, since there are *no* rotational DOFs, these DOFs will *always* be zero in the **LM** matrix. Hence rows 3 and 6 will always be zero. Using the figure above, the **LM** matrix is defined as

$$
\mathbf{LM} =
\begin{bmatrix}
0 & 0 & 0 \\
1 & 0 & 1 \\
0 & 0 & 0 \\
0 & 2 & 2 \\
0 & 3 & 3 \\
0 & 0 & 0
\end{bmatrix}
$$

Using the standard procedure for forming a structural stiffness, we can create the required CAL-90 input file:

```
Example: Truss Using Two-Part a Matrix
Q13
zero k r=3 c=3        : Zero global stiffness
loadi lm r=6 c=3      : Load integer LM matrix
0 0 0
1 0 1
0 0 0
0 2 2
0 3 3
0 0 0
frame k1 t1 x=0,0 y=0,48 a=3 e=29000 i=0
addk k k1 lm n=1
p k1
```

```
frame k2 t2 x=0,36 y=48,0 a=3 e=29000 i=0
addk k k2 lm n=2
p k2
frame k3 t3 x=0,36 y=0,0 a=3 e=29000 i=0
addk k k3 lm n=3
p k3
p k                   : Print final stiffness
load rhs r=3 c=1      : Load load vector
2
0
6
solve k rhs
p rhs                 : Print displacements
memfrc t1 rhs lm f1 n=1
p f1
memfrc t2 rhs lm f2 n=2
p f2
memfrc t3 rhs lm f3 n=3
p f3
return
```

To analyze the structure, CAL-90 needs to be started and the following command given:

CAL-90> submit q13

The final stiffness can be found in the output file. The portion of the output giving the printed version of the global stiffness matrix is

```
PRINT OF ARRAY NAMED ''K       ''

COL# =            1            2            3
ROW   1  1812.50000      .00000      .00000
ROW   2      .00000  2938.66667  -696.00000
ROW   3      .00000  -696.00000   928.00000
```

This is identical to the stiffness found using the methods in the preceding two examples.

5.11 Three-Dimensional Beam Elements

The extension of the frame and truss elements into full three-dimensional elements is a fairly straightforward manipulation. The method of derivation is identical for three dimensions as two dimensions. The stiffness matrix definition still requires application of a unit displacement at one of the DOFs and calculation of the forces required to maintain equilibrium. Let's look at the three-dimensional bending element. In three dimensions we need to consider six DOFs at each node. The three translations and three rotations at each end of the beam are shown in Figure 5.34.

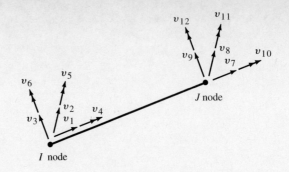

Figure 5.34 Local DOFs for three-dimensional beam element.

This results in a 12 × 12 stiffness matrix. We will then need to consider how to rotate this 12 × 12 into global coordinates. First let's look at the stiffness. We will derive the stiffness in the local coordinate system defined by $X'-Y'-Z'$. If we apply the required unit deformations, we see that the member acts like two uncoupled two-dimensional bending members. The deformed shapes and restraining forces are identical for the member in the local $X'-Y'$ and $Y'-Z'$ planes. The only difference is a change in axes about which the properties are calculated. From the deformed shapes we can form the stiffness matrix. The local element coordinate system and the element DOFs are defined by Figure 5.35. We can consider the two planes independently for which the element stiffness needs to be derived. On pp. 204 and 205, Figures 5.36 and 5.37 show the two-dimensional bending elements, their DOFs, and the deformed shapes required, together with the stiffness terms for the three-dimensional beam element in the $X'-Y'$ and $X'-Z'$ planes, respectively.

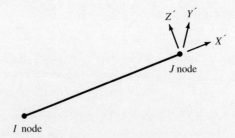

Figure 5.35 Local element coordinate system.

Also needed is the stiffness for the torsional rotation or torque stiffness. The stiffness for the torsion is the familiar mechanics value:

$$\mathbf{K}_{torsion} = \frac{JG}{L} \tag{5.70}$$

From the deformed shapes and forces, the forces given with their correct sign for the coordinate system used (in both planes), we can form the three-dimensional beam stiffness matter. The final stiffness matrix is

$$
\begin{bmatrix}
\frac{AE}{L} & 0 & 0 & 0 & 0 & 0 & -\frac{AE}{L} & 0 & 0 & 0 & 0 & 0 \\[2mm]
0 & \frac{12EI_z}{L^3} & 0 & 0 & 0 & \frac{6EI_z}{L^2} & 0 & -\frac{12EI_z}{L^3} & 0 & 0 & 0 & \frac{6EI_z}{L^2} \\[2mm]
0 & 0 & \frac{12EI_y}{L^3} & 0 & -\frac{6EI_y}{L^2} & 0 & 0 & 0 & -\frac{12EI_y}{L^3} & 0 & -\frac{6EI_y}{L^2} & 0 \\[2mm]
0 & 0 & 0 & \frac{JG}{L} & 0 & 0 & 0 & 0 & 0 & -\frac{JG}{L} & 0 & 0 \\[2mm]
0 & 0 & -\frac{6EI_y}{L^2} & 0 & \frac{4EI_y}{L} & 0 & 0 & 0 & \frac{6EI_y}{L^2} & 0 & \frac{2EI_y}{L} & 0 \\[2mm]
0 & \frac{6EI_z}{L^2} & 0 & 0 & 0 & \frac{4EI_z}{L} & 0 & -\frac{6EI_z}{L^2} & 0 & 0 & 0 & \frac{2EI_z}{L} \\[2mm]
-\frac{AE}{L} & 0 & 0 & 0 & 0 & 0 & \frac{AE}{L} & 0 & 0 & 0 & 0 & 0 \\[2mm]
0 & -\frac{12EI_z}{L^3} & 0 & 0 & 0 & -\frac{6EI_z}{L^2} & 0 & \frac{12EI_z}{L^3} & 0 & 0 & 0 & -\frac{6EI_z}{L^2} \\[2mm]
0 & 0 & -\frac{12EI_y}{L^3} & 0 & \frac{6EI_y}{L^2} & 0 & 0 & 0 & \frac{12EI_y}{L^3} & 0 & \frac{6EI_y}{L^2} & 0 \\[2mm]
0 & 0 & 0 & -\frac{JG}{L} & 0 & 0 & 0 & 0 & 0 & \frac{JG}{L} & 0 & 0 \\[2mm]
0 & 0 & -\frac{6EI_y}{L^2} & 0 & \frac{2EI_y}{L} & 0 & 0 & 0 & \frac{6EI_y}{L^2} & 0 & \frac{4EI_y}{L} & 0 \\[2mm]
0 & \frac{6EI_z}{L^2} & 0 & 0 & 0 & \frac{2EI_z}{L} & 0 & -\frac{6EI_z}{L^2} & 0 & 0 & 0 & \frac{4EI_z}{L}
\end{bmatrix}
\tag{5.71}
$$

The subscripts on the moment of inertia indicate about which axis bending takes place. For example, I_z is the property used when bending takes place in the X'–Y' plane or movement in DOFs 2, 6, 8, or 12.

5.11.1 Rotational Transformation in Three Dimensions

Following the standard procedure, any local element needs to be rotated to global directions before assembly. This process is identical for a three-dimensional element. What is required is a rotation matrix that rotates the three coordinate axes, not just the two axes as in two dimensions. By the usual definition, we need to convert the global coordinates to local coordinates. Just as the two-dimensional case, we need the projections of a unit vector in each of the local coordinate vectors onto the global system. If the vector components of these unit vectors along each of the local directions are placed in the rows of the matrix, one for each local coordinate, X', Y', and Z', we have the required rotation matrix. The required projections for the X' coordinate are shown in Figure 5.38.

The global components of the local $\mathbf{X}'$ unit vector become row 1 of the transformation matrix. Following the same procedure as that used for the $\mathbf{Y}'$ and $\mathbf{Z}'$ vectors, we get the

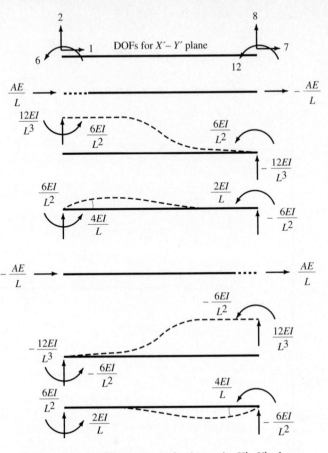

Figure 5.36 Stiffness terms for beam in $X'-Y'$ plane.

final rotation matrix:

$$\mathbf{v} = \begin{bmatrix} \alpha_{x'x} & \alpha_{x'y} & \alpha_{x'z} \\ \alpha_{y'x} & \alpha_{y'y} & \alpha_{y'z} \\ \alpha_{z'x} & \alpha_{z'y} & \alpha_{z'z} \end{bmatrix} \mathbf{r} = \boldsymbol{\alpha}\mathbf{r} \qquad (5.72)$$

where the subscript indicates the projection of the first vector onto the second. For example, $\alpha_{x'x}$ is the component in the $\mathbf{X}$ global coordinate of the unit vector along the $\mathbf{X}'$ local axis. This matrix is also referred to as the *matrix of direction cosines* since it can also be defined as the cosine of the angle between the local and global vectors. The $\alpha_{x'x}$ value is also the cosine of the angle between the $\mathbf{X}$ and $\mathbf{X}'$ vectors. While direction cosines are usually given as the definition of the transformation matrix, vector components are generally used to create the matrix, due to the simplicity of formation.

To use vector methods to form the rotation matrix, the vector components of the local coordinate axes are required. In most analysis programs, we only want the user to specify the end nodes of a member and some easy way to define the orientation of the

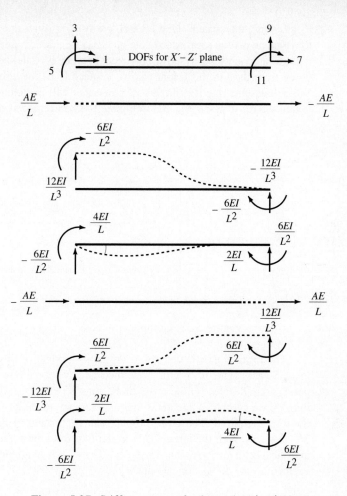

Figure 5.37 Stiffness terms for beam in X'–Z' plane.

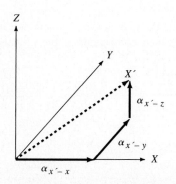

Figure 5.38 Components of **X**$'$ vector in global coordinates.

local coordinate system. One common way is through the use of a third point in space to define one of the local coordinate planes. Since three points define a plane, this uniquely defines the coordinate system. The usual practice is to define the local X' axis as a vector going from the I to the J node. Now, we need to define either the $X'-Y'$ or $X'-Z'$ plane. We will assume that it is the $X'-Z'$ plane that will be defined. If we give any vector that lies in the $X'-Z'$plane, we can complete the coordinate system (Figure 5.39).

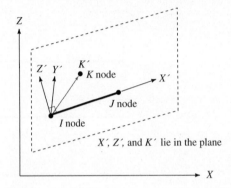

Figure 5.38 Components of $\mathbf{X}'$ vector in global coordinates.

The local Y' axis is defined as the cross product of this in-plane vector, $\mathbf{K}'$, into the local X' axis. The $\mathbf{K}'$ vector is defined as a vector going from the I node to the third point, node K. Clearly, the cross of $\mathbf{K}'$ into $\mathbf{X}'$ gives the vector $\mathbf{Y}'$. If we then cross $\mathbf{X}'$ into $\mathbf{Y}'$, we get the $\mathbf{Z}'$ vector. If these vectors are made into unit vectors, we have exactly the terms required. The cross product can be formed in the following manner. Form a 3×3 matrix with the terms i, j, k in the first row, the vector to cross in the second row, and the vector to cross into in the third row. For example, if we want to cross the $\mathbf{K}'$ vector into the $\mathbf{X}'$ vector, we have

$$\det \begin{vmatrix} i & j & k \\ K'_x & K'_y & K'_z \\ X'_x & X'_y & X'_z \end{vmatrix} = \mathbf{K}' \times \mathbf{X}' \tag{5.73}$$

where X'_x is the global X component of the unit vector along the X' axis. The subscript indicates which component in the global coordinate system is used. The K'_x is the X component of the $\mathbf{K}$ vector. Note that the $\mathbf{X}'$ vector is usually just the vector from the I to the J node.

If we take the determinate of the matrix above, the result is the vector resulting from the cross product of $\mathbf{K}'$ into $\mathbf{X}'$. The $i, j,$ and k values specify the $X, Y,$ and Z components of the vector. If we take the determinate, we get (in our case) the components for the $\mathbf{Y}'$ vector:

$$\begin{aligned} \mathbf{Y}'(\text{components}) &= (K'_y X'_z - K'_z X'_y)\mathbf{i} \\ &\quad - (K'_x X'_z - K'_z X'_x)\mathbf{j} \\ &\quad + (K'_x X'_y - K'_y X'_x)\mathbf{k} \end{aligned} \tag{5.74}$$

This vector needs to be converted to a unit vector before placing it into the transformation matrix.

The rotation matrix can be used to convert the three translations or three rotations at either end of the three-dimensional bending element. As before, this matrix needs to be put into the larger $\mathbf{a}_\alpha$ matrix needed to transform the stiffness. The $\mathbf{a}$ matrix for a three-dimensional beam is a 12×12 matrix composed of the above 3×3 matrix on the diagonal in blocks. If we represent the 3×3 matrix of direction cosines by the matrix α, we can show the final bending transformation matrix as

$$\mathbf{a}_\alpha = \begin{bmatrix} \alpha & 0 & 0 & 0 \\ 0 & \alpha & 0 & 0 \\ 0 & 0 & \alpha & 0 \\ 0 & 0 & 0 & \alpha \end{bmatrix} \qquad \alpha = 3 \times 3 \text{ direction cosine matrix} \qquad (5.75)$$

The $\mathbf{a}$ matrix above can also be used to transform any vector quantity, displacements, and forces, completely consistent with all equations derived in previous chapters.

5.12 Problems

5.1. Using the full (not two-part) $\mathbf{a}$ matrix method for stiffness formation, calculate the stiffness matrix for the following structures. Ignore axial deformations. Use CAL-90.

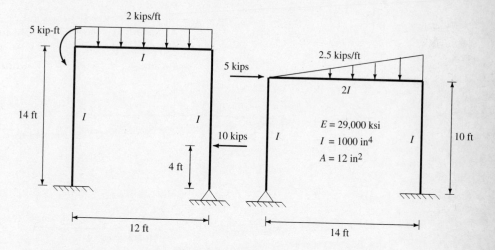

5.2. For the structures in Problem 5.1, calculate the load matrix, **R**.

5.3. Using the full (not two-part) **a** matrix method for stiffness formation, find the 2 × 2 global stiffness matrix for the two DOFs shown. Use CAL-90.

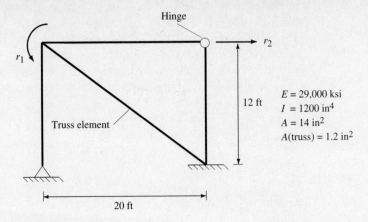

5.4. Given the following structure:
   (a) Using the full (not two-part) **a** matrix method for stiffness formation, develop the stiffness matrix for the two DOFs shown. Use $E = 29,000$ ksi, $I = 750$ in⁴, and $A = 18$ in² for the member properties. Assume small angles for the truss member and no axial deformations for the beams.
   (b) Develop the global load vector **R** for the load shown.

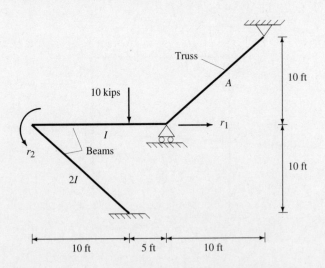

5.5. Given the following structure, find the stiffness matrix for the two DOFs shown. Use the full (not two-part) **a** matrix method for stiffness formation. Use Cal-90.

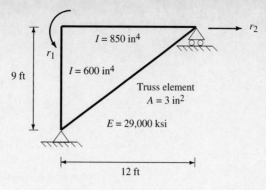

5.6. Use the full (not the two-part) **a** matrix to find the displacements and support reactions for the following structures.

(a)

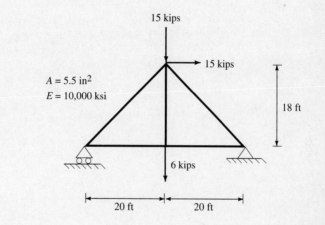

(b)

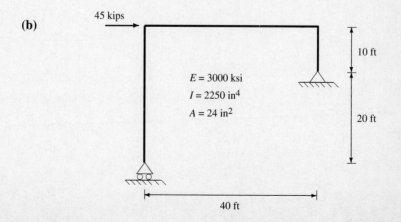

(c)

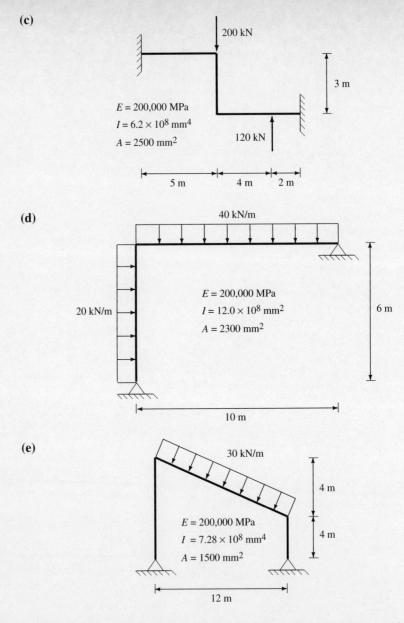

200 kN

3 m

$E = 200,000$ MPa
$I = 6.2 \times 10^8$ mm^4
$A = 2500$ mm^2

120 kN

5 m   4 m   2 m

(d)

40 kN/m

$E = 200,000$ MPa
$I = 12.0 \times 10^8$ mm^2
$A = 2300$ mm^2

6 m

20 kN/m

10 m

(e)

30 kN/m

4 m

$E = 200,000$ MPa
$I = 7.28 \times 10^8$ mm^4
$A = 1500$ mm^2

4 m

12 m

5.7. Analyze Problems 4.2 through 4.11 using the full (not the two-part) **a** matrix method.

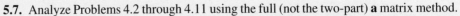

5.8. Using CAL-90 and the direct stiffness commands, analyze the following structures and find the displacements and the member forces. Check overall structural equilibrium using the support reactions.

(a)

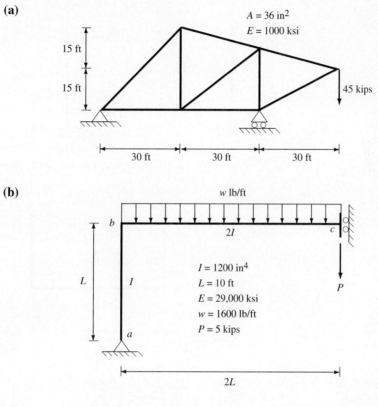

$A = 36$ in^2
$E = 1000$ ksi

15 ft

15 ft

45 kips

30 ft 30 ft 30 ft

(b)

w lb/ft

b

$2I$

c

L I

$I = 1200$ in^4
$L = 10$ ft
$E = 29{,}000$ ksi
$w = 1600$ lb/ft
$P = 5$ kips

P

a

$2L$

(c)

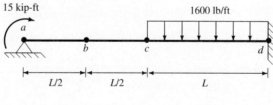

15 kip-ft

a

b c d

1600 lb/ft

$L/2$ $L/2$ L

$L = 10$ ft $E = 29{,}000$ ksi $I = 1200$ in^4

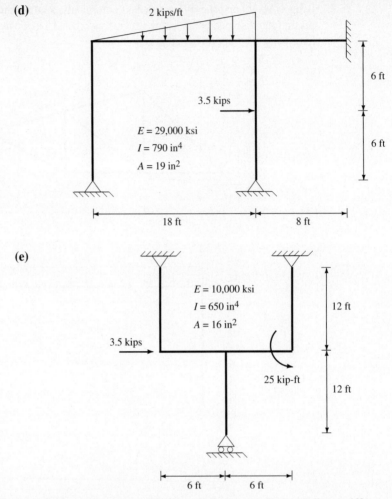

(d)

2 kips/ft

3.5 kips

$E = 29{,}000$ ksi

$I = 790$ in^4

$A = 19$ in^2

6 ft

6 ft

18 ft

8 ft

(e)

$E = 10{,}000$ ksi

$I = 650$ in^4

$A = 16$ in^2

12 ft

3.5 kips

25 kip-ft

12 ft

6 ft

6 ft

5.9. Reanalyze the structures in Problem 5.6 using the CAL-90 direct stiffness commands.

5.10. Reanalyze the structures in Problems 4.2 through 4.11 using the CAL-90 direct stiffness commands.

5.11. For the following structures, use CAL-90 and the direct stiffness commands to perform the analysis. Find the displacements and member forces.

(a)

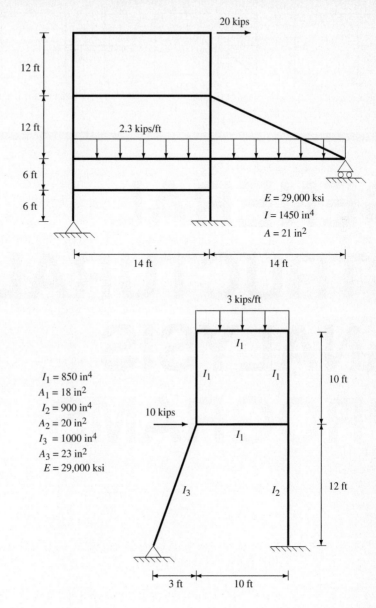

20 kips

12 ft

12 ft

2.3 kips/ft

6 ft

6 ft

$E = 29{,}000$ ksi
$I = 1450$ in^4
$A = 21$ in^2

14 ft

14 ft

(b)

3 kips/ft

I_1

I_1

I_1

I_1

10 ft

$I_1 = 850$ in^4
$A_1 = 18$ in^2
$I_2 = 900$ in^4
$A_2 = 20$ in^2
$I_3 = 1000$ in^4
$A_3 = 23$ in^2
$E = 29{,}000$ ksi

10 kips

I_1

I_3

I_2

12 ft

3 ft

10 ft

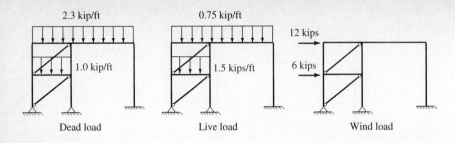

2.3 kip/ft

1.0 kip/ft

Dead load

0.75 kip/ft

1.5 kips/ft

Live load

12 kips

6 kips

Wind load

GENERAL STRUCTURAL ANALYSIS PROGRAMS

6.1 General Program Structure

General structural analysis programs perform in exactly the same steps as those given in Chapter 5. The differences lie in the order in which the steps are performed to increase computational efficiency. Since the procedure is basically the same, the required input for any finite-element analysis program is almost identical. Again, changes for efficiency, ease of use, and type of computer all play a role in choosing the details of implementation. Regardless of implementation details, the order discussed below is the most common.

6.1.1 Set Up the Problem Geometry

The problem geometry consists of defining points in space usually called *nodes*. The location of a node is defined by its coordinates (X and Y in two dimensions, X, Y, and Z in three dimensions). Each node is assigned boundary conditions that tells in what directions the node is allowed to move. In two dimensions, a node is allowed to move in the X and Y directions and rotate about the Z axis (assuming an X–Y plane). In three dimensions, a node is allowed to move in the X, Y, and Z directions and rotate about all three axes. Most computer programs require you to give the coordinates and boundary conditions of each node. The boundary conditions are generally specified by allowing or not allowing movement of the node in a particular direction. Most programs allow for three-dimensional analysis, and therefore special attention has to be given to a two-dimensional problem. All nodes are generally assumed to have the six DOFs shown in Figure 6.1 (p. 216).

For a two-dimensional problem in the X–Y plane, you need only consider the DOFs shown in Figure 6.2 (p. 216). When using a three-dimensional program to perform a two-dimensional analysis, it is mandatory that the additional DOFs not required in two dimensions be fixed (not allowed to move). This requirement is a specialized form of the rule that only the DOFs that have stiffness can be allowed to move. For example, our two-dimensional frame element in a general X–Y configuration has the DOFs shown in Figure 6.3 (p. 216). Therefore, these are the only DOFs to which the element can add stiffness in the global stiffness matrix. Since nodes in general three-dimensional programs are allowed to move and rotate in the X, Y, and Z directions, we need to prevent that movement where no stiffness exists. The same concepts are true for truss elements. A general two-dimensional version of a truss element would have DOFs as shown in Figure 6.4 (p. 217).

We would have to prevent the motion of the DOFs that do not have stiffness supplied by the truss. In this case, the rotation in the plane must also be fixed. This does not mean

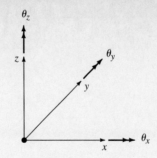

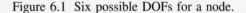

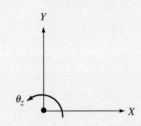

Figure 6.1 Six possible DOFs for a node.

Figure 6.2 Three possible DOFs in two dimensions for a node.

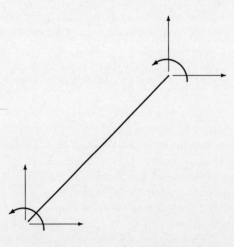

Figure 6.3 Actual DOFs for general two-dimensional frame element.

that truss bars do not rotate with respect to each other. It does mean that there is no bending rotation. As a result, a truss rotation can be calculated by the difference in end displacements and is not a DOF. Once all the DOFs have been designated as released or fixed, most programs assign numbers to the DOFs that are released. These numbers are the equation numbers for the unknown displacement at that released DOF. The order in which the equation numbers are assigned can greatly affect the efficiency of the solution process. Most of the more complete programs include a renumbering scheme that automatically numbers the equations (DOFs) in an optimal fashion.

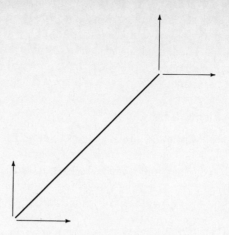

Figure 6.4 Actual DOFs for general two-dimensional truss element.

6.1.2   Form All Element Matrices

Formation of all element matrices consists of forming the element stiffness in global directions ($\mathbf{K}_{xy}$), the force transformation matrix ($\mathbf{FT}_\alpha = \mathbf{K}_e \mathbf{a}_\alpha$), the $\mathbf{LV}$ vector (the pointer form of the $\mathbf{a}_L$ matrix), and the member loads in the global and local directions ($\mathbf{S}^f$ and $\mathbf{a}_\alpha^T \mathbf{S}^f$). This step requires the numbers of the nodes to which the element is connected in order to acquire all the information to form the element matrices. By knowing the nodes, the coordinates at the ends of the member can be found and the rotation angle and member length calculated. The properties of the element and member load are the only additional information needed to form the required matrices.

6.1.3   Assemble the Global Stiffness and Loads

Now that all the element matrices have been formed, the pointer method of assembly can be used to form the global stiffness and load matrices. The load matrix has contributions from the element loads, which must also be assembled, and concentrated loads. It is at this point that the concentrated loads are usually added to the element contributions assembled into the global load vector.

6.1.4   Solve for the Unknown Displacements

To solve for the unknown displacements, any type of equation solver can be used that solves simultaneous equations. The most common forms are the banded and the profile or skyline solvers. Both of these are variations on the Gauss elimination technique discussed in Chapter 2. These are merely storage manipulations that can reduce memory and computational demands. The profile is generally better than the banded; however, both are orders of magnitude better than regular square matrix techniques.

Note that there are methods that combine the assembly and solution steps into one. One such method is called the *frontal method*. It works by assembling a portion of the global stiffness and then solves a portion of the equations. The method continues in this fashion until the solution is complete. The method is identical to the profile solution method in terms of computations but uses less memory during the process. It is also more restrictive for other advanced analysis techniques, such as dynamics and nonlinear. Iterative solution techniques such as the preconditioned conjugate gradient method are becoming more prevalent as computer technology changes to parallel architectures.

6.1.5 Recover the Element Forces

Using the solved-for displacements, we can recover the element forces. The **LV** pointer is again required to tell which DOFs are connected to each element. Once the correct DOFs are found, forces can be recovered using the force transformation matrix calculated earlier ($\mathbf{FT}_\alpha = \mathbf{K}_e\mathbf{a}_\alpha$) and the appropriate displacements. Again, the element load vector needs to be subtracted to correct the forces for element loadings ($\mathbf{S}^f$) for beam members.

6.2 Load Cases

There are two other concepts that occur when using most general analysis programs. The first is the definition of multiple-load cases. In design, the building codes usually require that you check a structure for different combinations of loadings to determine which combination is the most critical. For example, you might need to check dead plus live load, $\frac{3}{4}$(dead $+ \frac{1}{2}$ live $+$ wind), and so on. Since only the loading and not the structure changed, you can solve for these multiple-load cases simultaneously.

Most programs allow you to define multiple-load cases that consist of different sets of loads, both concentrated and distributed. In addition, some programs allow for these basic load cases to be combined to form load combinations. For example, in Figure 6.5, three load cases are defined. These three load cases could be combined in any fashion required to satisfy building code requirements. For example, you could combine load cases 1 and 2 (dead + live) to form a combination needed to perform a code design. Many programs allow you to specify combinations of these basic load cases. Programs

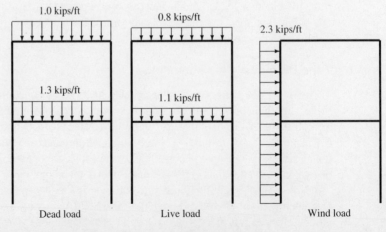

Figure 6.5 Basic load cases.

that do not have combination capability can perform the same analysis by defining the required combinations as one of the basic load cases. For example, the required combination of dead plus live load could be achieved by using the applied loads given in Figure 6.6 as basic load input.

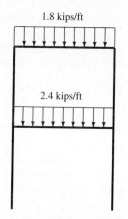

Figure 6.6 Dead and live load combined.

6.3 Element Types

General analysis programs usually have more than one type of element. Most usually have the standard truss and frame elements as well as a number of finite elements. The program sometimes needs to be informed about how many of these different types of elements are going to be used in an analysis. Therefore, the number of element types refers to how many different stiffness matrix types (truss, frame, plate, etc.) are to be used. This is not to be confused with the number of different cross sections and material properties specified for a particular type of element. For example, this is not the number of different steel wide-flange sections that are required for a particular analysis. The number of different sections is an element property, not an element type.

6.4 SSTAN Capabilities

The general analysis program included with this book is called SSTAN. It is a three-dimensional finite-element program that has a graphics postprocessor. It is written to run on a microcomputer and designed to be used as an educational tool. It is a batch program which assumes that the structural information is contained in a file named ''INPUT.'' The results are put into a file named ''OUTPUT.'' SSTAN uses the same free-form input as CAL-90. Therefore, comment lines, arithmetic statements, and header lines are all valid. We will explain what input is required and its form and show an example later in this chapter. See Appendix B for more details on the required input. Since SSTAN is a batch program, all of the input needs to be prepared in advance and then the program runs a complete analysis without additional user interaction. The input is prepared and stored in the INPUT file. Then the program is run by typing

C:\> sstan

SSTAN can be stored in any directory on the PC; however, it is an overlaid program and therefore needs to have a path to the directory in which it is stored or it must be in the current directory. The INPUT file must always be in the current directory. A summary of the required input follows. The first line in the file is an analysis title. *It must be the first line.* It is also important that it does not begin with any of the header names, such as COORDINATE, FRAME, and so on. The second line is the analysis control information line. It is of the form

N1, N2, N3, N4

where N1 is the number of nodes in the structure and N2 is the number of different element types in the structure. Again, this means that if you are using both truss and beams, there are two, regardless of the number of member properties. N3 is the number of load cases and N4 is the number of load combinations.

The rest of the INPUT is on a free-formatted header basis. All data are arranged by groups and signified by a header. For example, nodal coordinates are signified by the header COORDINATE. These data are order independent; that is, they can be placed in the INPUT file in any order. All data groups must end with a blank line. The following headers are available:

1. *COORDINATE.* This specifies the nodal coordinates of the structure to be analyzed. Note that the program assumes that the structure is three-dimensional.

2. *BOUNDARY.* This specifies the boundary conditions or nodal DOFs. Nodes can be either released or fixed and only in the global $X-Y-Z$ coordinate system.

3. *TRUSS.* This specifies the truss element data. Trusses can have initial tension, rigid-end offsets, and the ability not to take either compression or tension (not both). The zero-compression option is useful for slender bracing members. The zero-tension option is useful for gap elements or uplift problems.

4. *BEAM.* This specifies the bending member data. These elements can be used for beams as well as columns. Beams can have uniform loads applied to the member, rigid-end offsets, and can include $P-\Delta$ effects.

5. *LOADS.* This specifies the concentrated loads applied to the structure. They can be concentrated loads or moments.

6. *COMBINATIONS.* This section allows linear combinations of the basic load cases. This option is not valid when using noncompression/tension trusses or $P-\Delta$ effects.

7. *PLATE, MEMBRANE, SHELL, BRICK, and AXISYMMETRIC.* These additional types of members, which constitute what are usually called finite elements, are described in later chapters. Several examples will be given to demonstrate the use of SSTAN.

The coordinate system assumed in SSTAN is a right-hand-rule system. All displacements, forces, and moments are given in this right-handed system. The results for element

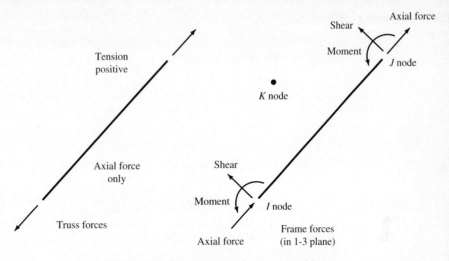

Figure 6.7 Force output for truss and beam.

forces are given at the nodes of a member in the local coordinate system. The local coordinate system for the truss and beam members is given in Figure 6.7.

6.5 Use of STANPLOT

STANPLOT is a graphics postprocessor for the SSTAN program. It is a mouse-based program that allows the user to view the structure and results of an analysis. STANPLOT can plot any of the SSTAN element types, displaced shapes, axial and moment results for beams and truss members, and stress contours for membranes, plates, shells, and solid elements. The program allows windowing and changing of the structure view as well as many other useful options.

To use the program, the file PLOT.FON must be in the root directory. SSTAN creates a number of output files, some of which are required by STANPLOT. The data files that are created by running the SSTAN program must be in the current directory. Some of the files are temporary for the plotting program, and some are output files. STANPLOT requires the following files:

PLTDB   Structure geometry and displacement results

STRESS   Stress information for membrane, plate, and solid elements (if any exist)

MOMENT   Moment results for beam elements (if any exist)

AXIAL   Axial results for truss and beam elements (if any exist)

VMDB   Shear and bending moment diagram information for beams (if any exist)

STANPLOT creates two additional temporary files, AXIDA and MOMDA. These files will be created by every run of STANPLOT and do not need to be saved.

6.5.1 Running STANPLOT

When STANPLOT is run, the screen will display the main menu items (Figure 6.8). A description of each menu item follows.

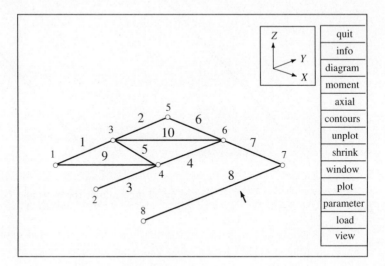

Figure 6.8

QUIT

This option quits the STANPLOT program.

INFO

This option allows you to get the nodal properties for any node in the structure. When it is chosen, a small menu box will be displayed where the nodal information will be given. By moving the mouse to a node and clicking the left mouse button, the node's information will be given. If a load case has been chosen, the node's displacements will also be given.

DIAGRAM

This option allows plotting of the shear and moment diagrams for any beam member. Again, a small menu box will be displayed asking for a member to be chosen. Once a member is chosen, the diagrams can be plotted for either local axis.

MOMENT

This option allows plotting of the maximum moment found in each beam member for the entire structure. Each member will be color coded as to its maximum moment.

AXIAL

This option allows plotting of the axial stress in each beam and truss member. Again, the members are color coded as to their magnitude.

CONTOURS

This option allows the plotting of stress or displacement contours for membrane, shell, plate, and brick elements. A pop-up menu box will give you the choice of stress or displacement. A second menu will give a further choice of which result to plot (e.g., X displacement, stress in the X direction, etc.). Windowing works with this option.

UNPLOT

This option allows selective elements to be turned off, hence not plotted. Individual or entire groups of elements can be turned off. A new menu is given from which the type of element to turn off can be chosen. A second menu is given to select ALL or SOME elements. If SOME is chosen, a final box requests the element numbers to turn off.

SHRINK

This option shrinks all the elements to help check connectivity. This is useful for membrane, plate, and solid elements, where it is difficult to see the connectivity on adjacent elements.

WINDOW

This option offers an explanation of how to use the window option. It also allows you to return to the default viewing window, which is the entire structure. Windowing works by clicking the left mouse button on two opposite corners of the area you wish to zoom in on. This process can continue for up to 40 levels. The right mouse button causes the zoom to back out one level. Pressing both the left and right mouse buttons causes the system to return to the default window.

PLOT

This causes the structure to be replotted. This allows many different options to be chosen before the structure is replotted.

PARAMETER

This option produces another menu that allows many setup options to be changed. The main item of interest is the SIZE suboption. This choice allows the element and node numbers to be turned on and off. Other options, such as number of contour colors and node markers, can also be turned on and off or changed.

LOAD

This option allows the choice of load case to be used for result plotting. If the analysis failed, no load case will be available but the structure can be plotted. Load combination results are added to the result set of load cases. Hence, if there are three load cases and two load combinations, there will be a total of five result sets. The first three are the load cases and the last two are from the load combinations. This is true for displacements as well as moments, axial loads, and stresses.

VIEW

This option allows the view direction for the structure to be changed. A menu box will be displayed with the current view of the coordinate system and choices for changing. The choices are:

Menu	Returns to the main plot menu.
Default	Gives the default view.
X–Y plane	Gives a view of the X–Y plane.
Y–Z plane	Gives a view of the Y–Z plane.
Z–X plane	Gives a view of the Z–X plane.
FWD or REV	Toggle that chooses the direction of rotation when using the X, Y, or Z choices.
X	Rotates the structure about the screen X direction (horizontal axis).
Y	Rotates the structure about the screen Y direction (vertical axis).
Z	Rotates the structure about the screen Z direction (axis pointing out of the screen).
10 deg	Changes the amount the structure is rotated when using the X, Y, and Z options. Clicking this cycles between different rotation increments.

The plots shown in Figures 6.9 through 6.15 demonstrate some of the capabilities of STANPLOT. The sequence of the plots is as follows. The first plot is the default view

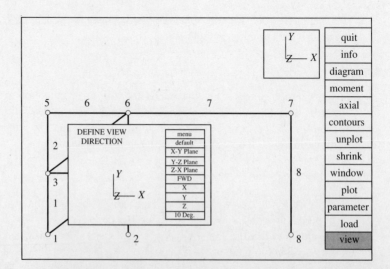

Figure 6.9

(perspective) of the structure shown in Figure 6.8. Next, the view option is chosen and then the X–Y plane direction is chosen (Figure 6.9). Now the LOAD option is chosen to pick the desired load case. Here the first load case is chosen (Figure 6.10). By choosing the PLOT option, the view and load case chosen are displayed (Figure 6.11). The load case is then switched to the second load case and then the INFO option is chosen. Node 6 is chosen by clicking on the original position of the node. Its information is displayed in the information box (Figure 6.12).

Next, LOAD is again chosen and the NONE option for no loads is chosen. The SHRINK option is chosen to see the connectivity. Finally, the windowing option is selected

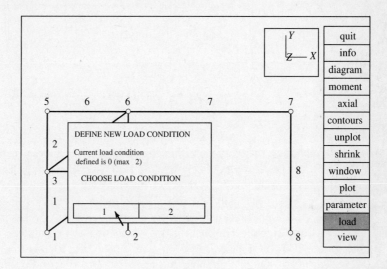

Figure 6.10

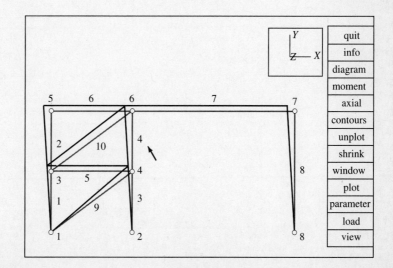

Figure 6.11

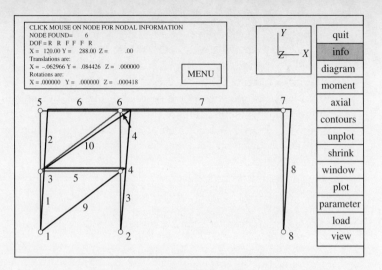

Figure 6.12

by choosing two corners within the structure. The plot is displayed as in Figure 6.13. Now the LOAD option is chosen and load case 2 is again selected. Then the structure is PLOTted. Again INFO is selected, but this time node 5 is selected. The result is shown in Figure 6.14. Finally, the DIAGRAM option is selected and then a beam member is chosen. Members are chosen by clicking the left mouse button at the element center (near the element number). Then the X–Y plane results are plotted as shown in Figure 6.15.

STANPLOT has many additional capabilities. It is an essential tool in checking the structure to see that its input and analysis are correct. Using the plotting program, such things as connectivity, node coordinates, symmetry in displacements, and many other

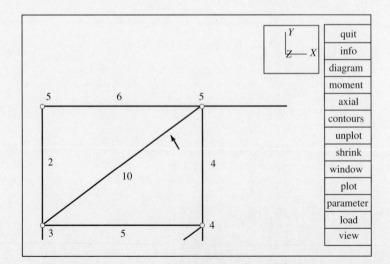

Figure 6.13

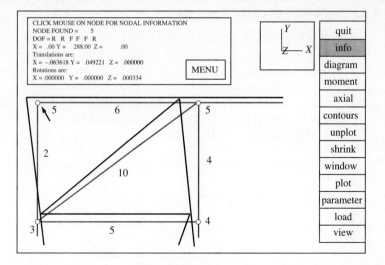

Figure 6.14

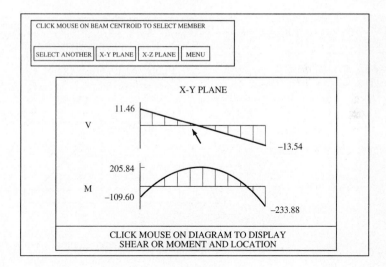

Figure 6.15

things can be checked. Verifying the structure and results is the most important part of a structural analysis.

6.6   SSTAN Examples

To better explain how to use general analysis programs, and in particular the SSTAN program, four examples will be given. These examples demonstrate the basic concepts as well as the data preparation required for truss and frame analysis.

6.6.1   Two-Dimensional Truss Example

The first example we will look at is a two-dimensional truss structure. The truss structure will be analyzed in the X–Y plane and is subjected to concentrated loads. Figure 6.16 shows the structure to be analyzed. There is only a single load case to be analyzed. The top cord members all have the same cross-sectional area. The bottom cord members also have the same properties. All remaining members, verticals and diagonals, have a third cross-sectional property. The node and element numbering used for the example is given in Figure 6.17.

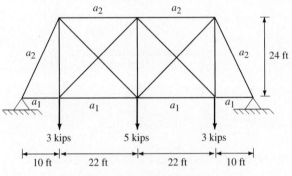

Verticals and bracing members use a_3

Figure 6.16 Truss example.

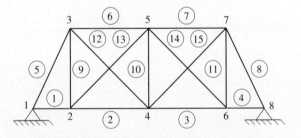

Figure 6.17 Node and element numbering.

The INPUT file required for this structure is as follows:

```
Example 1: Truss
8,1,1
:
Coordinates
1 X=0 Y=0 Z=0
2 X=10*12
3 Y=24*12
4 X=32*12 Y=0
5 Y=24*12
```

```
6 X=54*12 Y=0
7 Y=24*12
8 X=64*12 Y=0
:
Boundary
1,8 F=R,R,F,F,F,F    :  Release all X and Y displacements
1,8,7 F=F,F,F,F,F,F  :  Pin (fully fix) supports
:
TRUSS
15,3
1 A=.12 E=29000      : Bottom cord
2 A=.25 E=29000      : Top cord
3 A=.18 E=29000      : Bracing
1 1,2 M=1            : Bottom cord members
2 2,4 G=2,1,2,2
5 1,3 M=2 G=2,1,2,2  : Top cord members
8 7,8
9 2,3 M=3 G=2,1,2,2  : Vertical members
12 3,4 M=3 G=1,2,2,2 : Diagonals
13 2,5 G=1,2,2,2
:
Loads
2,6,4 L=1 F=0,-3
4 L=1 F=0 -5
:
```

Note how the boundary conditions are specified. First, all nodes are given the two-dimensional truss DOFs for a structure in the X–Y plane, released only to translate in the X and Y directions. Then the two pinned nodes, 1 and 8, are fully fixed. Remember, this acts as a pin since a truss member has no rotational stiffness. Note also how the generation is used to specify the truss members.

A copy of the OUTPUT file is given below.

```
*************************************************************
****  SSTAN  -  Simple Structural Analysis Program  ****
Copyright 1993  By  Dr. Marc Hoit, University of Florida
*************************************************************

Example 1 - Truss
NUMBER OF JOINTS                  =   8
NUMBER OF DIFFERENT ELEMENT TYPES =   1
NUMBER OF LOAD CONDITIONS         =   1
NUMBER OF LOAD COMBINATIONS       =   0
```

NODE	BOUNDARY CONDITION CODES						NODAL POINT COORDINATES		
NUMBER	X	Y	Z	XX	YY	ZZ	X	Y	Z
1	F	F	F	F	F	F	.000	.000	.000
2	R	R	F	F	F	F	120.000	.000	.000
3	R	R	F	F	F	F	120.000	288.000	.000
4	R	R	F	F	F	F	384.000	.000	.000
5	R	R	F	F	F	F	384.000	288.000	.000
6	R	R	F	F	F	F	648.000	.000	.000
7	R	R	F	F	F	F	648.000	288.000	.000
8	F	F	F	F	F	F	768.000	.000	.000

```
EQUATION NUMBERS
  N    X    Y    Z    XX   YY   ZZ
  1    0    0    0    0    0    0
  2    1    2    0    0    0    0
  3    3    4    0    0    0    0
  4    5    6    0    0    0    0
  5    7    8    0    0    0    0
  6    9   10    0    0    0    0
  7   11   12    0    0    0    0
  8    0    0    0    0    0    0

***** TRUSS MEMBERS *****

NUMBER OF DIFFERENT MEMBER PROPERTIES =    3

MEMBER PROPERTY NUMBER--=    1
AXIAL AREA--------------=         .1200
MODULUS OF ELASTICITY, E=    29000.0000

MEMBER PROPERTY NUMBER--=    2
AXIAL AREA--------------=         .2500
MODULUS OF ELASTICITY, E=    29000.0000

MEMBER PROPERTY NUMBER--=    3
AXIAL AREA--------------=         .1800
MODULUS OF ELASTICITY, E=    29000.0000
MEMB   CONNECTIVITY   MATERIAL              END ECCENTRICITIES
NUM.                    SET     *****  I-END   ****  ****  J-END    *****
         I    J                 X-VALUE Y-VALUE Z-VALUE X-VALUE Y-VALUE Z-VALUE
    1    1    2          1         .00     .00     .00     .00     .00     .00
    2    2    4          1         .00     .00     .00     .00     .00     .00
    3    4    6          1         .00     .00     .00     .00     .00     .00
    4    6    8          1         .00     .00     .00     .00     .00     .00
    5    1    3          2         .00     .00     .00     .00     .00     .00
    6    3    5          2         .00     .00     .00     .00     .00     .00
    7    5    7          2         .00     .00     .00     .00     .00     .00
    8    7    8          2         .00     .00     .00     .00     .00     .00
    9    2    3          3         .00     .00     .00     .00     .00     .00
   10    4    5          3         .00     .00     .00     .00     .00     .00
   11    6    7          3         .00     .00     .00     .00     .00     .00
   12    3    4          3         .00     .00     .00     .00     .00     .00
   14    5    6          3         .00     .00     .00     .00     .00     .00
   13    2    5          3         .00     .00     .00     .00     .00     .00
   15    4    7          3         .00     .00     .00     .00     .00     .00

THE NODE NUMBERING USED PRODUCED A HALF BANDWIDTH OF      5
TOTAL STORAGE REQUIRED  =       134
TOTAL STORAGE AVAILABLE =     38000

*** CONCENTRATED NODAL LOADS ***
 NODE  LOAD       X          Y          Z         XX         YY         ZZ
    2    1    .00E+00   -.30E+01   .00E+00    .00E+00    .00E+00    .00E+00
    6    1    .00E+00   -.30E+01   .00E+00    .00E+00    .00E+00    .00E+00
    4    1    .00E+00   -.50E+01   .00E+00    .00E+00    .00E+00    .00E+00

SOLUTION CONVERGED IN   1 ITERATION(S)
*** PRINT OF FINAL DISPLACEMENTS ***
DISPLACEMENTS FOR LOAD CONDITION  1
NODE        X            Y            Z           XX           YY           ZZ
   1  .00000E+00   .00000E+00   .00000E+00   .00000E+00   .00000E+00   .00000E+00
   2 -.14671E-01  -.54070E+00   .00000E+00   .00000E+00   .00000E+00   .00000E+00
   3  .14436E+00  -.33793E+00   .00000E+00   .00000E+00   .00000E+00   .00000E+00
   4 -.15897E-16  -.72162E+00   .00000E+00   .00000E+00   .00000E+00   .00000E+00
   5  .98063E-17  -.64713E+00   .00000E+00   .00000E+00   .00000E+00   .00000E+00
   6  .14671E-01  -.54070E+00   .00000E+00   .00000E+00   .00000E+00   .00000E+00
   7 -.14436E+00  -.33793E+00   .00000E+00   .00000E+00   .00000E+00   .00000E+00
   8  .00000E+00   .00000E+00   .00000E+00   .00000E+00   .00000E+00   .00000E+00
```

```
-------------FORCE'S TRUSS MEMBERS--------------
ASTERISK (*) MEANS COMPRESSION NOT ALLOWED IN MEMBER
MEMBER  LOAD  IEND  JEND       AXIAL           AXIAL
  #      #     #     #         FORCES          STRESS
  1      1     1     2        -.4254          -3.5454
  2      1     2     4         .1934           1.6115
  3      1     4     6         .1934           1.6115
  4      1     6     8        -.4254          -3.5454
  5      1     1     3       -5.9583         -23.8333
  6      1     3     5       -3.9645         -15.8580
  7      1     5     7       -3.9645         -15.8580
  8      1     7     8       -5.9583         -23.8333
  9      1     2     3        3.6751          20.4172
 10      1     4     5        1.3502           7.5010
 11      1     6     7        3.6751          20.4172
 12      1     3     4        2.4756          13.7534
 14      1     5     6        -.9158          -5.0878
 13      1     2     5        -.9158          -5.0878
 15      1     4     7        2.4756          13.7534
```

The first thing to do is plot the structure and verify the connectivity. Also while in the plot program, the displacements, symmetry, and axial stresses can be checked. Looking at the output, we can see the symmetry of displacements at nodes 2/6 and 3/7. We should also check the symmetry of forces and stresses. Again we can see that members 9 and 11 have the same force. Finally, notice that there is no rotation given for any of the nodes (ZZ). This is true for a truss analysis in that the nodes themselves do not rotate. There is, however, a difference in angle between the truss members, as can be seen in the plot.

6.6.2 Continuous Beam Example

The second example we look at is a continuous beam with load cases and combinations. The structure shown in Figure 6.18 will be analyzed. The load cases for dead load and live load are to be analyzed separately. The single combination of dead plus live load will be analyzed. Figure 6.19 gives the analysis model with the node and element number-

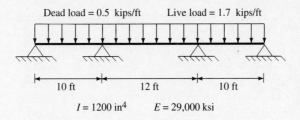

Figure 6.18 Continuous beam example.

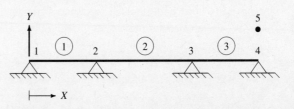

Figure 6.19 Numbered nodes and elements.

ing. The INPUT file required for this structure is

```
Three-Span Continuous Beam
5,1,2,1 : 5 nodes, 1 type element, 2 load cases, 1 comb.
:
COORDINATES
1 X=0.0 Y=0.0 Z=0.0
2 X=10*12
3 X=22*12
4 X=32*12
5 Y=1                    : Used for defining strong axis of beams
:
BOUNDARY
1,4 DOF=F,F,F,F,F,R  : Only rotations allowed
:
BEAM
3,1                      : 3 spans, one section type
1 I=1200 E=29000
1 1,2,5 M=1  L=0.5/12, 1.7/12  : DL, LL
2 2,3,5      L=0.5/12, 1.7/12  : DL, LL
3 3,4,5      L=0.5/12, 1.7/12  : DL, LL
:
COMBINATION
1 C=1.0, 1.0  : Load Case 1* 1.0 + Case 2 * 1.0
```

There are a couple of special things to notice about this structure. First, an extra node is required to be used for specifying the "strong and weak" axis orientation of the beams. This is because the program is three-dimensional and beams can be rotated to any special orientation, such as when used in columns.

All three-dimensional beams have two axes of bending, usually referred to as the *strong axis* and the *weak axis*. All three-dimensional finite-element programs need to define these two axes in a reference coordinate system called the *local coordinate system*. This system coincides with the strong and weak axes. In a general structure, the strong axis can be oriented in any direction. For example, the plan view of a simple three-dimensional frame gives two possible orientations for the columns (Figure 6.20). To specify which of these orientations is desired, the relation of the local to the global axes must be specified. This is needed to form the three-dimensional rotation matrix. In SSTAN, this is done by specifying a third node to define the local three-axis direction for a frame element. This third node is used to define a plane in which the local 1 and 3 axes lie. Therefore, the local coordinate system is defined as follows (Figure 6.21):

1. The local 1 axis goes from node *I* to *J*.

2. The local 3 axis is perpendicular to the local 1 axis and goes toward the third or *K* node.

3. The local 2 axis completes the right-hand rule.

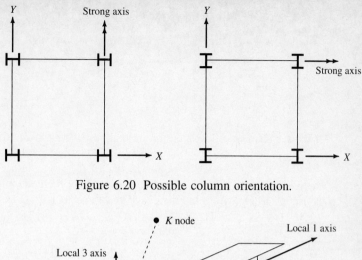

Figure 6.20 Possible column orientation.

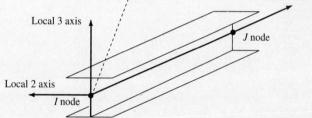

Figure 6.21 Definition of local beam coordinates.

In the example shown in Figure 6.18, the third node is in the X–Y plane and above the structure; therefore, the 3 axis points up in the global Y direction. It is important to understand the local coordinate conventions for an element since this is the coordinate system in which the final element forces will be given. In our example, the K node (node 5) was specified above the beams. This causes the local 3 axis to point upward and the moments to be positive when clockwise. If node 5 had been given a negative Y value (below the structure), this would have made the local 2 axis point downward and in addition, reversed the sign on the end moments (Figure 6.22). The choice of K node is also important for the distributed loading on the beams. The distributed load is assumed positive-acting in the *negative* local 3-axis direction (Figure 6.22).

Another thing to notice about the input file is that the designation for the boundary condition of node 5 is omitted. This is because all nodes not specified are defaulted to fixed (no movement allowed). For the BEAM specification, the member property and distributed load values are given only for the first beam; they will be continued for all subsequent beams as the default. On the property specification line, notice that only I_{22} is specified, as well as Young's modulus. This is because only bending about the 2 axis is allowed and the only stiffness matrix terms used have EI_{22}/L. Finally, note that there is no LOAD header. This is because no concentrated loads are applied, only distributed loads.

A copy of the OUTPUT file is given below. Note that a complete copy of the input is repeated to the output. This is useful for checking potential input errors. Also notice

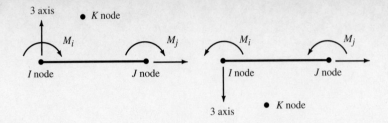

Figure 6.22 Local 3-axis definition and positive moments for different position of the K node.

that a copy of the equation numbers is given after the node information. This information is useful to help determine the cause of an unstable structure or zeros on the diagonal.

```
*************************************************************
****   SSTAN  -  Simple Structural Analysis Program  ****
Copyright 1993  By  Dr. Marc Hoit, University of Florida
*************************************************************

Three-Span Continuous Beam
NUMBER OF JOINTS                        =    5
NUMBER OF DIFFERENT ELEMENT TYPES       =    1
NUMBER OF LOAD CONDITIONS               =    2
NUMBER OF LOAD COMBINATIONS             =    1

NODE    BOUNDARY CONDITION CODES              NODAL POINT COORDINATES
NUMBER  X    Y    Z    XX   YY   ZZ          X            Y            Z
   1    F    F    F    F    F    R         .000         .000         .000
   2    F    F    F    F    F    R     120.000         .000         .000
   3    F    F    F    F    F    R     264.000         .000         .000
   4    F    F    F    F    F    R     384.000         .000         .000
   5    F    F    F    F    F    F     384.000        1.000         .000

EQUATION NUMBERS
    N    X    Y    Z    XX   YY   ZZ
    1    0    0    0    0    0    1
    2    0    0    0    0    0    2
    3    0    0    0    0    0    3
    4    0    0    0    0    0    4
    5    0    0    0    0    0    0

***** FRAME MEMBERS *****

NUMBER OF MEMBER PROPERTIES =    1

MEMBER PROPERTY NUMBER --------- =    1
AXIAL AREA, A ------------------ =         .0000
SHEAR AREA, 3 Axis ------------- =         .0000
SHEAR AREA, 2 Axis ------------- =         .0000
TORSIONAL MOMENT OF INERTIA, J  =         .0000
MOMENT OF INERTIA, I(2) -------- =     1200.0000
MOMENT OF INERTIA, I(3) -------- =         .0000
MODULUS OF ELASTICITY, E ------- =   .290000E+05
SHEAR MODULUS, G --------------- =   .111538E+05(USED FOR JG/L CAL.)

MEMB   CONNECTIVITY   MATERIAL             END ECCENTRICITIES
NUM. K=0 FOR Z-AXIS    SET   *****  I-END  ****   ****  J-END   *****  PIN
      I    J    K            X-VALUE Y-VALUE Z-VALUE X-VALUE Y-VALUE Z-VALUE OPT

   1   1    2    5      1      .00     .00     .00     .00     .00     .00   0
UNIFORM LOAD =    .042    .142

   2   2    3    5      1      .00     .00     .00     .00     .00     .00   0
UNIFORM LOAD =    .042    .142
```

```
      3   3   4   5   1      .00      .00      .00      .00      .00      .00   0
UNIFORM LOAD =    .042    .142

THE NODE NUMBERING USED PRODUCED A HALF BANDWIDTH OF    1

TOTAL STORAGE REQUIRED  =       35
TOTAL STORAGE AVAILABLE =    35000

SOLUTION CONVERGED IN    1 ITERATION(S)

TOTAL STORAGE REQUIRED  =       31
TOTAL STORAGE AVAILABLE =    35000

LOAD LOAD        LOAD CASE MULTIPLIERS
COMB CASE    1       2
  1   3    1.00    1.00

*** PRINT OF FINAL DISPLACEMENTS ***

DISPLACEMENTS FOR LOAD CONDITION  1
NODE        X            Y            Z            XX           YY           ZZ
  1   .00000E+00   .00000E+00   .00000E+00   .00000E+00   .00000E+00  -.44212E-04
  2   .00000E+00   .00000E+00   .00000E+00   .00000E+00   .00000E+00   .22167E-05
  3   .00000E+00   .00000E+00   .00000E+00   .00000E+00   .00000E+00  -.22167E-05
  4   .00000E+00   .00000E+00   .00000E+00   .00000E+00   .00000E+00   .44212E-04
  5   .00000E+00   .00000E+00   .00000E+00   .00000E+00   .00000E+00   .00000E+00

DISPLACEMENTS FOR LOAD CONDITION  2
NODE        X            Y            Z            XX           YY           ZZ
  1   .00000E+00   .00000E+00   .00000E+00   .00000E+00   .00000E+00  -.15032E-03
  2   .00000E+00   .00000E+00   .00000E+00   .00000E+00   .00000E+00   .75369E-05
  3   .00000E+00   .00000E+00   .00000E+00   .00000E+00   .00000E+00  -.75369E-05
  4   .00000E+00   .00000E+00   .00000E+00   .00000E+00   .00000E+00   .15032E-03
  5   .00000E+00   .00000E+00   .00000E+00   .00000E+00   .00000E+00   .00000E+00

DISPLACEMENTS FOR COMBINATION NUMBER  1
NODE        X            Y            Z            XX           YY           ZZ
  1   .00000E+00   .00000E+00   .00000E+00   .00000E+00   .00000E+00  -.19453E-03
  2   .00000E+00   .00000E+00   .00000E+00   .00000E+00   .00000E+00   .97537E-05
  3   .00000E+00   .00000E+00   .00000E+00   .00000E+00   .00000E+00  -.97537E-05
  4   .00000E+00   .00000E+00   .00000E+00   .00000E+00   .00000E+00   .19453E-03
  5   .00000E+00   .00000E+00   .00000E+00   .00000E+00   .00000E+00   .00000E+00

---------------------- FRAME MEMBER RESULTS ----------------
MEM LOAD NODE         1-2 PLANE                   1-3 PLANE        AXIAL FORCE
 #    #    #      MOMENT      SHEAR        MOMENT      SHEAR
 1    1    1   .00000E+00  .00000E+00   .00000E+00  .18911E+01   .00000E+00
           2   .00000E+00  .00000E+00   .73071E+02  .31089E+01   .00000E+00
                                                 AXIAL TORQUE =   .00000E+00
** MAXIMUM MIDSPAN MOMENT =    .429E+02 AT DISTANCE    45.39 FROM NODE     1
 1    2    1   .00000E+00  .00000E+00   .00000E+00  .64296E+01   .00000E+00
           2   .00000E+00  .00000E+00   .24844E+03  .10570E+02   .00000E+00
                                                 AXIAL TORQUE =   .00000E+00
** MAXIMUM MIDSPAN MOMENT =    .146E+03 AT DISTANCE    45.39 FROM NODE     1
 1    3    1   .00000E+00  .00000E+00   .00000E+00  .83207E+01   .00000E+00
           2   .00000E+00  .00000E+00   .32151E+03  .13679E+02   .00000E+00
                                                 AXIAL TORQUE =   .00000E+00
** MAXIMUM MIDSPAN MOMENT =    .189E+03 AT DISTANCE    45.39 FROM NODE     1
 2    1    2   .00000E+00  .00000E+00  -.73071E+02  .30000E+01   .00000E+00
           3   .00000E+00  .00000E+00   .73071E+02  .30000E+01   .00000E+00
                                                 AXIAL TORQUE =   .00000E+00
** MAXIMUM MIDSPAN MOMENT =    .181E+03 AT DISTANCE    72.00 FROM NODE     2
 2    2    2   .00000E+00  .00000E+00  -.24844E+03  .10200E+02   .00000E+00
           3   .00000E+00  .00000E+00   .24844E+03  .10200E+02   .00000E+00
                                                 AXIAL TORQUE =   .00000E+00
** MAXIMUM MIDSPAN MOMENT =    .616E+03 AT DISTANCE    72.00 FROM NODE     2
 2    3    2   .00000E+00  .00000E+00  -.32151E+03  .13200E+02   .00000E+00
           3   .00000E+00  .00000E+00   .32151E+03  .13200E+02   .00000E+00
                                                 AXIAL TORQUE =   .00000E+00
```

```
**  MAXIMUM MIDSPAN MOMENT =    .797E+03  AT DISTANCE      72.00  FROM NODE      2
     3    1    3   .00000E+00   .00000E+00  -.73071E+02   .31089E+01  .00000E+00
               4   .00000E+00   .00000E+00   .00000E+00   .18911E+01  .00000E+00
                                             AXIAL TORQUE =       .00000E+00
**  MAXIMUM MIDSPAN MOMENT =    .189E+03  AT DISTANCE      74.61  FROM NODE      3
     3    2    3   .00000E+00   .00000E+00  -.24844E+03   .10570E+02  .00000E+00
               4   .00000E+00   .00000E+00   .00000E+00   .64296E+01  .00000E+00
                                             AXIAL TORQUE =       .00000E+00
**  MAXIMUM MIDSPAN MOMENT =    .643E+03  AT DISTANCE      74.61  FROM NODE      3
     3    3    3   .00000E+00   .00000E+00  -.32151E+03   .13679E+02  .00000E+00
               4   .00000E+00   .00000E+00   .00000E+00   .83207E+01  .00000E+00
                                             AXIAL TORQUE =       .00000E+00
**  MAXIMUM MIDSPAN MOMENT =    .832E+03  AT DISTANCE      74.61  FROM NODE      3
```

From the output above, we can see that the displacements for both the load cases and the single combination are printed. Also, the forces from these loads and combination are given as results. On members where a uniform load is applied, the maximum moment in the span is printed.

SSTAN Solution Errors

If a structure is unstable, the stiffness matrix generated is singular. As a result, during the equation-solving process an error will be generated specifying a singular matrix or that there is a negative on the diagonal. Some typical causes of a singular matrix are as follows:

1. The structure is unstable. The cause of this generally is improper specification of boundary conditions. As in the example above, all out-of-plane DOFs were fixed; if this were not the case, an error would have occurred giving a singular matrix. Another possible cause is if the structure is placed on rollers. Then at least one horizontal fixity must be supplied for stability. In the direct stiffness method, all active DOFs must have some associated stiffness. Even if no load is applied to the horizontal direction, the structure must be stabilized.

2. The elements are not connected correctly, causing an unstable structure. Usually, the real structure is stable but the model is unstable, due to a typographical error in element connectivity. Check to be sure that the elements are connected to the proper nodes. STANPLOT can be very helpful for this type of error.

3. The element properties are specified incorrectly. If an element property is omitted or read in as zero, this can cause the structure to be unstable. This is common in three-dimensional structures where the torsional property is left out.

One final point to note is that when distributed loads are used, the maximum moment in the span is given along with the beam member forces. This is useful for design where the maximum moment is required. A good way to check the structure is through a graphical display of the structure and its results. SSTAN has a graphics postprocessor that allows display of the structure, displaced shapes, and stress and displacement contours for finite elements. To use the graphics postprocessor, just run the program (STANPLOT) in the directory where the result data are contained.

6.6.3 Two-Story Braced Frame

The third example is a two-story frame with truss bars to prevent sidesway. This example is also a two-dimensional structure. Load combinations will be handled by using the combination option. The structure to be analyzed is shown in Figure 6.23. All bending members are assumed to be bending about their strong axes. All members are steel and Young's modulus is 29,000 ksi. The member cross-sectional properties are

$$A = 4 \text{ in}^2 \text{ for truss members}$$
$$I_1 = \ \ 850 \text{ in}^4 \quad A_1 = \ \ 3 \text{ in}^2$$
$$I_2 = 1000 \text{ in}^4 \quad A_2 = \ \ 6 \text{ in}^2$$
$$I_3 = 1200 \text{ in}^4 \quad A_3 = \ \ 8 \text{ in}^2$$
$$I_4 = 2400 \text{ in}^4 \quad A_4 = 10 \text{ in}^2$$

The base load cases are dead load, live load, and wind load (Figure 6.24).

The following load combinations are to be analyzed: DL + LL and $\frac{3}{4}$(DL + $\frac{1}{2}$LL + WL). The first step is to number the nodes and members. The arbitrary numbering used

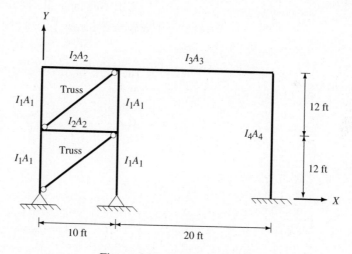

Figure 6.23 Frame example.

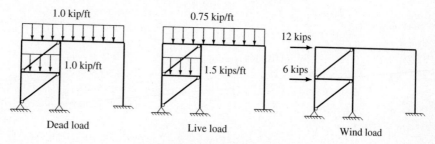

Figure 6.24 Base load cases.

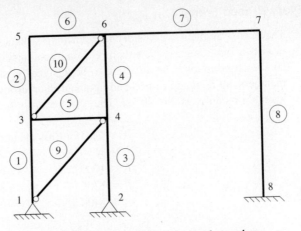

Figure 6.25 Numbered nodes and members.

is shown in Figure 6.25. Using the given dimensions, node and element numbers, and load combinations, we compile the following input file. Notice that the load combinations are handled using load combinations of the three base load cases.

```
Two-Dimensional Braced Frame
8,2,3,2 : 8 nodes, truss & beam, 3 load cases, 2 comb.
:
COORDINATES
1 X=0 Y=0 Z=0
2 X=10*12
3 X=0 Y=12*12
4 X=10*12
5 X=0 Y=24*12
6 X=10*12
7 X=30*12
8 Y=0
:
BOUNDARY
1,2 F=F,F,F,F,F,R
3,7 F=R,R,F,F,F,R
:
TRUSS
2,1 : 2 members, 1 property
1 A=4 E=29000
9 1,4 M=1
10 3,6
:
BEAM
8,4
1 I=850 A=3 E=29000
```

```
2 I=1000 A=6
3 I=1200 A=8
4 I=2400 A=10
C------ Single-story columns
1 1,3,2, M=1
2 3,5,2
3 2,4,8
4 4,6,8
C------ Beams -----
5 3,4,1 M=2 L=-1/12, -1.5/12
6 5,6,1     L=-1/12, -.75/12
7 6,7,1 M=3 L=-1/12, -.75/12
C------ 2-story column
8 8,7,2 M=4 L=0
:
LOADS
3 L=3 F=6
5 L=3 f=12
:
COMBINATION
1 c=1,1,0        : (DL + LL)
2 c=3/4,3/8,3/4  : 3/4 (DL + 1/2 LL + WL)
:
```

Notice how the trusses are used as bracing against sidesway. Although this is effective for the applied load, if the wind load were reversed, the real bracing would not be effective. This is since bracing is usually supplied by very slender members. They therefore cannot take compression. However, the numerical model using truss members will allow them to take compression. This is not correct modeling. Two things can be done to solve this problem. The first is to use bracing in both directions. This makes sure that load can be applied in either direction and is what would need to be done in a real structure. However, using truss members in both directions will again give false results if they are slender and cannot take compression. As a result, the correct thing to do is to use the given model for the applied load. When checking loads from the other direction, remove the existing trusses and use members going the other direction (from nodes 3–2 and 5–4). Then, when the structure is built, include both sets of braces. SSTAN is capable of handling this process automatically by specifying the trusses as noncompression members. In this way the load will determine what the member does. This option is not used in this example.

The local coordinate system for truss members is simpler than beam members since they are two force members. Trusses effectively only have one DOF, tension or compression. As a result, only a single force result is given. This result is defined as positive in tension. Let's look at the DOFs at the nodes where the trusses and beams connect. The trusses' and beams' individual DOFs in global directions are given in Figure 6.26. When they are attached together, one of each type of DOF needs to be released at the node. All other DOFs need to be fixed. This is because each DOF released in a model needs to have stiffness contributed to it from some member to stabilize that DOF. This means that

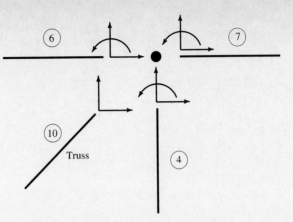

Figure 6.26 Stiffness applied to the node from each member.

the *X* and *Y* displacements and the *Z* rotation need to be released. While the truss does not need the rotational DOF, the beams and columns do. If the rotation is fixed, the beams are prevented from rotating and apply additional constraints to the structure. Also see that the loads are applied as a combination of distributed load on the beams and concentrated loads simulating wind loading. The OUTPUT file from this analysis is as follows:

```
***********************************************************
****   SSTAN  -  Simple Structural Analysis Program  ****
Copyright 1993  By  Dr. Marc Hoit, University of Florida
***********************************************************

Two-Dimensional Braced Frame
NUMBER OF JOINTS                      =    8
NUMBER OF DIFFERENT ELEMENT TYPES  =    2
NUMBER OF LOAD CONDITIONS            =    3
NUMBER OF LOAD COMBINATIONS          =    2

NODE    BOUNDARY CONDITION CODES          NODAL POINT COORDINATES
NUMBER  X    Y    Z   XX   YY   ZZ          X            Y            Z
     1  F    F    F    F    F    R        .000         .000         .000
     2  F    F    F    F    F    R     120.000         .000         .000
     3  R    R    F    F    F    R        .000      144.000         .000
     4  R    R    F    F    F    R     120.000      144.000         .000
     5  R    R    F    F    F    R        .000      288.000         .000
     6  R    R    F    F    F    R     120.000      288.000         .000
     7  R    R    F    F    F    R     360.000      288.000         .000
     8  F    F    F    F    F    F     360.000         .000         .000

***** TRUSS MEMBERS *****

NUMBER OF DIFFERENT MEMBER PROPERTIES =    1

MEMBER PROPERTY NUMBER--=    1
AXIAL AREA---------------=         4.0000
MODULUS OF ELASTICITY, E=     29000.0000

MEMB  CONNECTIVITY  MATERIAL              END ECCENTRICITIES
NUM.                 SET     *****   I-END   ****  ****   J-END   *****
        I    J              X-VALUE Y-VALUE Z-VALUE X-VALUE Y-VALUE Z-VALUE
     9   1    4        1        .00     .00     .00     .00     .00     .00
    10   3    6        1        .00     .00     .00     .00     .00     .00

***** FRAME MEMBERS *****

NUMBER OF MEMBER PROPERTIES =    4
```

```
MEMBER PROPERTY NUMBER --------- =    1
AXIAL AREA, A ----------------- =        3.0000
SHEAR AREA, 3 Axis ------------- =         .0000
SHEAR AREA, 2 Axis ------------- =         .0000
TORSIONAL MOMENT OF INERTIA, J   =         .0000
MOMENT OF INERTIA, I(2) -------- =      850.0000
MOMENT OF INERTIA, I(3) -------- =         .0000
MODULUS OF ELASTICITY, E ------- =   .290000E+05
SHEAR MODULUS, G --------------- =   .111538E+05(USED FOR JG/L CAL.)

MEMBER PROPERTY NUMBER --------- =    2
AXIAL AREA, A ----------------- =        6.0000
SHEAR AREA, 3 Axis ------------- =         .0000
SHEAR AREA, 2 Axis ------------- =         .0000
TORSIONAL MOMENT OF INERTIA, J   =         .0000
MOMENT OF INERTIA, I(2) -------- =     1000.0000
MOMENT OF INERTIA, I(3) -------- =         .0000
MODULUS OF ELASTICITY, E ------- =   .290000E+05
SHEAR MODULUS, G --------------- =   .111538E+05(USED FOR JG/L CAL.)

MEMBER PROPERTY NUMBER --------- =    3
AXIAL AREA, A ----------------- =        8.0000
SHEAR AREA, 3 Axis ------------- =         .0000
SHEAR AREA, 2 Axis ------------- =         .0000
TORSIONAL MOMENT OF INERTIA, J   =         .0000
MOMENT OF INERTIA, I(2) -------- =     1200.0000
MOMENT OF INERTIA, I(3) -------- =         .0000
MODULUS OF ELASTICITY, E ------- =   .290000E+05
SHEAR MODULUS, G --------------- =   .111538E+05(USED FOR JG/L CAL.)

MEMBER PROPERTY NUMBER --------- =    4
AXIAL AREA, A ----------------- =       10.0000
SHEAR AREA, 3 Axis ------------- =         .0000
SHEAR AREA, 2 Axis ------------- =         .0000
TORSIONAL MOMENT OF INERTIA, J   =         .0000
MOMENT OF INERTIA, I(2) -------- =     2400.0000
MOMENT OF INERTIA, I(3) -------- =         .0000
MODULUS OF ELASTICITY, E ------- =   .290000E+05
SHEAR MODULUS, G --------------- =   .111538E+05(USED FOR JG/L CAL.)

MEMB   CONNECTIVITY  MATERIAL                END ECCENTRICITIES
NUM.  K=0 FOR Z-AXIS   SET  *****  I-END  ****  ****  J-END  *****  PIN
       I    J    K         X-VALUE Y-VALUE Z-VALUE X-VALUE Y-VALUE Z-VALUE OPT

   1    1    3    2    1     .00     .00     .00     .00     .00     .00   0
UNIFORM LOAD =    .000    .000    .000
   2    3    5    2    1     .00     .00     .00     .00     .00     .00   0
UNIFORM LOAD =    .000    .000    .000
   3    2    4    8    1     .00     .00     .00     .00     .00     .00   0
UNIFORM LOAD =    .000    .000    .000
   4    4    6    8    1     .00     .00     .00     .00     .00     .00   0
UNIFORM LOAD =    .000    .000    .000
   5    3    4    1    2     .00     .00     .00     .00     .00     .00   0
UNIFORM LOAD =   -.083   -.125    .000
   6    5    6    1    2     .00     .00     .00     .00     .00     .00   0
UNIFORM LOAD =   -.083   -.063    .000
   7    6    7    1    3     .00     .00     .00     .00     .00     .00   0
UNIFORM LOAD =   -.083   -.063    .000
   8    8    7    2    4     .00     .00     .00     .00     .00     .00   0
UNIFORM LOAD =    .000    .000    .000

THE NODE NUMBERING USED PRODUCED A HALF BANDWIDTH OF     6
TOTAL STORAGE REQUIRED   =     203
TOTAL STORAGE AVAILABLE  =   38000

*** CONCENTRATED NODAL LOADS ***
 NODE  LOAD     X        Y        Z        XX       YY       ZZ
   3     3   .60E+01  .00E+00  .00E+00  .00E+00  .00E+00  .00E+00
   5     3   .12E+02  .00E+00  .00E+00  .003+00  .00E+00  .00E+00
```

```
SOLUTION CONVERGED IN   1 ITERATION(S)
TOTAL STORAGE REQUIRED  =      126
TOTAL STORAGE AVAILABLE =    38000

LOAD LOAD      LOAD CASE MULTIPLIERS
COMB CASE    1       2       3
 1    4    1.00    1.00     .00
 2    5     .75     .38     .75

*** PRINT OF FINAL DISPLACEMENTS ***
DISPLACEMENTS FOR LOAD CONDITION  1
NODE        X          Y          Z          XX         YY         ZZ
  1    .00000E+00 .00000E+00 .00000E+00 .00000E+00 .00000E+00 -.94116E-04
  2    .00000E+00 .00000E+00 .00000E+00 .00000E+00 .00000E+00 -.17884E-03
  3    .15958E-01 -.18194E-01 .00000E+00 .00000E+00 .00000E+00 -.14424E-03
  4    .17785E-01 -.24230E-01 .00000E+00 .00000E+00 .00000E+00 -.12834E-04
  5    .38230E-01 -.24313E-01 .00000E+00 .00000E+00 .00000E+00 -.20514E-03
  6    .38033E-01 -.44744E-01 .00000E+00 .00000E+00 .00000E+00 -.24325E-03
  7    .35355E-01 -.10857E-01 .00000E+00 .00000E+00 .00000E+00  .26857E-03
  8    .00000E+00 .00000E+00 .00000E+00 .00000E+00 .00000E+00  .00000E+00

DISPLACEMENTS FOR LOAD CONDITION  2
NODE        X          Y          Z          XX         YY         ZZ
  1    .00000E+00 .00000E+00 .00000E+00 .00000E+00 .00000E+00 -.97658E-04
  2    .00000E+00 .00000E+00 .00000E+00 .00000E+00 .00000E+00 -.20988E-03
  3    .16878E-01 -.19777E-01 .00000E+00 .00000E+00 .00000E+00 -.15630E-03
  4    .18440E-01 -.23568E-01 .00000E+00 .00000E+00 .00000E+00  .35594E-04
  5    .32311E-01 -.24548E-01 .00000E+00 .00000E+00 .00000E+00 -.13114E-03
  6    .31951E-01 -.38529E-01 .00000E+00 .00000E+00 .00000E+00 -.17365E-03
  7    .29829E-01 -.82602E-02 .00000E+00 .00000E+00 .00000E+00  .20037E-03
  8    .00000E+00 .00000E+00 .00000E+00 .00000E+00 .00000E+00  .00000E+00

DISPLACEMENTS FOR LOAD CONDITION  3
NODE        X          Y          Z          XX         YY         ZZ
  1    .00000E+00 .00000E+00 .00000E+00 .00000E+00 .00000E+00 -.79112E-03
  2    .00000E+00 .00000E+00 .00000E+00 .00000E+00 .00000E+00 -.68904E-03
  3    .99699E-01 .13212E-01 .00000E+00 .00000E+00 .00000E+00 -.49483E-03
  4    .92378E-01 -.33743E-01 .00000E+00 .00000E+00 .00000E+00 -.54647E-03
  5    .18413E+00 .12435E-01 .00000E+00 .00000E+00 .00000E+00 -.55422E-03
  6    .17646E+00 -.36174E-01 .00000E+00 .00000E+00 .00000E+00 -.21710E-03
  7    .17253E+00 -.33644E-02 .00000E+00 .00000E+00 .00000E+00 -.44403E-03
  8    .00000E+00 .00000E+00 .00000E+00 .00000E+00 .00000E+00  .00000E+00

DISPLACEMENTS FOR COMBINATION NUMBER  1
NODE        X          Y          Z          XX         YY         ZZ
  1    .00000E+00 .00000E+00 .00000E+00 .00000E+00 .00000E+00 -.19177E-03
  2    .00000E+00 .00000E+00 .00000E+00 .00000E+00 .00000E+00 -.38873E-03
  3    .32836E-01 -.37971E-01 .00000E+00 .00000E+00 .00000E+00 -.30054E-03
  4    .36225E-01 -.47798E-01 .00000E+00 .00000E+00 .00000E+00  .22760E-04
  5    .70541E-01 -.48861E-01 .00000E+00 .00000E+00 .00000E+00 -.33629E-03
  6    .69984E-01 -.83272E-01 .00000E+00 .00000E+00 .00000E+00 -.41690E-03
  7    .65184E-01 -.19117E-01 .00000E+00 .00000E+00 .00000E+00  .46893E-03
  8    .00000E+00 .00000E+00 .00000E+00 .00000E+00 .00000E+00  .00000E+00

DISPLACEMENTS FOR COMBINATION NUMBER  2
NODE        X          Y          Z          XX         YY         ZZ
  1    .00000E+00 .00000E+00 .00000E+00 .00000E+00 .00000E+00 -.70055E-03
  2    .00000E+00 .00000E+00 .00000E+00 .00000E+00 .00000E+00 -.72962E-03
  3    .93073E-01 -.11153E-01 .00000E+00 .00000E+00 .00000E+00 -.53791E-03
  4    .89538E-01 -.52318E-01 .00000E+00 .00000E+00 .00000E+00 -.40613E-03
  5    .17889E+00 -.18113E-01 .00000E+00 .00000E+00 .00000E+00 -.61870E-03
  6    .17285E+00 -.75137E-01 .00000E+00 .00000E+00 .00000E+00 -.41038E-03
  7    .16710E+00 -.13763E-01 .00000E+00 .00000E+00 .00000E+00 -.56462E-04
  8    .00000E+00 .00000E+00 .00000E+00 .00000E+00 .00000E+00  .00000E+00

-----------FORCE'S TRUSS MEMBERS--------------
ASTERISK (*) MEANS COMPRESSION NOT ALLOWED IN MEMBER
MEMBER  LOAD  IEND  JEND      AXIAL         AXIAL
  #      #     #     #       FORCES         STRESS

    9     1     1     4     -4.4734        -1.1184
    9     2     1     4     -3.8987         -.9747
```

9	3	1	4	20.5562	5.1391
9	4	1	4	-8.3722	-2.0930
9	5	1	4	10.6001	2.6500
10	1	3	6	-3.8767	-.9692
10	2	3	6	-2.9431	-.7358
10	3	3	6	6.9327	1.7332
10	4	3	6	-6.8198	-1.7049
10	5	3	6	1.1883	.2971

```
---------------------- FRAME MEMBER RESULTS ----------------
MEM LOAD NODE       1-2 PLANE              1-3 PLANE      AXIAL FORCE
 #    #    #     MOMENT      SHEAR      MOMENT      SHEAR
 1    1    1  .00000E+00  .00000E+00 -.95618E-14  .11916E+00  .10992E+02
           3  .00000E+00  .00000E+00 -.17159E+02 -.11916E+00 -.10992E+02
                                     AXIAL TORQUE =    .00000E+00
 1    2    1  .00000E+00  .00000E+00 -.11800E-13  .13943E+00  .11948E+02
           3  .00000E+00  .00000E+00 -.20077E+02 -.13943E+00 -.11948E+02
                                     AXIAL TORQUE =    .00000E+00
 1    3    1  .00000E+00  .00000E+00 -.83988E-13 -.70444E+00 -.79824E+01
           3  .00000E+00  .00000E+00  .10144E+03  .70444E+00  .79824E+01
                                     AXIAL TORQUE =    .00000E+00
 1    4    1  .00000E+00  .00000E+00 -.30649E-13  .25859E+00  .22941E+02
           3  .00000E+00  .00000E+00 -.37236E+02 -.25859E+00 -.22941E+02
                                     AXIAL TORQUE =    .00000E+00
 1    5    1  .00000E+00  .00000E+00 -.72276E-13 -.38668E+00  .67381E+01
           3  .00000E+00  .00000E+00  .55681E+02  .38668E+00 -.67381E+01
                                     AXIAL TORQUE =    .00000E+00
 2    1    3  .00000E+00  .00000E+00 -.10146E+02  .28573E+00  .36965E+01
           5  .00000E+00  .00000E+00 -.30999E+02 -.28573E+00 -.36965E+01
                                     AXIAL TORQUE =    .00000E+00
 2    2    3  .00000E+00  .00000E+00 -.41843E+02  .52135E+00  .28826E+01
           5  .00000E+00  .00000E+00 -.33230E+02 -.52135E+00 -.28826E+01
                                     AXIAL TORQUE =    .00000E+00
 2    3    3  .00000E+00  .00000E+00  .73641E+02 -.88159E+00  .46937E+00
           5  .00000E+00  .00000E+00  .53308E+02  .88159E+00 -.46937E+00
                                     AXIAL TORQUE =    .00000E+00
 2    4    3  .00000E+00  .00000E+00 -.51990E+02  .80707E+00  .65791E+01
           5  .00000E+00  .00000E+00 -.64229E+02 -.80707E+00 -.65791E+01
                                     AXIAL TORQUE =    .00000E+00
 2    5    3  .00000E+00  .00000E+00  .31930E+02 -.25139E+00  .42054E+01
           5  .00000E+00  .00000E+00  .42709E+01  .25139E+00 -.42054E+01
                                     AXIAL TORQUE =    .00000E+00
 3    1    2  .00000E+00  .00000E+00 -.51764E-14 -.39469E+00  .14639E+02
           4  .00000E+00  .00000E+00  .56835E+02  .39469E+00 -.14639E+02
                                     AXIAL TORQUE =    .00000E+00
 3    2    2  .00000E+00  .00000E+00 -.33454E-14 -.58363E+00  .14239E+02
           4  .00000E+00  .00000E+00  .84042E+02  .58363E+00 -.14239E+02
                                     AXIAL TORQUE =    .00000E+00
 3    3    2  .00000E+00  .00000E+00 -.68792E-13 -.33898E+00  .20386E+02
           4  .00000E+00  .00000E+00  .48813E+02  .33898E+00 -.20386E+02
                                     AXIAL TORQUE =    .00000E+00
 3    4    2  .00000E+00  .00000E+00 -.26510E-13 -.97831E+00  .28878E+02
           4  .00000E+00  .00000E+00  .14088E+03  .97831E+00 -.28878E+02
                                     AXIAL TORQUE =    .00000E+00
 3    5    2  .00000E+00  .00000E+00  .97422E-14 -.76911E+00  .31609E+02
           4  .00000E+00  .00000E+00  .11075E+03  .76911E+00 -.31609E+02
                                     AXIAL TORQUE =    .00000E+00
 4    1    4  .00000E+00  .00000E+00  .52349E+02 -.17925E+00  .12393E+02
           6  .00000E+00  .00000E+00 -.26537E+02  .17925E+00 -.12393E+02
                                     AXIAL TORQUE =    .00000E+00
 4    2    4  .00000E+00  .00000E+00  .61288E+02 -.35373E+00  .90389E+01
           6  .00000E+00  .00000E+00 -.10350E+02  .35373E+00 -.90389E+01
                                     AXIAL TORQUE =    .00000E+00
 4    3    4  .00000E+00  .00000E+00  .15123E+03 -.28834E+01  .14687E+02
           6  .00000E+00  .00000E+00  .26399E+03  .28834E+01 -.14687E+02
                                     AXIAL TORQUE =    .00000E+00
 4    4    4  .00000E+00  .00000E+00  .11364E+03 -.53298E+00  .21432E+02
           6  .00000E+00  .00000E+00 -.36887E+02  .53298E+00 -.21432E+02
                                     AXIAL TORQUE =    .00000E+00
```

```
   4    5    4   .00000E+00   .00000E+00   .17566E+03  -.24297E+01   .13786E+02
             6   .00000E+00   .00000E+00   .17421E+03   .24297E+01  -.13786E+02
                                               AXIAL TORQUE =    .00000E+00
   5    1    3   .00000E+00   .00000E+00   .27305E+02  -.43177E+01  -.26484E+01
             4   .00000E+00   .00000E+00  -.10918E+03  -.56823E+01   .26484E+01
                                               AXIAL TORQUE =    .00000E+00
** MAXIMUM MIDSPAN MOMENT =    .845E+02 AT DISTANCE    51.81 FROM NODE     3
   5    2    3   .00000E+00   .845E+00   .61921E+02  -.68049E+01  -.22660E+01
             4   .00000E+00   .00000E+00  -.14533E+03  -.81951E+01   .22660E+01
                                               AXIAL TORQUE =    .00000E+00
** MAXIMUM MIDSPAN MOMENT =    .123E+03 AT DISTANCE    54.44 FROM NODE     3
   5    3    3   .00000E+00   .00000E+00  -.17508E+03   .31260E+01   .10615E+02
             4   .00000E+00   .00000E+00  -.20004E+03  -.31260E+01  -.10615E+02
                                               AXIAL TORQUE =    .00000E+00
   5    4    3   .00000E+00   .00000E+00   .89226E+02  -.11123E+02  -.49144E+01
             4   .00000E+00   .00000E+00  -.25451E+03  -.13877E+02   .49144E+01
                                               AXIAL TORQUE =    .00000E+00
** MAXIMUM MIDSPAN MOMENT =    .208E+03 AT DISTANCE    53.39 FROM NODE     3
   5    5    3   .00000E+00   .00000E+00  -.87611E+02  -.34456E+01   .51255E+01
             4   .00000E+00   .00000E+00  -.28642E+03  -.96794E+01  -.51255E+01
                                               AXIAL TORQUE =    .00000E+00
** MAXIMUM MIDSPAN MOMENT =    .142E+03 AT DISTANCE    31.50 FROM NODE     3
   6    1    5   .00000E+00   .00000E+00   .30999E+02  -.36965E+01   .28573E+00
             6   .00000E+00   .00000E+00  -.18742E+03  -.63035E+01  -.28573E+00
                                               AXIAL TORQUE =    .00000E+00
** MAXIMUM MIDSPAN MOMENT =    .510E+02 AT DISTANCE    44.36 FROM NODE     5
   6    2    5   .00000E+00   .00000E+00   .33230E+02  -.28826E+01   .52135E+00
             6   .00000E+00   .00000E+00  -.13731E+03  -.46174E+01  -.52135E+00
                                               AXIAL TORQUE =    .00000E+00
** MAXIMUM MIDSPAN MOMENT =    .332E+02 AT DISTANCE    46.12 FROM NODE     5
   6    3    5   .00000E+00   .00000E+00  -.53308E+02  -.46937E+00   .11118E+02
             6   .00000E+00   .00000E+00   .10963E+03  -.46937E+00  -.11118E+02
                                               AXIAL TORQUE =    .00000E+00
   6    4    5   .00000E+00   .00000E+00   .64229E+02  -.65791E+01   .80707E+00
             6   .00000E+00   .00000E+00  -.32473E+03  -.10921E+02  -.80707E+00
                                               AXIAL TORQUE =    .00000E+00
** MAXIMUM MIDSPAN MOMENT =    .842E+02 AT DISTANCE    45.11 FROM NODE     5
   6    5    5   .00000E+00   .00000E+00  -.42709E+01  -.42054E+01   .87486E+01
             6   .00000E+00   .00000E+00  -.10983E+03  -.61071E+01  -.87486E+01
                                               AXIAL TORQUE =    .00000E+00
** MAXIMUM MIDSPAN MOMENT =    .107E+03 AT DISTANCE    48.94 FROM NODE     5
   7    1    6   .00000E+00   .00000E+00   .21396E+03  -.90681E+01   .25883E+01
             7   .00000E+00   .00000E+00  -.43762E+03  -.10932E+02  -.25883E+01
                                               AXIAL TORQUE =    .00000E+00
** MAXIMUM MIDSPAN MOMENT =    .279E+03 AT DISTANCE   108.82 FROM NODE     6
   7    2    6   .00000E+00   .00000E+00   .14766E+03  -.66825E+01   .20517E+01
             7   .00000E+00   .00000E+00  -.34387E+03  -.83175E+01  -.20517E+01
                                               AXIAL TORQUE =    .00000E+00
** MAXIMUM MIDSPAN MOMENT =    .210E+03 AT DISTANCE   106.92 FROM NODE     6
   7    3    6   .00000E+00   .00000E+00  -.37362E+03   .33877E+01   .37968E+01
             7   .00000E+00   .00000E+00  -.43943E+03  -.33877E+01  -.37968E+01
                                               AXIAL TORQUE =    .00000E+00
   7    4    6   .00000E+00   .00000E+00   .36162E+03  -.15751E+02   .46400E+01
             7   .00000E+00   .00000E+00  -.78149E+03  -.19249E+02  -.46400E+01
                                               AXIAL TORQUE =    .00000E+00
** MAXIMUM MIDSPAN MOMENT =    .489E+03 AT DISTANCE   108.00 FROM NODE     6
   7    5    6   .00000E+00   .00000E+00  -.64373E+02  -.67662E+01   .55582E+01
             7   .00000E+00   .00000E+00  -.78674E+03  -.13859E+02  -.55582E+01
                                               AXIAL TORQUE =    .00000E+00
** MAXIMUM MIDSPAN MOMENT =    .331E+03 AT DISTANCE    78.73 FROM NODE     6
   8    1    8   .00000E+00   .00000E+00  -.30781E+03   .25883E+01   .10932E+02
             7   .00000E+00   .00000E+00  -.43762E+03  -.25883E+01  -.10932E+02
                                               AXIAL TORQUE =    .00000E+00
   8    2    8   .00000E+00   .00000E+00  -.24702E+03   .20517E+01   .83175E+01
             7   .00000E+00   .00000E+00  -.34387E+03  -.20517E+01  -.83175E+01
                                               AXIAL TORQUE =    .00000E+00
   8    3    8   .00000E+00   .00000E+00  -.65405E+03   .37968E+01   .33877E+01
             7   .00000E+00   .00000E+00  -.43943E+03  -.37968E+01  -.33877E+01
                                               AXIAL TORQUE =    .00000E+00
```

```
8   4   8  .00000E+00  .00000E+00 -.55483E+03  .46400E+01  .19249E+02
        7  .00000E+00  .00000E+00 -.78149E+03 -.46400E+01 -.19249E+02
                                   AXIAL TORQUE =    .00000E+00
8   5   8  .00000E+00  .00000E+00 -.81403E+03  .55582E+01  .13859E+02
        7  .00000E+00  .00000E+00 -.78674E+03 -.55582E+01 -.13859E+02
                                   AXIAL TORQUE =    .00000E+00
```

From the output it can be seen that the truss members are in tension (positive). This means that the model follows the standard design of slender members only taking tension. If this were not true, the structure would need to be rerun with the members in compression removed.

6.6.4 Three-Dimensional Example

The final example of SSTAN is a three-dimensional structure. The structure is a square single-story frame. The load on the structure will be a uniform load on the roof of 0.8 kip/ft². It is assumed that the roof load can be converted into a line load on each beam. Each beam will take one-fourth of the total roof load. Although the use of a line load is not completely accurate, it will be sufficient for demonstration purposes. The structure to be analyzed is shown in Figure 6.27. The properties of the members are
Columns:

$$I_x = 650 \text{ in}^4$$
$$I_y = 54 \text{ in}^4$$
$$A = 4 \text{ in}^2$$
$$J = 60 \text{ in}^4$$

Beams:

$$I_x = 450 \text{ in}^4$$
$$I_y = 32 \text{ in}^4$$
$$A = 3.2 \text{ in}^2$$
$$J = 43 \text{ in}^4$$

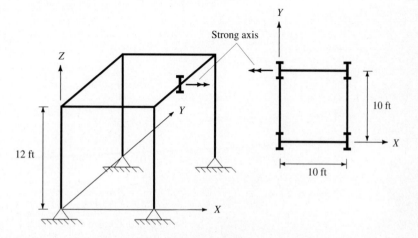

Figure 6.27 Three-dimensional example.

For all beams, the strong axis lies along the *X* or *Y* axis. This means that the roof load will cause bending about the strong or local 2 axis. The first step is to number the nodes and members. The scheme shown in Figure 6.28 is used.

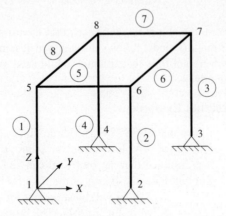

Figure 6.28 Node and element numbering.

The resulting input file is as follows:

```
Three-Dimensional Structure
8,1,1 : 8 nodes, beams only, one load case
:
COORDINATES
1 X=0 Y=0 Z=0
2 X=10*12
3 Y=10*12
4 X=0
5 Y=0 Z=12*12
6 X=10*12
7 Y=10*12
8 X=0
:
BOUNDARY
1,4 F=F,F,F,R,R,R
5,8 F=R,R,R,R,R,R
:
BEAMS
8,2
C-----Properties
1 A=4 E=29000 I=650,54 J=60
2 A=3.2 E=29000 I=450,32 J=43
C-----Columns -----
1 1,5,4 M=1
2 2,6,3
3 3,7,2
4 4,8,1
```

```
C-----Beams -----
5 5,6,1 M=2 L=-2/12 : 2 kip/ft = 1/4 total load
6 6,7,2      L=-2/12
7 7,8,3      L=-2/12
8 8,5,4      L=-2/12
:
```

Notice how the third node is used to specify that the strong direction of the columns and beams is used. For the beams, a node in the vertical plane is chosen (just the base node under the *I* node). The columns' strong axis is defined following Figure 6.26. In this case a node from another column is chosen. Note that any node can be used to define the *K* node for a beam. It does not matter whether the node is part of the structure or an extra node with no active DOFs. Nodes can be included for the purpose of defining a beam's *K* node (as was done in Example 6.6.1) and then given *no* released DOFs.

The output file is given below. Again, watch the output force sign convention since it is dependent on the choice of the third node for axis definition. Here notice that forces exist in both planes for the beam members.

```
************************************************************
****   SSTAN  -   Simple Structural Analysis Program   ****
Copyright 1993  By  Dr. Marc Hoit, University of Florida
************************************************************

Three-Dimensional Structure
NUMBER OF JOINTS                        =    8
NUMBER OF DIFFERENT ELEMENT TYPES       =    1
NUMBER OF LOAD CONDITIONS               =    1
NUMBER OF LOAD COMBINATIONS             =    0

NODE    BOUNDARY CONDITION CODES            NODAL POINT COORDINATES
NUMBER  X    Y    Z    XX   YY   ZZ           X            Y            Z
   1    F    F    F    R    R    R          .000         .000         .000
   2    F    F    F    R    R    R       120.000         .000         .000
   3    F    F    F    R    R    R       120.000      120.000         .000
   4    F    F    F    R    R    R          .000      120.000         .000
   5    R    R    R    R    R    R          .000         .000      144.000
   6    R    R    R    R    R    R       120.000         .000      144.000
   7    R    R    R    R    R    R       120.000      120.000      144.000
   8    R    R    R    R    R    R          .000      120.000      144.000

EQUATION NUMBERS
   N    X    Y    Z    XX   YY   ZZ
   1    0    0    0     1    2    3
   2    0    0    0     4    5    6
   3    0    0    0     7    8    9
   4    0    0    0    10   11   12
   5   13   14   15    16   17   18
   6   19   20   21    22   23   24
   7   25   26   27    28   29   30
   8   31   32   33    34   35   36

***** FRAME MEMBERS *****

NUMBER OF MEMBER PROPERTIES =    2

MEMBER PROPERTY NUMBER --------- =    1
AXIAL AREA, A ------------------ =          4.0000
SHEAR AREA, 3 Axis ------------- =           .0000
SHEAR AREA, 2 Axis ------------- =           .0000
TORSIONAL MOMENT OF INERTIA, J  =         60.0000
MOMENT OF INERTIA, I(2) -------- =        650.0000
MOMENT OF INERTIA, I(3) -------- =         54.0000
MODULUS OF ELASTICITY, E ------- =    .290000E+05
SHEAR MODULUS, G --------------- =    .111538E+05(USED FOR JG/L CAL.)
```

```
MEMBER PROPERTY NUMBER --------- =    2
AXIAL AREA, A ------------------ =        3.2000
SHEAR AREA, 3 Axis ------------- =         .0000
SHEAR AREA, 2 Axis ------------- =         .0000
TORSIONAL MOMENT OF INERTIA, J   =       43.0000
MOMENT OF INERTIA, I(2) -------- =      450.0000
MOMENT OF INERTIA, I(3) -------- =       32.0000
MODULUS OF ELASTICITY, E ------- =   .290000E+05
SHEAR MODULUS, G --------------- =   .111538E+05(USED FOR JG/L CAL.)


MEMB  CONNECTIVITY  MATERIAL              END ECCENTRICITIES
NUM. K=0 FOR Z-AXIS  SET  *****    I-END    ****    ****   J-END    *****  PIN
       I   J   K         X-VALUE Y-VALUE Z-VALUE X-VALUE Y-VALUE Z-VALUE OPT

  1    1   5   4      1     .00     .00     .00     .00     .00     .00    0
UNIFORM LOAD =    .000
  2    2   6   3      1     .00     .00     .00     .00     .00     .00    0
UNIFORM LOAD =    .000
  3    3   7   2      1     .00     .00     .00     .00     .00     .00    0
UNIFORM LOAD =    .000
  4    4   8   1      1     .00     .00     .00     .00     .00     .00    0
UNIFORM LOAD =    .000
  5    5   6   1      2     .00     .00     .00     .00     .00     .00    0
UNIFORM LOAD =   -.167
  6    6   7   2      2     .00     .00     .00     .00     .00     .00    0
UNIFORM LOAD =   -.167
  7    7   8   3      2     .00     .00     .00     .00     .00     .00    0
UNIFORM LOAD =   -.167
  8    8   5   4      2     .00     .00     .00     .00     .00     .00    0
UNIFORM LOAD =   -.167


THE NODE NUMBERING USED PRODUCED A HALF BANDWIDTH OF    14
TOTAL STORAGE REQUIRED  =        590
TOTAL STORAGE AVAILABLE =      35000

SOLUTION CONVERGED IN   1 ITERATION(S)
*** PRINT OF FINAL DISPLACEMENTS ***

DISPLACEMENTS FOR LOAD CONDITION  1
NODE        X           Y           Z          XX          YY          ZZ
  1   .00000E+00  .00000E+00  .00000E+00  .15917E-03 -.39863E-03  .42948E-19
  2   .00000E+00  .00000E+00  .00000E+00  .15917E-03  .39863E-03  .14883E-19
  3   .00000E+00  .00000E+00  .00000E+00 -.15917E-03  .39863E-03  .34747E-19
  4   .00000E+00  .00000E+00  .00000E+00 -.15917E-03 -.39863E-03 -.10867E-19
  5   .11703E-03  .57540E-03 -.24828E-01 -.33033E-03  .79971E-03  .42948E-19
  6  -.11703E-03  .57540E-03 -.24828E-01 -.33033E-03 -.79971E-03  .14883E-19
  7  -.11703E-03 -.57540E-03 -.24828E-01  .33033E-03 -.79971E-03  .34747E-19
  8   .11703E-03 -.57540E-03 -.24828E-01  .33033E-03  .79971E-03 -.10867E-19


---------------------- FRAME MEMBER RESULTS ----------------
MEM LOAD NODE         1-2 PLANE              1-3 PLANE        AXIAL FORCE
 #   #    #      MOMENT      SHEAR      MOMENT      SHEAR
 1   1    1  -.53256E-15  .18100E+00  .16660E-13  .88996E+00  .20000E+02
          5   .25064E+02 -.18100E+00 -.12815E+03 -.88996E+00  .20000E+02
                                             AXIAL TORQUE =   .00000E+00
 2   1    2   .35735E-15 -.18100E+00  .11935E-13  .88996E+00  .20000E+02
          6  -.26064E+02  .18100E+00 -.12815E+03 -.88996E+00 -.20000E+02
                                             AXIAL TORQUE =   .00000E+00
 3   1    3   .83961E-15  .18100E+00  .89789E-14  .88996E+00  .20000E+02
          7   .26064E+02 -.18100E+00 -.12815E+03 -.88996E+00 -.20000E+02
                                             AXIAL TORQUE =   .00000E+00
 4   1    4  -.60889E-15 -.18100E+00  .16667E-13  .88996E+00  .20000E+02
          8  -.26064E+02  .18100E+00 -.12815E+03 -.88996E+00 -.20000E+02
                                             AXIAL TORQUE =   .00000E+00
 5   1    5   .11819E-15  .55870E-17  .26064E+02 -.10000E+02  .18100E+00
          6   .55226E-15 -.55870E-17 -.26064E+02 -.10000E+02 -.18100E+00
                                             AXIAL TORQUE =  -.43336E-15
** MAXIMUM MIDSPAN MOMENT =  -.326E+03 AT DISTANCE    60.00 FROM NODE     5
 6   1    6  -.73785E-17 -.26832E-17  .12815E+03 -.10000E+02  .88996E+00
          7  -.31460E-15  .26832E-17 -.12815E+03 -.10000E+02 -.88996E+00
                                             AXIAL TORQUE =   .00000E+00
```

```
** MAXIMUM MIDSPAN MOMENT =  -.428E+03 AT DISTANCE    60.00 FROM NODE    6
   7   1   7  .64439E-15  .16619E-16  .26064E+02 -.10000E+02  .18100E+00
               8  .13499E-14 -.16618E-16 -.26064E+02 -.10000E+02 -.18100E+00
                                          AXIAL TORQUE = -.21673E-15
** MAXIMUM MIDSPAN MOMENT =  -.326E+03 AT DISTANCE    60.00 FROM NODE    7
   8   1   8  .64658E-15  .38404E-17  .12815E+03 -.10000E+02  .88996E+00
               5 -.18574E-15 -.38404E-17 -.12815E+03 -.10000E+02 -.88996E+00
                                          AXIAL TORQUE =  .86671E-15
** MAXIMUM MIDSPAN MOMENT =  -.428E+03 AT DISTANCE    60.00 FROM NODE    8
```

6.7 Solution Errors and Model Correctness

When using any general analysis program, it is always the responsibility of the engineer performing the analysis to check the correctness and accuracy of the analysis. Errors can occur due to bad input, bad modeling, and errors within the analysis program itself. When using a different computer program, an engineer should always check the program and his or her understanding of its use on a few simple problems that can be checked using other methods, including hand methods and other analysis programs.

6.7.1 Solution Errors

A number of common mistakes can occur when using computer programs to analyze a structure, the most common being an error in input to the program. Such an error can be the result of a misunderstanding about what is actually required as input, a typographical error, or bad modeling. Whatever the cause, some method of determining the correctness of solution is required.

The easiest error to handle is not understanding the required input. When an engineer is unsure of the required input, a simple model should be tried to test the input convention. For example, if the direction of applied member loading is confusing, try a single-element model, for which the results should be obvious.

Typographical errors and bad modeling are more difficult to check. If the errors are significant, the analysis will fail during some portion of the process. For example, if the structure is unstable, the equation solver will fail and give the message of a singular matrix. A good program will check the stiffness matrix for zeros on the diagonal and report the DOFs or nodes that have no stiffness connected. The engineer can then check the model to see why the DOF has no stiffness.

Smaller, more subtle problems are much harder to catch since the solution will usually finish but the answers may be in error by a small amount. Whatever the cause, there are some common methods of finding the errors. One common cause of errors in incorrect connectivity, connecting the members to the wrong nodes. Another is incorrect model geometry, with bad nodal coordinate specifications. Usually, the use of a graphics program that displays the structural configuration will help in determining these errors. It is usually obvious when a member is connected incorrectly. Incorrect connectivity can cause a structure to be unstable when the real structure is not. Some finite elements will give an error if they are not defined correctly. If they give a Jacobian of less than zero, this means that the element was connected incorrectly or has bad nodal coordinates.

Another common cause of errors is member properties. When incorrect values or directions are given for member properties, a solution can fail or be in error. This is a more subtle error and requires careful hand checking of the output. A missing property, specified as zero, can cause an unstable structure. An error in direction or value will cause a change in the results. Finding a small error in properties is very difficult and requires careful hand checking. It is always important to use the output for checking the structure since what you type in the input file and what the computer reads may be entirely different.

Bad boundary conditions will also cause solution problems. If insufficient restraints exist, the solution will fail since the structure is unstable. Solutions will change when the type of fixity used is incorrect. Simple checks on the support forces and equilibrium should help determine these errors. For example, pinned supports should have no moment result on the connected member.

6.7.2 Model Correctness

Once a model has been developed, the initial errors removed and the analysis run, the results need to be checked to verify the correctness of the model. Errors that cause the program not to complete the analysis are easier to find. Errors that cause the results to be off by 10 pecent are much more difficult. Thorough checking of an analysis is the responsibility of the engineer.

The first method of checking is to plot the structure. Plotting the structure allows a visual check of the nodal coordinates and element connectivity. Viewing the structure from different planes makes sure that nodes are directly lined up and plot on top of one another. Using shrunken elements allows connectivity checks of more complicated elements. Many errors are caused by input errors that result in an improper model. Some of this can be caught through visual inspection.

The next check is to use simple models to perform basic results checking. This is done to verify the range of the answers. All complicated structures can be represented by simplified models to get ''ballpark'' answers. As an example, a multistory frame structure can be modeled by a simple cantilever beam to check the drift at the top of the structure (Figure 6.29). The approximate moment of inertia is the Ad^2 effect of the columns.

Another example is a complicated bridge structure of which a simple representation is given in Figure 6.30. This complex structure can always be represented by a simple beam structure. Again, Ad^2 of the top and bottom chords gives the approximate moment

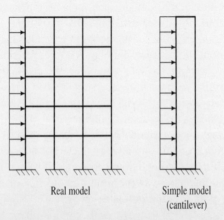

Real model   Simple model
(cantilever)

Figure 6.29 Simple verification model.

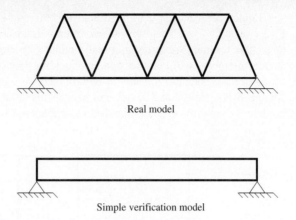

Real model

Simple verification model

Figure 6.30 Verification with simple models.

of inertia. Often, a simple structural representation can be used that requires only hand calculations to check the finite-element model. Other times, a simpler finite-element model needs to be used to verify a complex model. Either way, some method of checking must be performed.

The next way of checking a complicated structure is through the use of symmetry. In general, most structures have some type of symmetry. This symmetry can be used as a check on structural modeling. Even if the boundary conditions or the loads on a structure are not symmetric, a symmetric structure can be checked. The first step is to temporarily force the structure to be symmetric, including both boundary conditions and loads if necessary. Then run a preliminary analysis and check the symmetry of the results.

The important result to check is displacements. Remember, we are using a displacement-based method. Therefore, displacements are the primary unknowns. If the displacements are not correct, nothing else can be. Make sure that the displacements are symmetrical. It may even be necessary to make the element properties the same in this preliminary analysis to force symmetry. Once symmetry has been checked, the boundary conditions, properties, and correct loads can be put back on with more confidence in the model correctness.

Many other techniques can be used to check the results from a computer program. Whatever methods used, remember, the engineer is responsible for the results. Be sure you are confident that the results are correct before you sign the calculations.

6.8 Result Interpretation

Once the structural model has been verified, the results from the final analysis need to be interpreted. This mainly involves understanding the element results and their coordinate systems. For SSTAN we will look at the truss and beam members. The truss is a two-force member; its result can only be an axial tension or compression. In SSTAN, a tension is positive, compression is negative. SSTAN also prints the member stress (force divided by area).

Remember, SSTAN is a linear analysis program unless zero compression/tension or P–Δ effects are used. Therefore, slender members are assumed to take compression equally as well as tension. We know that real members cannot do this because they buckle. If a slender member is in compression, its buckling load must be checked. If a member is found to have buckled, it needs to be removed from the structure and the analysis rerun (Figure 6.31). Note that SSTAN can handle this automatically as one of its two limited nonlinear-capability, zero-compression members.

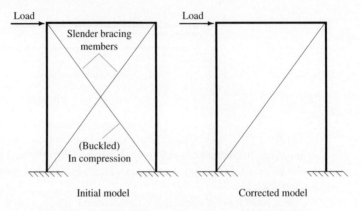

Figure 6.31 Model correction for buckled members.

Condensed Example: Zero-Compression Members

The following structure needs to be designed for the given load. Note that the horizontal load can come from either side. Also note that the bracing members are slender and can take no compression.

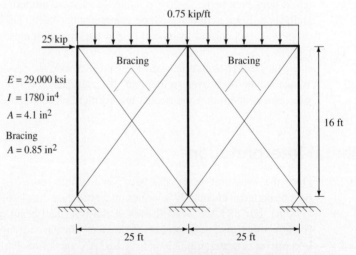

Example using zero-compression members.

Since the structure is symmetric, we can analyze the system for the horizontal load on one side and the results for the other are just flipped. However, if the bracing members are slender, they can take no compression. The general stiffness model allows any member to take equal tension and compression. Allowing the members to go into compression would give additional bracing that is not found in the real structure. Therefore, we must use the noncompression truss option.

To analyze the structure, we first need to number the nodes and elements.

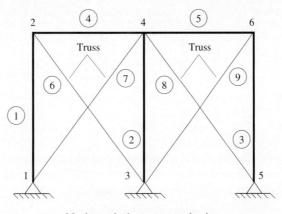

Node and element numbering.

From the preceding two figures, we can create the required SSTAN input:

```
Example: Zero-Compression Members
6,2,1  c=1   : 6 nodes, 2 types (truss, beam), 1 load
:
Coordinates
C---- Analyze in X-Z plane !
1 x=0 y=0 z=0
2 z=16*12
3 x=25*12 z=0
4 z=16*12
5 x=50*12 z=0
6 z=16*12
:
Boundary
1,6    DOF=r,f,r,f,r,f    : Set all to 2 dimen. (X-Z system)
1,5,2  DOF=f,f,f,f,r,f    : Pinned supports
:
Truss
4,1
1 A=0.85 E=29000          : Bracing members
7 1,4 c=1 M=1             : C=1, take no compression
6 2,3 c=1
8 4,5 c=1
9 3,6 c=1
:
Beam
5,1
1 A=4.1 i=1780 e=29000
C--- Columns
1 1,2,3 m=1
2 3,4,5
3 5,6,1
C--- Beams
```

```
4 2,4,1 1=-0.75               : K node is below, negative load
5 4,6,3 1=-0.75
:
Loads
2 1=1 f=25
:
```

After running SSTAN we can look at the output file for the results. The truss member output is as follows:

```
-----------------FORCE'S TRUSS MEMBERS----------------
ASTERISK (*C or *T) MEANS COMPRESSION OR TENSION SET TO ZERO FOR THIS MEMBER
MEMBER  LOAD  IEND  JEND        AXIAL          AXIAL
   #      #     #     #        FORCES         STRESS

   7      1     1     4        4.8499         5.7057

   6      1     2     3         .0000          .0000 **C

   8      1     4     5         .0000          .0000 **C

   9      1     3     6        8.9005        10.4712
```

As can be seen from the above, two of the members go into compression and therefore take zero force. If the horizontal load had come from the other side, the other two members would have become zero. Therefore, all four bracing members are required, but the true required tension for each is known.

The beam member in SSTAN is a three-dimensional bending member. Therefore, it has six force results at each end. The coordinate system for these results is dictated by the K-node location. Figures 6.32 and 6.33 show the two possible results. Note that SSTAN gives the results in terms of local coordinate planes.

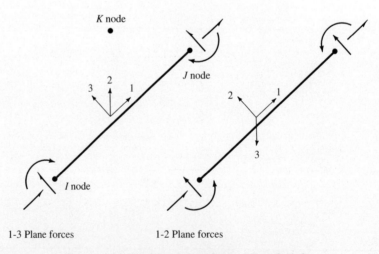

1-3 Plane forces   1-2 Plane forces

Figure 6.32 Positive beam forces, K node left.

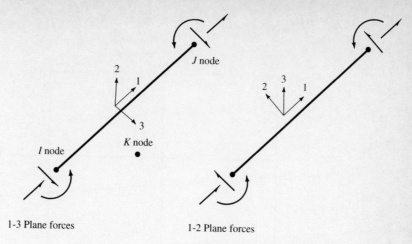

1-3 Plane forces   1-2 Plane forces

Figure 6.33 Positive beam forces, K node right.

6.9 Problems

6.1. Use SSTAN to analyze the following structures. Use $E = 29{,}000$ ksi for all structures unless otherwise specified. Check global equilibrium of the entire structure using the support reactions.

(a)

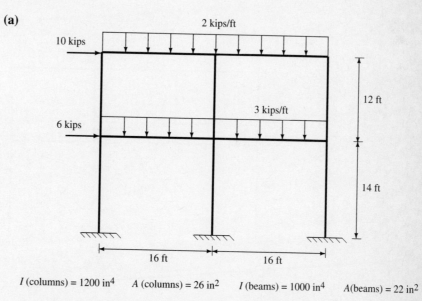

I (columns) $= 1200$ in^4 A (columns) $= 26$ in^2 I (beams) $= 1000$ in^4 A(beams) $= 22$ in^2

Draw the moment diagram for the first floor, left span beam.

(b)

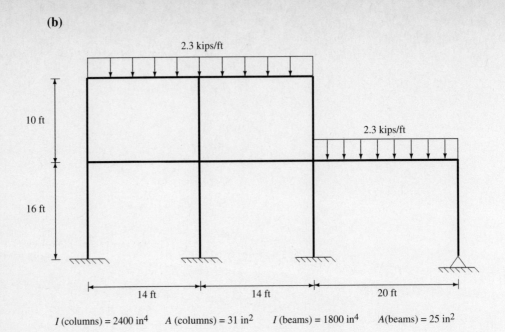

I (columns) = 2400 in^4 A (columns) = 31 in^2 I (beams) = 1800 in^4 A(beams) = 25 in^2

Show the final joint equilibrium where the one-story roof meets the rightmost two-story column.

(c)

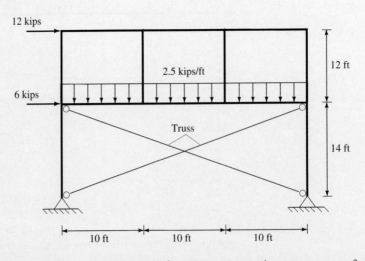

I (columns) = 1200 in^4 A (columns) = 27 in^2 I (beams) = 900 in^4 A(beams) = 20 in^2 A(truss) = 8 in^2

Show the final joint equilibrium for the joint where the 6-kip load is applied.

6.2. For the following structures, use SSTAN to analyze the displacements and the maximum force member. For beams, plot the moment diagram for the member with the maximum moment. For the truss structures, find the member with the maximum force and give the support reactions.

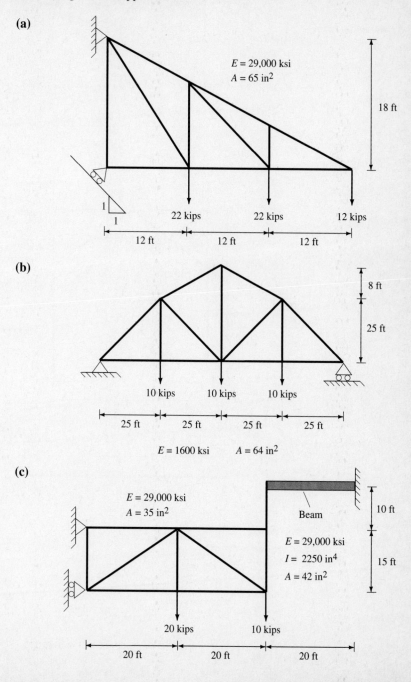

(d)

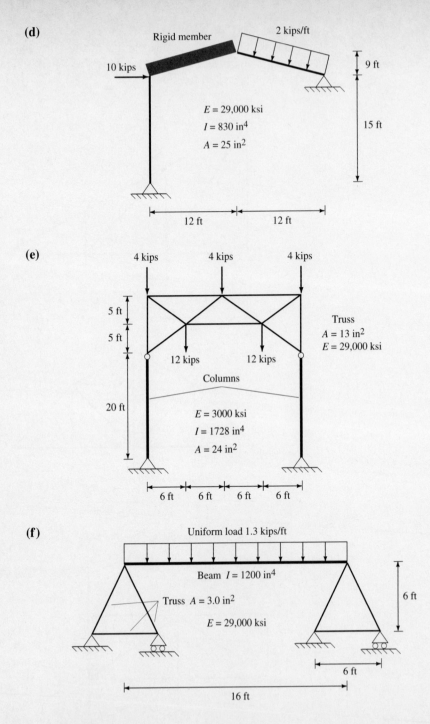

(e)

(f)

(g)

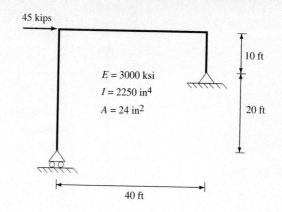

45 kips

$E = 3000$ ksi
$I = 2250$ in^4
$A = 24$ in^2

10 ft

20 ft

40 ft

(h)

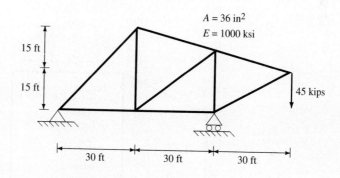

$A = 36$ in^2
$E = 1000$ ksi

15 ft

15 ft

45 kips

30 ft   30 ft   30 ft

(i)

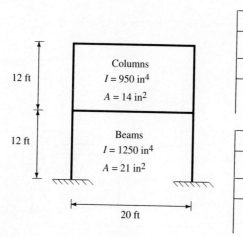

Columns
$I = 950$ in^4
$A = 14$ in^2

Beams
$I = 1250$ in^4
$A = 21$ in^2

12 ft

12 ft

20 ft

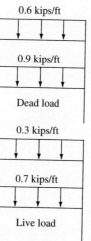

0.6 kips/ft

0.9 kips/ft

Dead load

0.3 kips/ft

0.7 kips/ft

Live load

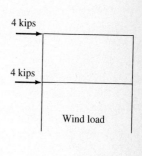

4 kips

4 kips

Wind load

(j)

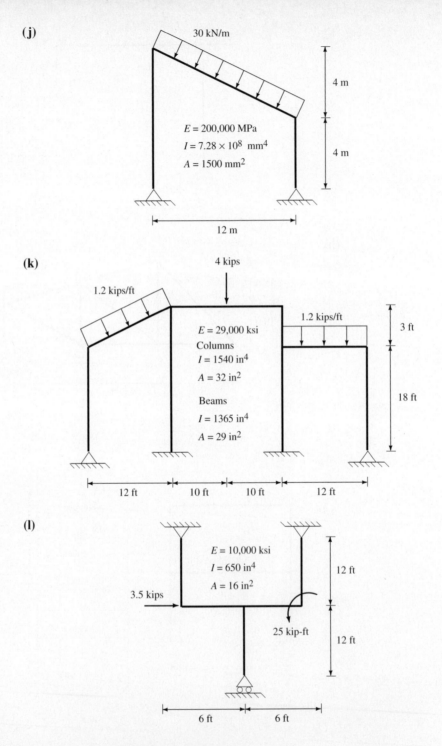

30 kN/m

4 m

$E = 200,000$ MPa

$I = 7.28 \times 10^8$ mm^4

$A = 1500$ mm^2

4 m

12 m

(k)

4 kips

1.2 kips/ft

$E = 29,000$ ksi

Columns
$I = 1540$ in^4
$A = 32$ in^2

Beams
$I = 1365$ in^4
$A = 29$ in^2

1.2 kips/ft

3 ft

18 ft

12 ft   10 ft   10 ft   12 ft

(l)

$E = 10,000$ ksi
$I = 650$ in^4
$A = 16$ in^2

12 ft

3.5 kips

25 kip-ft

12 ft

6 ft   6 ft

(m)

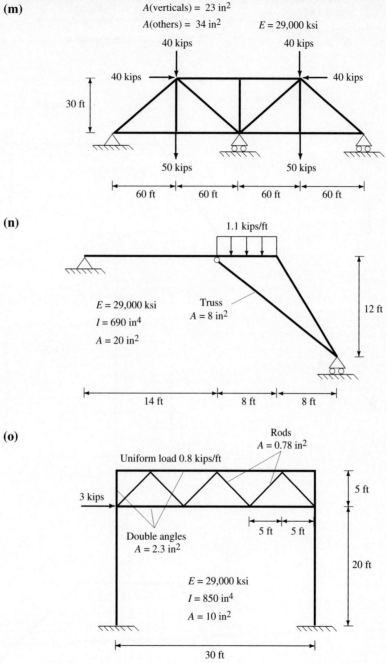

A(verticals) = 23 in^2
A(others) = 34 in^2 E = 29,000 ksi

40 kips   40 kips

40 kips →   ← 40 kips

30 ft

↓ 50 kips   ↓ 50 kips

|← 60 ft →|← 60 ft →|← 60 ft →|← 60 ft →|

(n)

1.1 kips/ft

E = 29,000 ksi
I = 690 in^4
A = 20 in^2

Truss
A = 8 in^2

12 ft

|← 14 ft →|← 8 ft →|← 8 ft →|

(o)

Rods
A = 0.78 in^2

Uniform load 0.8 kips/ft

3 kips →

5 ft

|← 5 ft →|← 5 ft →|

Double angles
A = 2.3 in^2

20 ft

E = 29,000 ksi
I = 850 in^4
A = 10 in^2

|← 30 ft →|

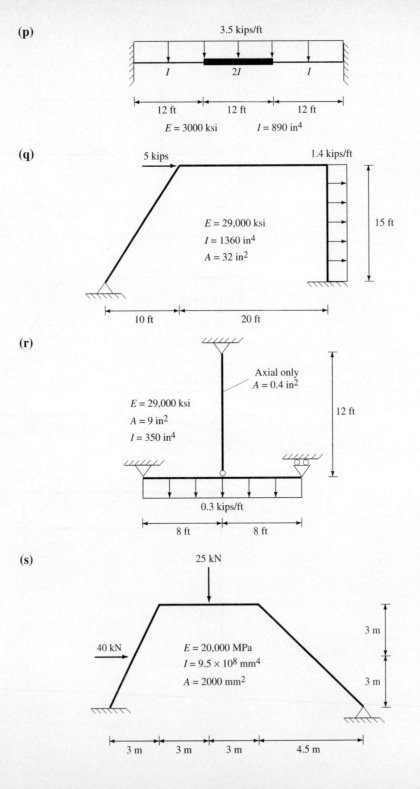

(p)

3.5 kips/ft

I $2I$ I

12 ft   12 ft   12 ft

$E = 3000$ ksi $I = 890$ in^4

(q)

5 kips   1.4 kips/ft

$E = 29,000$ ksi
$I = 1360$ in^4
$A = 32$ in^2

15 ft

10 ft   20 ft

(r)

Axial only
$A = 0.4$ in^2

$E = 29,000$ ksi
$A = 9$ in^2
$I = 350$ in^4

12 ft

0.3 kips/ft

8 ft   8 ft

(s)

25 kN

40 kN $E = 20,000$ MPa
$I = 9.5 \times 10^8$ mm^4
$A = 2000$ mm^2

3 m

3 m

3 m   3 m   3 m   4.5 m

(t)

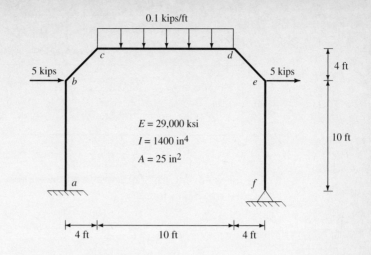

6.3. Analyze the following three-dimensional structures using SSTAN. Find the member with the maximum moment and plot its moment diagram.

(a)

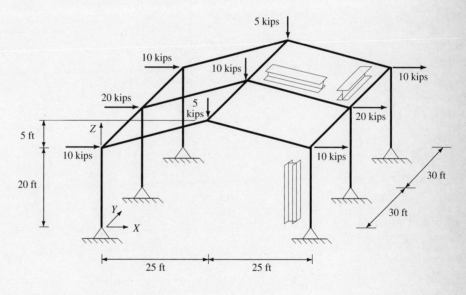

	E (ksi)	A (in²)	I_x (in⁴)	I_y (in⁴)	J (in⁴)
Columns	29,000	31.1	1910	220	7.48
Beams	29,000	24.7	2370	94	3.70

I_x is the strong axis and I_y is the weak axis.

(b) For the following truss structure, find the support reactions and which member has the maximum force.

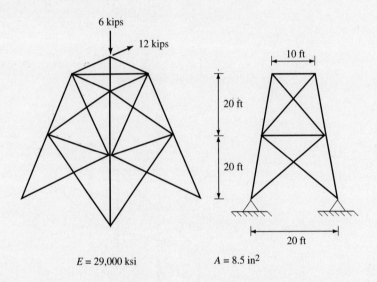

$E = 29,000$ ksi $A = 8.5$ in²

(c) The loading is a uniform load on both roofs. Assume that the load is taken by the "long" beams. Show the joint equilibrium for the joint where the lower roof meets the column.

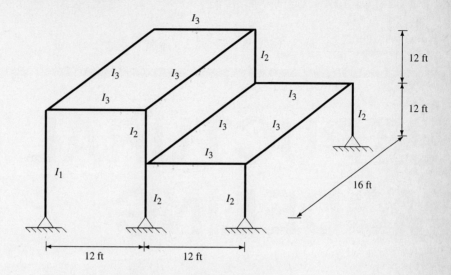

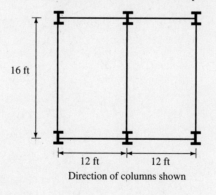

Uniform load on both roofs of 0.6 kips/ft^2

Direction of columns shown

$E = 29{,}000$ ksi $I_{X1} = 1850$ in^4 $I_{Y1} = 84$ in^4 $J_1 = 2$ in^4 $A_1 = 20.1$ in^2

$I_{X2} = 1350$ in^4 $I_{Y2} = 30$ in^4 $J_2 = 1.5$ in^4 $A_2 = 16.0$ in^2

$I_{X3} = 2100$ in^4 $I_{Y3} = 84$ in^4 $J_3 = 2$ in^4 $A_3 = 22.2$ in^2

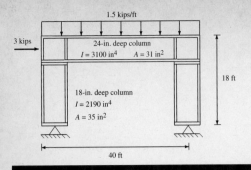

STRUCTURAL
MODELING

Understanding the methods for forming frame and truss stiffnesses, assembling them, and solving for the displacements and resulting forces is just the beginning of practical structural analysis. Very often, unique structures must be analyzed that require special loading or support modeling that does not fit the standard assumptions in matrix structural analysis. Such assumptions as that all boundaries act in the global X, Y, and Z directions and loads act perpendicular to the member are frequently not true for practical structures. In this chapter we discuss how to handle different modeling conditions to get numerical models that represent the real structure more accurately.

7.1 Load Projection

In design, distributed loads often need to be applied in directions other than perpendicular to a member. This is the case when following the building codes in applying snow loads. Snow loads are applied as vertical loads (direction of gravity) even on pitched roofs (Figure 7.1). Special consideration needs to be given when applying this projected load to a structure. To explain the effect, a simple slanted roof frame will be used (Figure 7.2). Notice how the load acts over the entire span but is not perpendicular to the member. Note that the total load applied to the structure is

$$P_{total} = wL = wL' \cos \theta \tag{7.1}$$

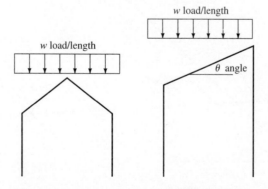

Figure 7.1 Snow loads on simple structures.

which is the distributed load times the span of the frame. By statics, it can be seen that there are no horizontal reactions and the vertical reactions are equal and have a value of

$$R_a = R_b = \frac{wL}{2} = \frac{wL' \cos \theta}{2} \tag{7.2}$$

To account for the member loads we need to resolve the load into its perpendicular and longitudinal components. Then we can form the required fixed-end forces (FEFs). The pure vector components of the load are shown in Figure 7.3. However, these components are valid for the span length and not the member length. Notice that the member length is longer than the span length and hence we would be adding too much load by using

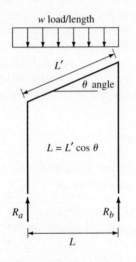

Figure 7.2 Projected load example.

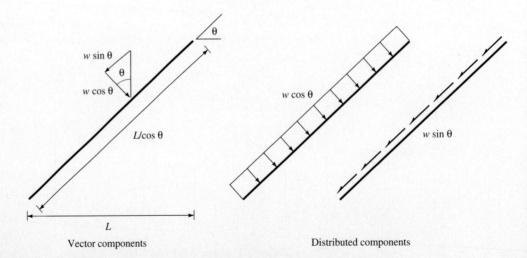

Figure 7.3 Load components on slanted member.

these components. This change can be accounted for by a change in integration when finding the fixed-end forces. The required change is to substitute $dl \cos \theta$ for dx during the integration. Since the $\cos \theta$ is a constant, this just becomes a multiplier times the FEFs of the distributed load. Another way to view this change is as a correction to the applied load. To correct for the real member length versus the span length over which the load acts, we need to reduce the load components. The reduction is clearly a fraction composed of the load length divided by the member length. As a result, the values needed for the calculation of equivalent member loads are shown in Figure 7.4.

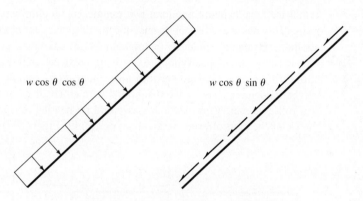

Figure 7.4 Final vector components.

The values to be used if the load were a horizontal projection are clearly similar. The reduction factor is always the load length divided by the member length. Hence, for wind loads, you would use the values shown in Figure 7.5. Applying these loads is now a matter of using these reduced loads and the standard fixed-end forces for the perpendicular portion. In an analysis program, the distributed load becomes this modified (reduced) value. The axial component is usually applied by splitting the total, giving half to each end of the beam. If no distributed load capabilities exist, the fixed-end forces and the axial component need to be resolved back into the global X and Y directions and then applied as concentrated loads. Remember, then, to remove them again from the final beam results.

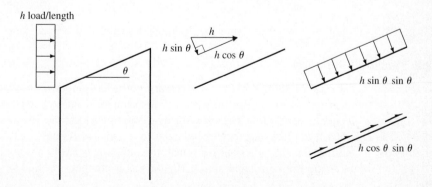

Figure 7.5 Horizontal loading components.

7.2 **Boundary Conditions**

Structural modeling requires that the analyst be able to approximate many different types of boundary conditions. While the standard boundary conditions of release or fixity in the global directions has already been covered by including or not including the DOFs, other support conditions are needed. The ability to model semirigid supports, rollers not in the global directions, and hinges is often required. These techniques are discussed next.

7.2.1 **Spring Supports**

So far we have handled fixed support conditions by eliminating a DOF at the location of fixity. This has the effect of specifying the displacement at that DOF as exactly zero. In addition, this method reduces the total number of unknowns to those that are not specified. Another method of specifying a fixity is to allow the DOF to exist and then place a very stiff spring at that DOF to restrain movement. For example, a simple single-bay frame can be modeled as having fixed supports with either of the methods shown in Figure 7.6. If a roller support were needed, this could be handled by placing a very stiff spring at the roller location offering vertical support only (Figure 7.7).

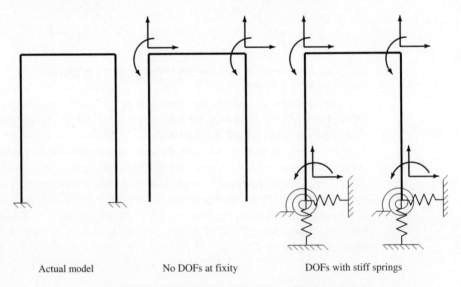

Actual model   No DOFs at fixity   DOFs with stiff springs

Figure 7.6 Methods for modeling a fixed support.

7.2.2 **Inclined Supports**

The use of springs becomes necessary when the support is needed in a direction other than one of the global directions. For example, if an inclined roller were needed, this could be handled by placing a spring at the roller location. However, the spring must be rotated to offer support perpendicular to the incline (Figure 7.8). Since the rotation of a stiffness matrix is a standard transformation, this is easily accomplished.

The use of rotated spring supports is a common method of specifying nonglobal

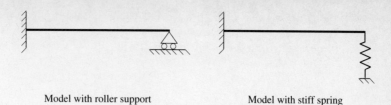

Model with roller support   Model with stiff spring

Figure 7.7 Methods for modeling a roller support.

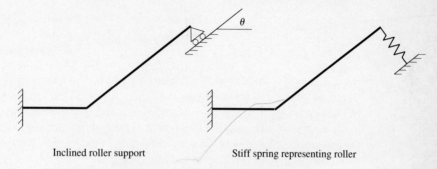

Inclined roller support   Stiff spring representing roller

Figure 7.8 Model for inclined support.

boundary conditions. When using a spring element for nonglobal boundary conditions, some care must be exercised. If the spring is situated in the global directions, the stiffness adds only to the diagonal of the equations and causes no problems when solving the equations. If, however, you need the support at some direction other than the global directions, the springs need to be rotated. If the values of the springs are very large, when rotated the off-diagonal values also become very large. These off-diagonal values can cause numerical instability during the solution of the equilibrium (stiffness) equations. The solution to this problem is to use spring stiffness values appropriate to the problem.

The rule of thumb for choosing support stiffness values is that the value should be approximately 1000 times greater than the structure to which they are connected. For example, when using the structure in Figure 7.8, which requires an inclined roller, we need to check the stiffness of the structure that the spring is to support. As can be seen, the support is to restrain displacements of the type drawn in Figure 7.9. Since stiffness

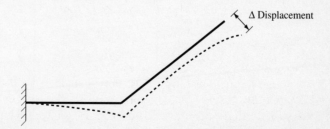

Figure 7.9 Unit displacement in the direction of the restraint.

is defined as force divided by displacement, we need to find the approximate stiffness of the structure for the displacement. This is approximately the stiffness of a straight cantilever subjected to a tip displacement. The stiffness in this case is

$$K \approx \frac{3EI}{L^3} \tag{7.3}$$

where L is the total length of the cantilever. Using this as a guide, the approximate stiffness of the spring should be

$$K_{\text{spring}} \approx \frac{3000EI}{L^3} \tag{7.4}$$

Let's look at another example of trying to find the approximate structural stiffness. Using the frame structure in Figure 7.10, let's find what the support spring values should be. As before, the first step is to draw the approximate deflected shape for a displacement in the direction of the required support. Now looking at this deflected shape (Figure 7.11), we can see that the member that absorbs most of the vertical displacement is the beam. Also, this beam is deformed primarily in a perpendicular displacement mode. This means that the approximate vertical structural stiffness to use as a fixed spring support would be

$$K_{\text{spring}} \approx \frac{12,000EI}{L^3} \tag{7.5}$$

where the length and properties used are those for the beam. The approximate horizontal displacement and resulting stiffness are also shown. It is important to note that these methods are approximate. As such, it is not crucial that the structure's stiffness be evaluated exactly. The usual method for checking whether the choice was correct is to look at the results to see if the displacement in the direction of the support is sufficiently small as

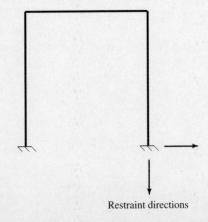

Restraint directions

Figure 7.10 Frame example for spring supports.

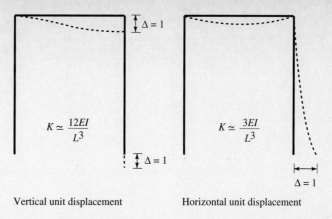

Figure 7.11 Approximate stiffness values for spring supports.

to be negligible. If it is not, increase the magnitude of the spring stiffness and rerun the problem. It is important to remember that modeling is not an exact process and often requires a trial-and-error process to develop a satisfactory model.

Often, programs do not have a rotated spring boundary condition option. If this is the case, there is a simple alternative. Translational springs can be modeled using truss or beam members. This process is exact since a truss or beam can act as an axial member. Therefore, a truss bar can be placed so that the axis is parallel to the line of support. The values for the area, Young's modulus, and length need to be chosen so that the truss has an equivalent stiffness of the required spring. For example, if a truss bar is to be used for the inclined cantilever, we would model it as shown in Figure 7.12. Note that a length of 1 and a Young's modulus of 1 could be used. This is an arbitrary choice that causes the area to be equal to the desired stiffness of the spring since the stiffness is AE/L.

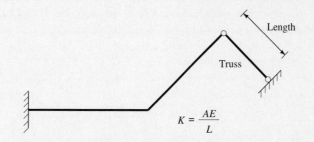

Figure 7.12 Truss bar used as equivalent support spring.

7.2.3 Hinges

In a similar fashion, rotational springs can be modeled using the torsional stiffness of a beam element. This assumes that the program can handle three-dimensional structures since the beam must be placed perpendicular to the structure. Figure 7.13 shows an example of torsional springs provided by beam elements. Again, when choosing the length

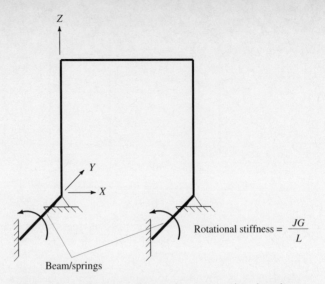

Figure 7.13 Beam torsion acting as rotational springs.

and rotational stiffness (torsional moment of inertia), the values need to be chosen so that the final rotational stiffness has the value of the desired spring. Recall that the torsional stiffness for a beam is *JG/L*.

The torsional property of a beam can also be used to form hinges within a structure. Many programs allow beams to have pins at the ends. However, a large number do not. A beam element can be used to model this condition. Suppose that the structure shown in Figure 7.14 is to be modeled. If the program has pinned-end beams, all that would be required is to place a pin at the right end of the beam. Note that a hinge must not be additionally placed at the end of the right-hand column if the rotational DOF of the corner is released. This is because one of the members (beam or column) must add stiffness to the node. If both are pinned, it is like connecting truss bars to a node where no rotational stiffness exists. So, in reality, the pin can exist at either the top of the column or the right end of the beam and get identical results. If pins exist at both, the rotation at the node must be fixed. This is shown in Figure 7.15. If the program does not offer pinned-end

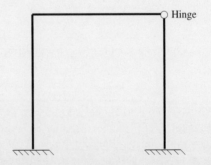

Figure 7.14 Example frame with a hinge.

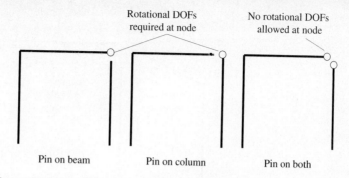

Figure 7.15 Possible pin locations on member ends to create a hinge.

beams, an alternative approach is to use a beam with zero torsional stiffness. Again, this creates a three-dimensional model, but it is exact. The structure shown in Figure 7.16 could be used.

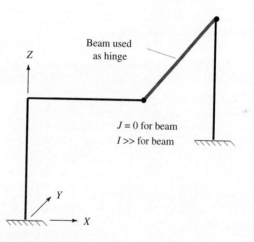

Figure 7.16 Hinge modeled by beam with $J = 0$.

Note that to maintain compatibility of the other displacements, large values for area and the two moments of inertia must be used. The torsional moment of inertia must be zero. Again, when choosing values for A and I to constrain the lateral and vertical DOFs to be the same, you must use relative values. The rule of thumb is to choose values 1000 times the connected stiffness values. In this case, just use 1000 times the moments of inertia of the beam and/or column.

Condensed Example: Hinged Member

The following single-bay structure is having a frame with a slanted roof attached. The attachment connection is to be considered as a hinge. Analyze the complete structure to find the moments in the original beam so that they can be checked later for adequacy.

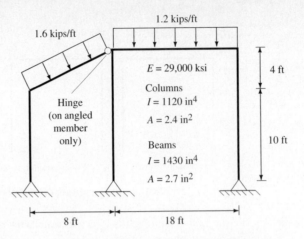

Example including a hinge on a member.

To create the SSTAN input, we need to number the nodes and elements. The following figure gives the numbering.

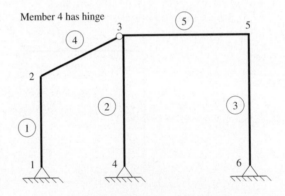

Node and element numbering.

Using the two figures, we create the following input file:

```
Example: Hinged Member
6,1,1               : 6 nodes, 1 type (beams), 1 load
:
Coordinates
1 x=0 y=0 z=0
2 y=10*12
3 x=8*12 y=14*12
4 y=0
5 x=26*12 y=14*12
6 y=0
:
```

```
Boundary
1,6     DOF= r,r,f,f,f,r    : Set all to two dimensions
1,4,3   DOF= f,f,f,f,f,r    : Pin supports
6       DOF= f,f,f,f,f,r    : Pin support
:
Beams
5,2
1 a=2.7 i=1430 e=29000      : Beams
2 a=2.4 i=1120              : Columns
C--- Columns
1 1,2,6 m=2
2 4,3,6
3 6,5,1
C--- Beams
4 2,3,1 m=1 l=-1.6/12 p=2   : Slanted, hinge at J (3) end
5 3,5,4     l=-1.2/12       : K node below, load negative
:
```

Note how the hinge is assigned to member 4. The *J* end (node 3) of the slanted member is hinged. Also notice that no concentrated loads are used. After running SSTAN, the two beam members' output is as follows:

```
---------------------- FRAME MEMBER RESULTS ----------------
MEM LOAD NODE        1-2 PLANE           1-3 PLANE       AXIAL FORCE
 #    #   #     MOMENT     SHEAR     MOMENT     SHEAR
 4    1   2   .00000E+00 .00000E+00 -.27956E+03 -.45508E+01 -.32921E+00
          3   .00000E+00 .00000E+00  .00000E+00 -.97600E+01  .32921E+00
                                        AXIAL TORQUE =   .00000E+00
** MAXIMUM MIDSPAN MOMENT =   .357E+03 AT DISTANCE   34.13 FROM NODE    2
 5    1   3   .00000E+00 .00000E+00 -.11441E+03 -.76342E+01  .33894E+01
          5   .00000E+00 .00000E+00 -.56941E+03 -.13966E+02 -.33894E+01
                                        AXIAL TORQUE =   .00000E+00
** MAXIMUM MIDSPAN MOMENT =   .406E+03 AT DISTANCE   76.34 FROM NODE    3
```

Notice how the moment at node 3 on member 4 has a value of zero. Also notice that both members have a positive moment in the span. In the case of member 4, the moment is larger than the end moment.

7.3 Imposed Displacements

When designing a structure, it is often necessary to include the effects of support settlements. The inclusion of imposed displacements is a fairly simple procedure. As an example, the structure shown in Figure 7.17 will be modeled. The solution method involves replacing the support with a spring to enforce the displacement boundary condition. This spring needs to act as any normal support spring. Next, a force is applied to the support that will cause the support to move the required amount (Figure 7.18). The value of the

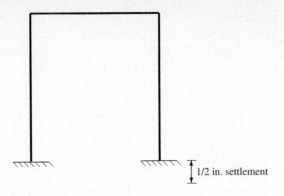

Figure 7.17 Applied support settlement example.

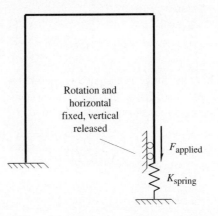

Figure 7.18 Model for imposed settlement.

required force is calculated from the force–displacement relationship of the spring. The equation for this relationship is

$$F_{\text{applied}} = K_{\text{spring}}\Delta(\text{settlement}) \tag{7.6}$$

In equation (7.6), Δ is the known value, in this case $\frac{1}{2}$ in. The stiffness value for the spring should follow the rule of thumb of 1000 times the connected stiffness, as discussed in Section 7.2. The required force can easily be calculated.

Since the spring used to impose the boundary condition is very stiff compared to the structure, the force required will be very large. However, this force will go directly into the spring support and not affect the rest of the structure except to impose the desired displacement. Again, as a check, the final displacement should be compared with the imposed settlement and the stiffness value and load increased if the accuracy is not sufficient.

Another method of imposing displacements on a structure is to remove the displacement DOFs from the equations. The displacement is a known value for some DOFs, yet it is found on the left-hand side of the stiffness equations with the other unknown displacements. As a result, it can be multiplied by the stiffness matrix and brought over to the right-hand side of the equations. As an example, the structure shown in Figure 7.19 is a model of the example frame with the DOFs shown. If we are given the imposed displacement at DOF r_4, we can remove it from the set of unknowns.

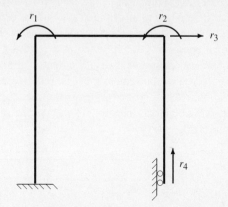

Figure 7.19 Model for settlement by removing DOF.

The equations for the structure shown in Figure 7.19 convert to

$$
\begin{bmatrix} K_{11} & K_{12} & K_{13} \\ K_{21} & K_{22} & K_{23} \\ K_{31} & K_{32} & K_{33} \end{bmatrix}
\begin{bmatrix} r_1 \\ r_2 \\ r_3 \end{bmatrix} =
\begin{bmatrix} R_1 - K_{14} \times \frac{1}{2} \\ R_2 - K_{24} \times \frac{1}{2} \\ R_3 - K_{34} \times \frac{1}{2} \end{bmatrix}
\tag{7.7}
$$

Note that the fourth equation is not necessary since the displacement for the fourth DOF is a known value. This process has the effect of converting the displacement to a load. This method is exact but complicates the equation-solving process. It requires the global stiffness to be formed and then modified with zeros for the column and row at which the displacement is applied. Next, the load must be modified with the stiffness times displacement terms. The stiffness must then be collapsed or the equation solver must skip zeros on the diagonal. This method is not found in many programs since skipping zeros on the diagonal can cause errors to be ignored and give results that are erroneous.

7.4 Symmetry and Antisymmetry

When modeling a structure, there is always a trade-off between accuracy and problem size. Usually, the more accurate the model, the larger it becomes. In the limit, the model becomes too large to analyze. It is the job of the analyst to create a model that is as accurate as required, yet reduces the computer demand to a minimum. Reducing the computer demand to a minimum is important because it allows additional time for "what if" analyses as well as result checking. It is very common to find that the model desired is too large for the computer or program that is being used for the analysis. This is often because a poor model was chosen; however, often the model is good but it is too large.

One way to reduce a model is to take advantage of symmetry. In this case, symmetry means symmetry of both the structure and the loads. For the structure to be symmetric, the members and the support conditions must be symmetric. A simple example of how symmetry works will demonstrate the concept.

Given the simple span beam in Figure 7.20 with a uniform distributed load, we can see that there is a line of symmetry about the center. If we draw the deflected shape, we can see some additional properties of symmetry (Figure 7.21). Here we can see that some additional constraints are enforced as a result of symmetry. The two constraints are that the rotation and horizontal displacement at the line of symmetry are zero. Also notice that the vertical displacement must be released and allowed to move. As a result of this observation, we could reduce the size of the structure to one of the symmetric parts and enforce these boundary constraints and get the same results. The new structure is shown in Figure 7.22.

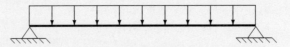

Figure 7.20 Simply supported beam with uniform load.

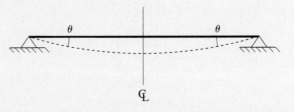

Figure 7.21 Displaced shape of uniformly loaded beam.

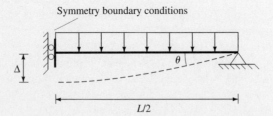

Figure 7.22 Symmetric portion of structure.

This concept can be extended to all three dimensions. The generalized process for three dimensions is as follows:

1. Cut the structure with a plane into its symmetric parts

2. Correct the boundary conditions at the plane of symmetry to:
   (a) All the in-plane DOFs are released.
   (b) All the out-of-plane DOFs are fixed.

In-plane DOFs are defined as the translations in the plane and the rotation vector perpendicular to the plane. Out-of-plane DOFs are those that are not in-plane. For example, if the cutting plane is the Y–Z plane, the Y and Z translations and the X rotation would be released. The others would be fixed. Remember, the decision of whether or not a DOF is to be fixed, based on there being stiffness to restrain it, still overrides the symmetry constraints. This means that a two-dimensional structure would still have the DOF perpendicular to the structure fixed to avoid an instability.

Some examples of symmetric structures and the symmetric boundary conditions are given below. The first structure is the two-dimensional truss structure shown in Figure 7.23. The total structure has only the horizontal and vertical translational DOFs as possible displacements. There are no rotations for truss members. The symmetric portion is also given in the figure. When a member is cut across its axis, the new member has the same properties as the original but it is shortened. Notice a special change for the horizontal truss member. If the symmetry boundary conditions are applied, the vertical displacement would be left free. However, this violates the principle that any DOF must have some associated stiffness. Truss members do not have stiffness perpendicular to their axis. In this case, the vertical DOF must be fixed so that the structure remains stable.

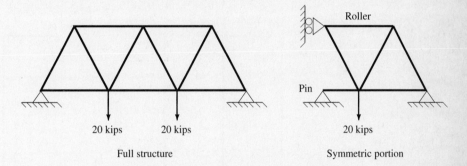

Full structure   Symmetric portion

Figure 7.23 Symmetry truss example.

The next structure is a three-bay frame in the X–Y plane, shown in Figure 7.24. In this case, the original DOFs allowed are the X and Y translations and the Z rotation. The symmetric portion of the structure shows the required symmetry boundary conditions. These are identical to the simple span beam. The vertical translation is released. The rotation and the horizontal translation are fixed. Notice how the center beams are cut in half but the load has the same magnitude and acts over the entire shortened length.

The next example is a two-bay frame, shown in Figure 7.25, in which a member as well as a load lies in the plane of symmetry. The center column and the vertical load on the center column needs to be split as a result of symmetry conditions. In the symmetric structure, the boundary conditions at the plane of symmetry are identical to the preceding example. In addition, the properties of the column are reduced by half and the concentrated load is also reduced by half. Notice that the use of half properties gives the correct force for the full member.

The final example is the three-dimensional structure shown in Figure 7.26. It is a simple rectangular frame structure. Both the full structure and symmetric portion, together

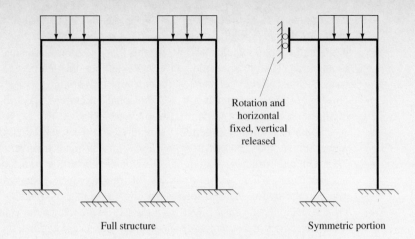

Full structure

Symmetric portion

Figure 7.24 Symmetry frame example.

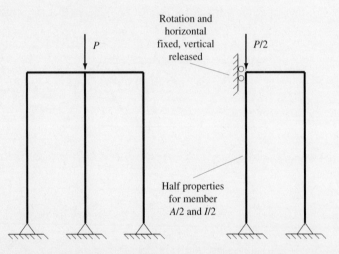

Figure 7.25 Symmetry conditions with member and load in plane.

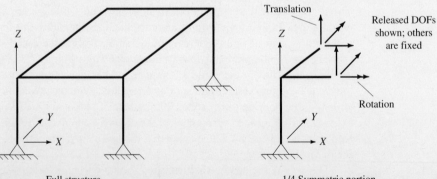

Full structure

1/4 Symmetric portion

Figure 7.26 Three-dimensional structure with symmetry boundary conditions.

with boundary conditions, are given. Since the structure is three-dimensional, all six DOFs exist at each node in the original structure. Notice that the three-dimensional structure has two planes of symmetry.

Finally, we need to discuss how to handle a symmetric structure when the loads are antisymmetric. First, we need to discuss what symmetric and antisymmetric loads are. Symmetric loads are loads that match. If the plane of symmetry were a mirror, the reflected half of the structure's loads are the same as the actual loads applied to that half. Antisymmetric loads are when the reflected loads are equal in magnitude but opposite in direction. As an example, the structures in Figure 7.27 are symmetric but have antisymmetric loading.

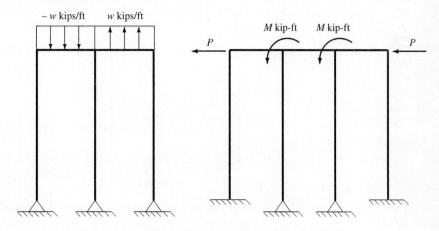

Figure 7.27 Symmetric structures with antisymmetric loading.

This type of structure can also be reduced through the use of boundary conditions. The rules for a symmetric structure with antisymmetric loads are:

1. Cut the structure with a plane into its symmetric parts.

2. Correct the boundary conditions at the plane of symmetry:
   (a) All the out-of-plane DOFs are released.
   (b) All the in-plane DOFs are fixed.

Notice that the releases and fixities are just the opposite of the symmetric structure/symmetric loading version. Again remember that out-of-plane DOFs for two-dimensional problems must still be fixed since no stiffness exists.

The advantages of using these rules to reduce the size of a structure are obvious. They can reduce the structure to a small portion of the original size. This may allow the analysis of a structure that could not be analyzed in its full size. In addition, using these rules can reduce the time it takes to create a structural model as well as to reduce the computer time to analyze the structure. All of this can be accomplished with no loss of accuracy since these methods are exact.

The only thing left to discuss is what to do when you have a symmetric structure but the loads are neither symmetric nor antisymmetric. Any load can always be converted into the sum of a symmetric and an antisymmetric load. Therefore, any symmetric structure can be handled by performing two analyses: one analysis for the symmetric portion of the load using symmetric boundary conditions and a second analysis on the same portion of the structure using the antisymmetric boundary conditions and load.

As an example, the loads on the structure in Figure 7.28 are broken up into its symmetric and antisymmetric parts. When a general load is converted into symmetric and antisymmetric parts, the structure will have to be analyzed twice, with totally separate results. These results now have to be added together to get the complete result. Usually, this addition must be done by hand. As a result, using this method is time effective only when a small portion of the results are needed or it is the only way to accomplish the task. This is because computer time is generally more effective than using a person to combine results by hand. Clearly, spending 1 hour of computer time to analyze the full structure as opposed to 15 minutes and then 2 hours of hand addition is a better approach.

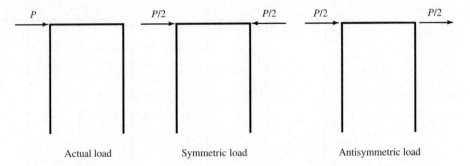

Figure 7.28 Conversion of load to symmetric and antisymmetric components.

7.5 Rigid-End Offsets

Up to now, we have used the centerline dimensions of the real structure to create our numerical models. Figure 7.29 shows an example of a real and numerical model of a

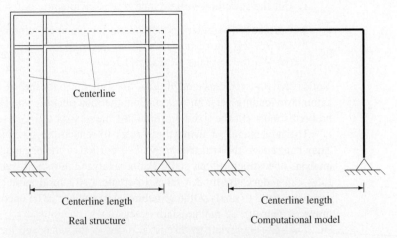

Figure 7.29 Centerline model assumption.

structure. This implies some assumptions about the structure that we have represented in our model. The main assumption is that the depth of the member is small enough not to affect the results. In reality, our beam or column members stop at the joint. The structural joint then connects the beams to the columns. Figure 7.30 shows the connection of the beams and columns to joint elements.

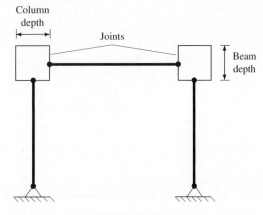

Figure 7.30 Computational model including joints.

When we use the centerline dimensions, we are assuming that the effect of the joint is negligible. This implies that the additional length of the beams and columns (half the beam or column depth) compensates for the effect of the joint. For most purposes, this is a reasonable assumption. However, as the depth of a member becomes large or the stiffness of a joint becomes large, this assumption fails.

In steel structures where 36-in. members and reinforced joints are used, the centerline dimensions may not be acceptable. In concrete structures, the extreme rigidity of the joint may cause the use of centerline dimensions to be inaccurate. The way to model this effect is through the use of rigid-end offsets when the flexibility of the joint can be ignored. Rigid-end offsets allow you to model the fact that the ends of a member are not considered to be part of the flexible portion. A model using rigid-end offsets can be represented schematically as shown in Figure 7.31.

Notice how the portion of the member that extends from the center of the joint to the end of the flexible portion of the member is replaced by a rigid piece. This effect can be handled in two ways: (1) through the use of additional nodes and rigid members, or (2) exactly through the use of a transformation matrix. The use of extra nodes and members is an easy-to-use method that works in all programs. The use of transformation matrices must be done at the program level and hence must be an option in the program. We will look at the modeling method using extra nodes and members first.

The most straightforward method of including rigid-end offsets is to convert the model in Figure 7.31 into an analysis model. This means replacing the rigid sections with small rigid beams. The model would look like the one given in Figure 7.32. Notice that each rigid end adds an additional node and member. The DOFs required for the rigid-

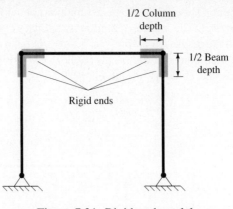

Figure 7.31 Rigid-end model.

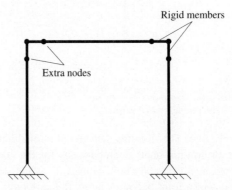

Figure 7.32 Joints modeled using additional rigid members.

end model are identical to the original X and Y translations and Z rotation. The choice of the properties for the rigid-beam segments needs to again follow the rule of thumb of 1000 times stiffer than the connected element. In this case, that means choosing a moment of inertia 1000 times that of the connected element.

In the case of concrete or reinforced steel joints, this model will give good results. However, the connection is still assumed to be rigid and the stiffness of the joint is neglected. It is possible to add flexibility to the joint by using two nodes and springs between them. This is identical to using a beam with a zero torsional moment of inertia to model a hinge. By making the torsional stiffness represent the flexibility of the joint, joint deformation can be included.

This same procedure of rigid members can be used to include the effect of a member connected to another member but having different neutral axes. The most common use is to include the effect of a prestressing tendon in a concrete beam. The concrete beam is modeled using a standard beam element. The prestressing element is modeled using a truss element. As an example, let's model a simple span beam with a straight prestressing tendon (Figure 7.33). Note that the tendon is located at a distance d from the neutral axis of the beam. This effect can be modeled using a rigid-end offset for the truss element.

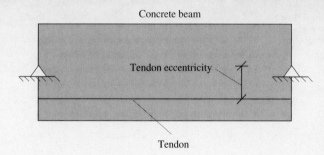

Figure 7.33 Prestressed concrete beam structure.

The model shown in Figure 7.34 can be used. Notice how rigid arms connect the beam to the truss. This has the effect of causing the truss to stiffen the rotational stiffness of the beam as well as its axial stiffness. This model can be created using either of the two methods above. If the truss element will allow initial forces, such as pretensioning or thermal effects, the pre- or posttensioning effects can be analyzed. The truss force at a distance from the neutral axis causes the required additional moment. The stiffening effect of the truss can be seen by looking at the deflected shape of the combined structure shown in Figure 7.35. Notice how the beam rotations cause the rigid arms to rotate, thus stretching the tendon (truss) member. If the program being used has the option of using rigid-end offsets, no rigid members or additional nodes are required. For the prestressed beam, the truss would be connected directly to the support nodes with the applied eccentricity.

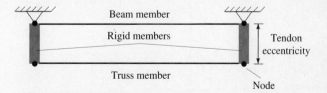

Figure 7.34 Computational model of prestressed beam.

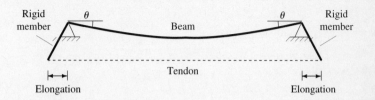

Figure 7.35 Tendon elongation due to beam rotations.

Condensed Example: Rigid-End Offsets

Given the following steel frame, we would like to find the end moments in the beam, including the effects of the joints.

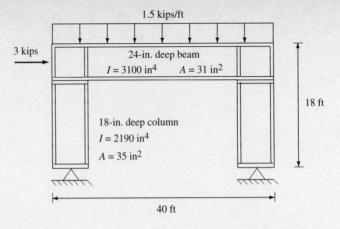

It is assumed that the joints are reinforced by welding additional plates in the panel section. Therefore, we can neglect any deformation of these joints and assume that they are rigid. Under this assumption we can use the rigid-end-offset capability of SSTAN. The centerline model of the structure with the nodes and members numbered is shown below.

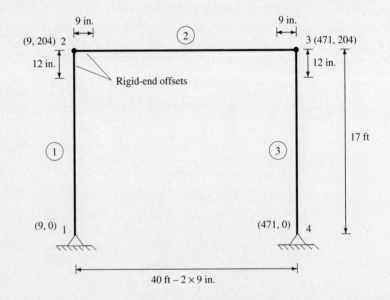

Using this figure we can create the required SSTAN input file. The file uses the centerline nodal locations and includes the rigid ends for the beams. Notice that the rigid ends are given in the global coordinate system. Hence the sign of the offset matches the global coordinate directions. For example, the tops of the columns have a negative Y offset since they are downward from the joint. Another point to notice is that the distributed load is applied to the beam but the beam is 18 in. shorter (due to the rigid ends). Therefore, 18 in. worth of the distributed load is neglected. The required input file is

```
Example: Rigid-End Offsets
4,1,1
:
Boundary
1,4 dof=r,r,f,f,f,r
1,4,3 dof=f,f,f,f,f,r              : Pinned supports
:
Coordinates
1 x=0+9   y=0 z=0                  : Add depth of column
2 y=18*12-12                      : Subtract half depth of beam
3 x=40*12-9                       : Subtract depth of beam
4 y=0
:
Beam
3,2
1 a=31  i=3100 e=29000            : Beam element
2 a=35  i=2190 e=29000            : Column element
1 1,2,3 m=2 j=0,-12               : Rigid-end offset at top col. - 12 in.
2 2,3,1 m=1 i=9 j=-9  l=-1.5/12   : Beam - offset both ends 9 in.
3 4,3,1 m=2 j=0,-12 i=0           : Column - offset top - 12 in.
:
Loads
2 l=1 f=3                         : 3-kip load at left joint
:
```

After running SSTAN, we can plot the structure to verify the input. Looking at the plot we can see the rigid-end offsets shown for each member. The maximum moment is found under the beam results. The output for the beam is as follows:

```
--------------------- FRAME MEMBER RESULTS ----------------
MEM LOAD NODE          1-2 PLANE                1-3 PLANE        AXIAL FORCE
 #   #    #     MOMENT      SHEAR      MOMENT       SHEAR
 2   1    2  .00000E+00  .00000E+00  .11444E+04 -.26425E+02  .97755E+01
          3  .00000E+00  .00000E+00 -.17325E+04 -.29075E+02 -.97755E+01
                                         AXIAL TORQUE =   .00000E+00
 ** MAXIMUM MIDSPAN MOMENT =   .165E+04 AT DISTANCE   211.40 FROM NODE    2
```

Note that the maximum positive moment is also given for loaded members.

7.6 Geometric Stiffness: P–Δ

Most structures are analyzed assuming small displacements and small strains. The small displacement assumption means that summing equilibrium about the undeformed configuration is sufficiently accurate as not to affect the results. Small strains assumes that the strains are small enough that the second-order terms in the strain tensor can be ignored. We also usually assume that the stress–strain relationship is linear.

It is often important to include the effects of a portion of the large displacement effect into the analysis without including the full effects. One such case is in the analysis of tall buildings, where the lateral displacement in conjunction with the height causes additional displacements that cannot be ignored. This is shown in Figure 7.36. These are called *P–Δ effects* and are due to the additional moment caused by the gravity loads acting through the drift.

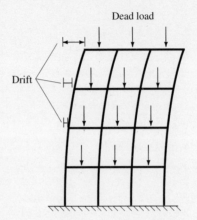

Figure 7.36 Additional *P–Δ* moments from drift.

P–Δ can be thought of as the additional load due to the axial load of a member not being collinear. This is as a result of considering the deformed system instead of the undeformed system for equilibrium calculations. A simple example will demonstrate this effect. Consider the structure in Figure 7.37, consisting of a rigid bar with a spring attached. Standard analysis says that **R** and **N** are independent (in the undeformed configuration) and the stiffness relationship is

$$\mathbf{Kr} = \mathbf{R} \tag{7.8}$$

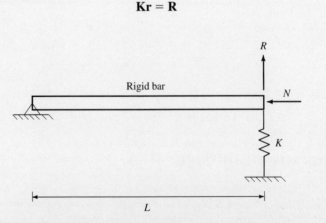

Figure 7.37 Rigid-bar example for geometric stiffness.

But if the structure is allowed to deform, we can calculate equilibrium in the deformed configuration as shown in Figure 7.38. Summing moments about the pinned end, we get

$$\mathbf{R}L + \mathbf{N}r = \mathbf{S}L \tag{7.9}$$

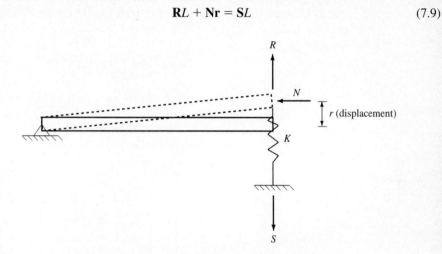

Figure 7.38 Deformed shape for rigid-bar example.

The equation for the response of the spring is

$$\mathbf{K}r = \mathbf{S} \tag{7.10}$$

Substituting **S** from equation (7.10) into (7.9), we get

$$\mathbf{R}L + \mathbf{N}r = (\mathbf{K}r)L \tag{7.11}$$

Dividing by L and grouping terms, we have

$$\left(\mathbf{K} - \frac{\mathbf{N}}{L}\right)r = \mathbf{R} \tag{7.12}$$

This can be rewritten further if we define the geometric stiffness as

$$\mathbf{K}_g = \frac{\mathbf{N}}{L} \tag{7.13}$$

giving the final form for equation (7.12) as

$$(\mathbf{K} - \mathbf{K}_g)r = \mathbf{R} \tag{7.14}$$

Note that the stiffness of the system is reduced as the axial load increases. A reduction in stiffness causes increased displacements. Hence the larger the axial load, the greater the displacements.

At some point, the axial load can become so large that the stiffness becomes zero or even negative. The point at which the stiffness becomes zero defines an instability point. At this point, there are an infinite number of solutions to the set of equations. When the stiffness becomes zero due to the axial load, we have reached the critical load or buckling load:

$$\mathbf{N}_{\text{critical}} = \mathbf{K}L = \text{buckling load} \qquad (7.15)$$

Using this form of the geometric stiffness, we can create a matrix version for use with bending elements. The matrix form causes the same effect of a reduction in stiffness. It causes an axial load to reduce the lateral stiffness of a bending element. For the two-dimensional element, the simplest form is

$$\mathbf{K}_g = \frac{\mathbf{N}}{L} \begin{bmatrix} 0 & 0 & 0 & 0 & 0 & 0 \\ 0 & 1 & 0 & 0 & -1 & 0 \\ 0 & 0 & 0 & 0 & 0 & 0 \\ 0 & 0 & 0 & 0 & 0 & 0 \\ 0 & -1 & 0 & 0 & 1 & 0 \\ 0 & 0 & 0 & 0 & 0 & 0 \end{bmatrix} \qquad (7.16)$$

This form assumes that the element is rigid for the definition of the geometric stiffness matrix, as in the example above.

7.6.1 Consistent Geometric Stiffness

A more complete form of the geometric stiffness assuming a flexible element can be derived using virtual work. We assume that a general bending element is in its deformed equilibrium configuration, as shown in Figure 7.39. The four end displacements (v_1, v_2, v_3, v_4) are used to define its position. There is an axial force in the element that can vary along its length, $N(X)$. As usual, we apply a small, compatible virtual displacement to the element, as shown in Figure 7.40.

The integration of the strain energy over the length of the element gives the internal

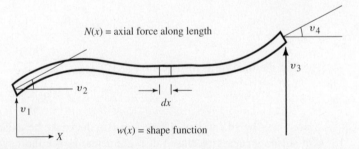

Figure 7.39 Deformed shape for consistent geometric stiffness.

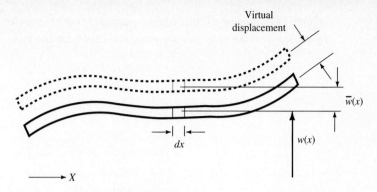

Figure 7.40 Virtual displacement for consistent geometric stiffness.

work in its normal form:

$$\delta W_i = \bar{\mathbf{v}}^{\mathrm{T}} \mathbf{K}_e \mathbf{v} \tag{7.17}$$

The external work is the sum of the work of the end forces through their respective displacements plus the work of the axial load $N(X)$. The end forces through their displacements again are in the standard form:

$$\delta W_e = \bar{\mathbf{v}}^{\mathrm{T}} \mathbf{S} + \int_0^L N(X) \times \text{displacements } dx \tag{7.18}$$

The only new portion is the contribution of the axial load to the external virtual work. To calculate this, we need to look more closely at the deformed shape of the element. If we take a small section of the element dx, we can show the equilibrium and virtual displacement in Figure 7.41. From this we can see that the axial force $N(X)$ goes through a small displacement $de(x)$. The vertical component of the virtual displacement for this section is shown as $dw(x)$. Since the section we are looking at is very small, we can assume that it is essentially straight. In addition, for a small enough length and small

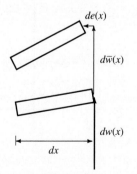

Figure 7.41 Displacements for dx segment.

virtual displacement, the displacements $de(x)$ and $dw(x)$ can be related by similar triangles to the slope of the small dx segment. Hence, by similar triangles,

$$\frac{d\bar{e}(x)}{d\bar{w}(x)} = \frac{dw(x)}{dx} \tag{7.19}$$

We can solve for $de(x)$ in the form

$$d\bar{e}(x) = \frac{d\bar{w}(x)}{dx}\frac{dw(x)}{dx}\,dx \tag{7.20}$$

We can substitute $\overline{de}(x)$ for the displacement associated with the axial force in equation (7.18) and integrate over the length of the element. This gives us

$$\delta W_e = \mathbf{v}^\mathrm{T}\mathbf{S} + \int_0^L \mathbf{N}(X)\,d\bar{e}(x)\,dx \tag{7.21}$$

Substituting for $\overline{de}(x)$ using equation (7.20), we get

$$\delta W_e = \mathbf{v}^\mathrm{T}\mathbf{S} + \int_0^L \mathbf{N}(X)\frac{d\bar{w}(x)}{dx}\frac{d\bar{w}(x)}{dx}\,dx \tag{7.22}$$

Now we need to define the displaced shape $w(x)$ in order to get $dw(x)/dx$.

The displaced shape must have a form that is consistent with the boundary conditions, that is, the end displacements v_1, v_2, v_3, and v_4. Since there are four boundary conditions, we can define a cubic function for the displaced shape. We can put these cubic functions into a matrix equation of the form

$$\mathbf{w(x)} = \mathbf{H(x)} \cdot \mathbf{v} \tag{7.23}$$

where $\mathbf{v}$ is the vector of end displacements for the element. $\mathbf{H(x)}$ is the matrix that converts these end displacements into a continuous displacement along the length. These functions are usually referred to as *shape functions*. The cubic form is

$$\mathbf{w(x)} = \begin{bmatrix} 1 - \dfrac{3X^2}{L^2} + \dfrac{2X^3}{L^3} \\[2mm] -X + \dfrac{2X^2}{L} - \dfrac{X^3}{L^2} \\[2mm] \dfrac{3X^2}{L^2} - \dfrac{2X^3}{L^3} \\[2mm] \dfrac{X^2}{L} - \dfrac{X^3}{L^2} \end{bmatrix}^{\mathrm{T}} \begin{bmatrix} v_1 \\ v_2 \\ v_3 \\ v_4 \end{bmatrix} \tag{7.24}$$

Note that the matrix is given in its transposed form. These are called the *cubic Hermitian polynomials* or *beam functions*. These functions can be used along with the finite-element method to derive the bending element stiffness matrix we have used in this book.

We take the derivative of these functions to form the matrix $dw(x)/dx$. We can then substitute into equation (7.22) and integrate to get the consistent form of the geometric stiffness. Note that when using the matrix form of virtual work, the virtual displacement term is transposed so that the work becomes a scaler value. As a result, the $d\overline{w}(x)/dx$ is replaced by $d\overline{w}(x)^{\mathrm{T}}/dx$ in equation (7.22). This final matrix has the form

$$
\mathbf{K}_g = \frac{\mathbf{N}L}{30}
\begin{bmatrix}
\dfrac{36}{L^2} & \dfrac{3}{L} & -\dfrac{36}{L^2} & \dfrac{3}{L} \\[2mm]
\dfrac{3}{L} & 4 & -\dfrac{3}{L} & -1 \\[2mm]
-\dfrac{36}{L^2} & -\dfrac{3}{L} & \dfrac{36}{L^2} & -\dfrac{3}{L} \\[2mm]
\dfrac{3}{L} & -1 & -\dfrac{3}{L} & 4
\end{bmatrix}
\tag{7.25}
$$

You could have derived the rigid member form using the same method and the linear shape functions:

$$
\mathbf{w(x)} = \left\langle 1 - \frac{X}{L}, \frac{X}{L} \right\rangle
\begin{bmatrix} v_1 \\ v_3 \end{bmatrix}
\tag{7.26}
$$

7.6.2   Geometric Stiffness Example

As an example, let's look at a simple cantilever beam. We will consider two cases: (1) no geometric stiffness and (2) geometric stiffness included. For this problem we will use two DOFs at the tip, lateral displacement, w, and rotation, θ. The loads on the structure will be a moment, R_0, and in the geometric case will include the axial load. The structure is shown in Figure 7.42.

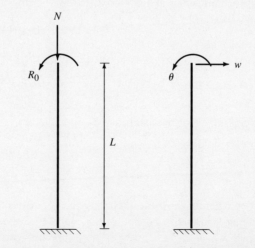

Figure 7.42 Geometric stiffness example for a cantilever beam.

Following standard procedure, the global stiffness will be a 2 × 2 matrix. This matrix can be assembled directly from the bending element stiffness matrix correcting for the 90° rotation. Since the only load is the applied moment, we generate the equilibrium equations in matrix form:

$$\frac{2EI}{L} \begin{bmatrix} \dfrac{6}{L^2} & \dfrac{3}{L} \\ \dfrac{3}{L} & 2 \end{bmatrix} \begin{bmatrix} w \\ \theta \end{bmatrix} = \begin{bmatrix} 0 \\ R_0 \end{bmatrix} \tag{7.27}$$

If we solve the set of equations symbolically, we get

$$\begin{bmatrix} w \\ \theta \end{bmatrix} = R_0 \begin{bmatrix} \dfrac{L}{EI} \\ -\dfrac{L^2}{2EI} \end{bmatrix} \tag{7.28}$$

The element end forces for the two DOFs can be recovered by multiplying the stiffness times the displacements and we get

$$\begin{bmatrix} S_w \\ S_\theta \end{bmatrix} = \begin{bmatrix} 0 \\ R_0 \end{bmatrix} \tag{7.29}$$

This solution clearly satisfies equilibrium. Now, let's include the effects of geometric stiffness. We will add an axial load, N, to the system. We will not include any additional DOFs. As a result, the axial load only has the effect of reducing the lateral stiffness. The global stiffness is identical to the first case. We need to assemble the geometric stiffness in the same manner as the global stiffness. Using the consistent stiffness matrix, we get

$$\mathbf{K}_g = \frac{NL}{30} \begin{bmatrix} \dfrac{36}{L^2} & \dfrac{3}{L} \\ \dfrac{3}{L} & 4 \end{bmatrix} \tag{7.30}$$

If we assume that the axial load has a value of

$$\mathbf{N} = \frac{2EI}{L} \tag{7.31}$$

we can form the stiffness, including the geometric effects. Note that for a cantilever, the buckling load is

$$\mathbf{N}_{\text{buckling}} = \frac{\pi^2 EI}{4L^2} \tag{7.32}$$

Compared to the buckling load, the choice for **N** in equation (7.31) is not unreasonable. Its expression in terms of the EI/L is useful so that the stiffness will be in a form that can easily be inverted. The final equilibrium equation is

$$(\mathbf{K} - \mathbf{K}_g)\mathbf{r} = \mathbf{R} \tag{7.33}$$

If we substitute in for the elastic stiffness and the geometric stiffness, we get the following matrix equations:

$$\frac{2EI}{L}\begin{bmatrix} \dfrac{4.8}{L^2} & \dfrac{2.9}{L} \\ \dfrac{2.9}{L} & 1.86\overline{6} \end{bmatrix}\begin{bmatrix} w \\ \theta \end{bmatrix} = \begin{bmatrix} 0 \\ R_0 \end{bmatrix} \tag{7.34}$$

Notice how the combined stiffness matrix has changed. All the terms have been reduced. Most notably, the 1,1 term has been reduced to one-fourth of its original value. This term directly reflects the lateral stiffness. Again, if we solve for the displacements, we get

$$\begin{bmatrix} w \\ \theta \end{bmatrix} = R_0 \begin{bmatrix} \dfrac{4.367L}{EI} \\ -\dfrac{2.644L^2}{2EI} \end{bmatrix} \tag{7.35}$$

Comparing these to the purely elastic case, we see that the displacements are more than four times their elastic value. In this case, the axial load caused a very large $P-\Delta$ effect. We can again recover the member end forces. We use the combined elastic and geometric stiffness matrix. The geometric matrix is required for the element to maintain equilibrium. We can look at the contributions to the member forces from each portion of the stiffness (elastic and geometric). The elastic portion of the member forces is

$$\begin{bmatrix} S_w \\ S_\theta \end{bmatrix} = \begin{bmatrix} \dfrac{5.45}{L}R_0 \\ 1.64R_0 \end{bmatrix} \tag{7.36}$$

Here notice the increase in the end moment and that there is now a shear at the end of the column. These are due to the added displacements caused by the inclusion of the geometric stiffness. Notice that the structure is not in equilibrium using the elastic forces alone. Looking at the structure as a whole, we have the free body shown in Figure 7.43.

To check equilibrium, we would need to include the forces from geometric stiffness. These forces must be combined with the elastic portion to get the true forces. The geometric forces are recovered in the same manner as the elastic forces: stiffness times displacements.

$$\begin{bmatrix} S_w \\ S_\theta \end{bmatrix}_{\text{geometric}} = \begin{bmatrix} -\dfrac{5.45}{L}R_0 \\ -0.64R_0 \end{bmatrix} \tag{7.37}$$

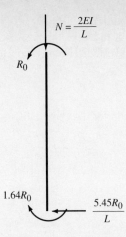

Figure 7.43 Free body of elastic results.

The geometric stiffness matrix can be used on large structures to include the effects of axial loads reducing the lateral stiffness. The process becomes more involved when indeterminate structures are to be analyzed. The difficulty arises because the axial force in a member is not known until the structure is analyzed. As a result, the process becomes iterative in nature.

For statically determinate structures, the axial loads in the members caused by the loading is known and does not depend on the deflection. Therefore, for any value of load, the axial load in all members is constant and the resulting force–displacement plot is always linear. A plot of the stiffness for different axial load values is given in Figure 7.44.

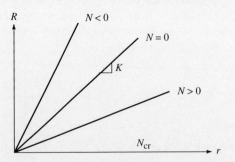

Figure 7.44 Geometric stiffness effects from axial loads.

For statically indeterminate structures, the axial force in a member is dependent on the applied loading and hence deflection. Therefore, the resulting force–displacement plot for the structure is nonlinear. This means that for an applied load you get a certain axial force. But this axial force creates a geometric stiffness that changes the deflections and hence the axial force. At some point this process reaches equilibrium or the structure becomes unstable. A plot of the stiffness for an indeterminate structure is given in Figure 7.45.

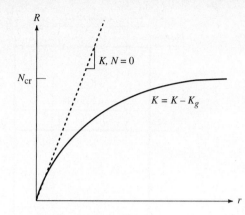

Figure 7.45 Geometric stiffness effects for indeterminate structures.

The process for including geometric effects in indeterminate structures is:

1. Form stiffness with *no* geometric stiffness.

2. Impose loads on the structure.

3. Solve for the axial forces in each member.

4. Use the axial force to form $K - K_g$.

5. Impose loads on the structure.

6. Solve for axial forces in each member.

7. Check if axial forces have changed since preceding cycle:
   (a) If the same, recover forces and quit.
   (b) If different, repeat steps 4 through 7.

Using this iterative process, $P-\Delta$ effects can be handled in a relatively simple analysis procedure. SSTAN has the capability to include $P-\Delta$ effects for all beam elements.

Condensed Example: $P-\Delta$ Analysis

Given the two-story single-bay frame shown at the top of page 300, find the maximum and end moments in the first-floor beam member. Show how the results change when $P-\Delta$ effects are included. To analyze this structure, we need to number the joints and members. The numbering shown in the second figure on page 300 will be used.

From the given numbering and the dimensions from the first figure, we can create the following SSTAN input file.

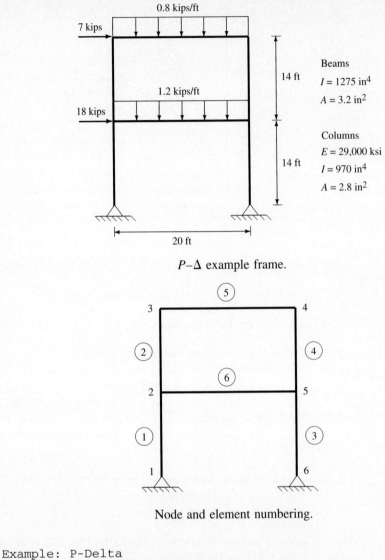

P–Δ example frame.

Node and element numbering.

```
Example: P-Delta
6,1,1  p=1    : 6 nodes, 1 type (beams), 1 load
:
Coordinates
1 x=0 y=0 z=0
3 y=28*12    g=1,3
6 x=20*12 y=0
4 y=28*12    g=4,6
:
Boundary
1,6    DOF=r,r,f,f,f,r    : Set all to 2 dimen. (X-Y plane)
1,6,5  DOF=f,f,f,f,f,r    : Pinned supports
:
```

```
Beam
6,2
1 a=3.2 i=1275 e=29000      : Beams
2 a=2.8 i=970               : Columns
C--- Columns
1 1,2,6 m=2
2 2,3,6
3 6,5,1
4 5,4,1
C--- Beams
5 3,4,2 m=1   l=-0.8/12     : Roof, K node below - neg. load
6 2,5,1       l=-1.2/12     : 1st floor
:
Loads
2 l=1 f=18
3     f=7
:
```

Notice that the second line of the input file has P=1 to turn on the P–Δ effects. To check the results without P–Δ, this should be removed.

The results for member 6, the first-floor beam, with no P–Δ effects are as follows:

```
----------------------- FRAME MEMBER RESULTS ----------------
MEM LOAD NODE           1-2 PLANE              1-3 PLANE        AXIAL FORCE
 #    #    #     MOMENT        SHEAR      MOMENT        SHEAR
 6    1    2   .00000E+00  .00000E+00  -.17340E+04  .57060E+01  .66168E+01
           5   .00000E+00  .00000E+00  -.25154E+04 -.29706E+02 -.66168E+01
                                             AXIAL TORQUE =   .00000E+00
```

The results from the same member when P–Δ effects are included are

```
----------------------- FRAME MEMBER RESULTS ----------------
MEM LOAD NODE           1-2 PLANE              1-3 PLANE        AXIAL FORCE
 #    #    #     MOMENT        SHEAR      MOMENT        SHEAR
 6    1    2   .00000E+00  .00000E+00  -.17557E+04  .58859E+01  .64798E+01
           5   .00000E+00  .00000E+00  -.25370E+04 -.29886E+02 -.64798E+01
                                             AXIAL TORQUE =   .00000E+00
```

Notice how when the P–Δ effects are included, the moments and shears increase. This is a result of the added load, which increases deflections.

7.7   Problems

7.1. Use SSTAN to analyze the following structures and find the maximum moment and shear in the structures.

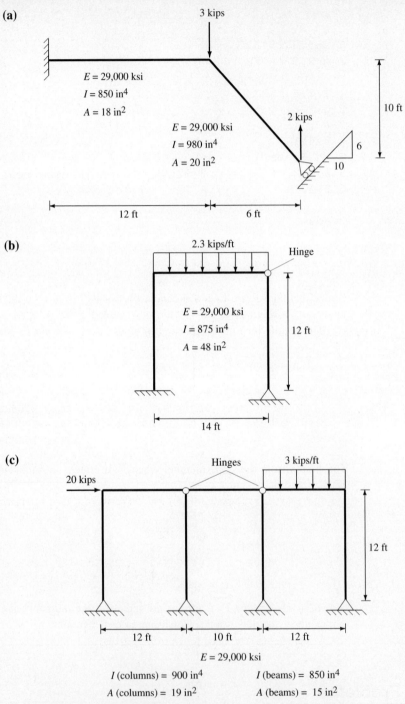

(a)

3 kips

$E = 29{,}000$ ksi
$I = 850$ in^4
$A = 18$ in^2

$E = 29{,}000$ ksi
$I = 980$ in^4
$A = 20$ in^2

2 kips

10 ft

6

10

12 ft

6 ft

(b)

2.3 kips/ft

Hinge

$E = 29{,}000$ ksi
$I = 875$ in^4
$A = 48$ in^2

12 ft

14 ft

(c)

Hinges

3 kips/ft

20 kips

12 ft

12 ft

10 ft

12 ft

$E = 29{,}000$ ksi

I (columns) $= 900$ in^4 I (beams) $= 850$ in^4

A (columns) $= 19$ in^2 A (beams) $= 15$ in^2

(d)

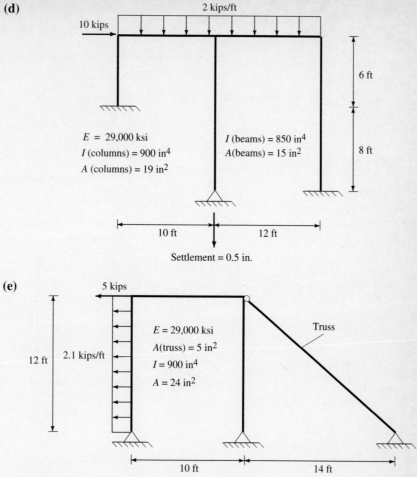

2 kips/ft

10 kips

$E = 29,000$ ksi
I (columns) = 900 in^4
A (columns) = 19 in^2

I (beams) = 850 in^4
A(beams) = 15 in^2

6 ft

8 ft

10 ft

12 ft

Settlement = 0.5 in.

(e)

5 kips

$E = 29,000$ ksi
A(truss) = 5 in^2
$I = 900$ in^4
$A = 24$ in^2

Truss

12 ft   2.1 kips/ft

10 ft

14 ft

Include a settlement of $\frac{1}{2}$ in. at the right column support and $\frac{1}{4}$ in. at the truss support.

(f)

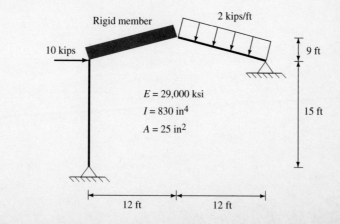

Rigid member

2 kips/ft

10 kips

9 ft

$E = 29,000$ ksi
$I = 830$ in^4
$A = 25$ in^2

15 ft

12 ft

12 ft

(g)

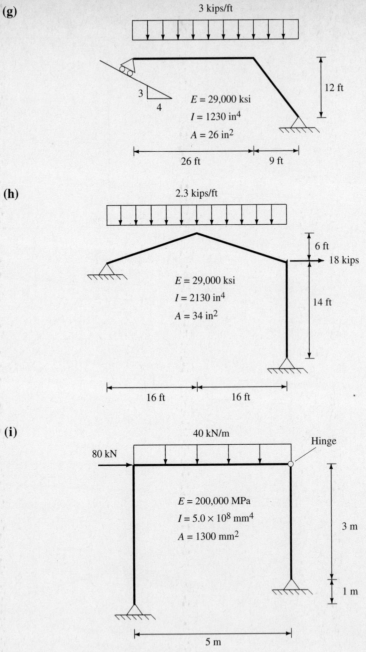

3 kips/ft

$E = 29{,}000$ ksi
$I = 1230$ in^4
$A = 26$ in^2

12 ft

3
4

26 ft   9 ft

(h)

2.3 kips/ft

6 ft
18 kips

$E = 29{,}000$ ksi
$I = 2130$ in^4
$A = 34$ in^2

14 ft

16 ft   16 ft

(i)

40 kN/m

Hinge

80 kN

$E = 200{,}000$ MPa
$I = 5.0 \times 10^8$ mm^4
$A = 1300$ mm^2

3 m

1 m

5 m

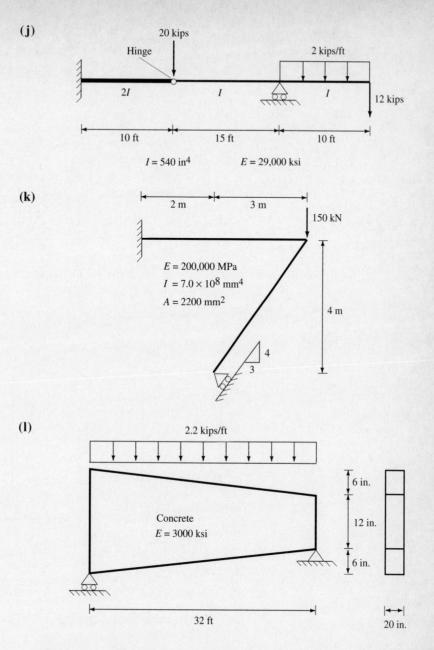

(j)

20 kips

Hinge

2 kips/ft

2*I*   *I*   *I*

12 kips

10 ft   15 ft   10 ft

I = 540 in⁴   *E* = 29,000 ksi

(k)

2 m   3 m

150 kN

E = 200,000 MPa

I = 7.0 × 10⁸ mm⁴

A = 2200 mm²

4 m

4
3

(l)

2.2 kips/ft

Concrete
E = 3000 ksi

6 in.

12 in.

6 in.

32 ft

20 in.

(***Hint:*** Break the element up into many smaller elements each with a constant
moment of inertia equal to the average at its center.)

(m)

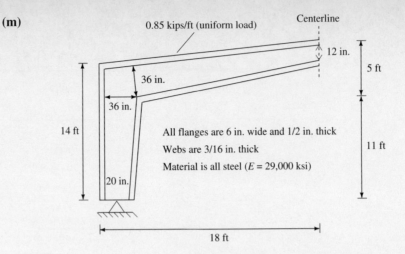

(*Hint:* Break the elements up into many smaller elements each with a constant moment of inertia equal to the average at its center.)

See the effects of including rigid ends to model the panel joint as rigid.

(n)

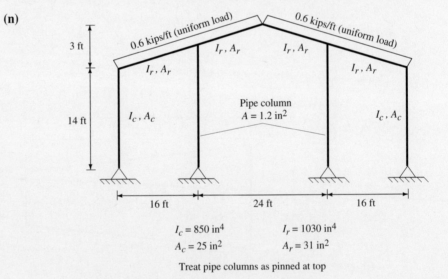

$$I_c = 850 \text{ in}^4 \qquad I_r = 1030 \text{ in}^4$$
$$A_c = 25 \text{ in}^2 \qquad A_r = 31 \text{ in}^2$$

Treat pipe columns as pinned at top

Include a settlement of $\frac{1}{2}$ in. at the left center column.

(o)

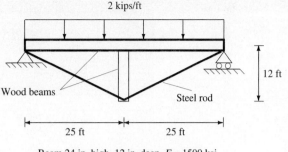

2 kips/ft

Wood beams

Steel rod

12 ft

25 ft   25 ft

Beam 24 in. high, 12 in. deep, $E = 1500$ ksi
Steel rod – $A = 1.5$ in^2 $E = 29,000$ ksi

Model first using all components as a combination of beams and trusses. Then, model using rigid ends on the truss (rods) and do not include the vertical wood beam. Notice the difference in the results.

(p)

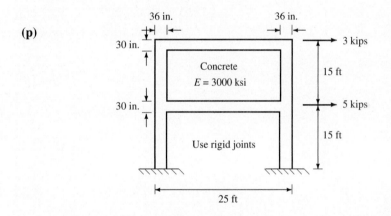

36 in.   36 in.

30 in.

Concrete
$E = 3000$ ksi

3 kips

15 ft

30 in.

5 kips

15 ft

Use rigid joints

25 ft

Analyze the structure above with and without the inclusion of rigid-end offsets. Show the difference in moment in the columns and the change in the maximum displacement.

7.2. Redo Problem 6.1(a), including the effects of geometric stiffness.

7.3. Analyze the following concrete beams, including the prestressed tendons. Assume that $E_c = 4000$ ksi for concrete and $E = 29,000$ for steel. The tension in the tendons is 1020 kips. The properties for the concrete portion of the beam are $f'_c = 5000$ psi, $A = 1000$ in^2, and $I = 520,000$ in^4. Multiple tendons (modeled as a single tendon) have a combined area: $A = 5$ in^2. The beams have a dead weight of 3.5 kips/ft. Check the response with the dead load and a live load consisting of a 63-kip point load at the center of the span. Assume that the beams are simply supported.

(a)

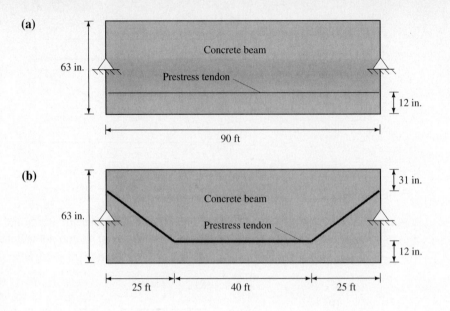

(b)

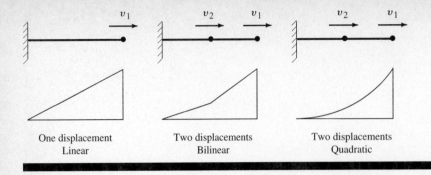

One displacement
Linear

Two displacements
Bilinear

Two displacements
Quadratic

FINITE ELEMENTS

8.1 Finite-Element Theory

The next step in structural analysis is to be able to model more complex structures. The ability to include the effects of shear walls, floor diaphragms, complex connections, and model shell structures is required. This is accomplished by using additional types of elements in the modeling process. The finite-element method (FEM) allows more complex element behavior to be modeled. The FEM was originally just an extension of matrix structural analysis, developed by structural engineers. It has since been used in just about every field where differential equations define the problem behavior. The result of the FEM is to create a stiffness matrix and a set of loads. After that, the solution process is identical to that covered in this book. There are many excellent books covering the FEM; this chapter is intended only as an introduction.

The basic idea of the finite-element method is to break up a continuum into a discrete number of smaller *elements*. The elements can be modeled mathematically by a stiffness matrix and are connected by nodes that have degrees of freedom. This is identical to what we have done with bending and truss elements. However, beams and trusses have natural locations at which to define nodes. In addition, the derivation of their stiffness matrices can be done on a physical basis.

8.1.1 Simple FEM Theory

More general finite elements require slightly more complicated procedures than those used for beams in order to derive the stiffness matrix. The basic procedure is to assume a shape function that describes how the nodal displacements are distributed throughout the element. From the differential equation, we form an operator matrix that will convert the displacements within the element into strains. Next the internal and external virtual work can be formed and equated to develop the stiffness matrix. The last step is identical to that used for truss and bending elements.

As an example, we will develop the stiffness matrix for a truss element, an axial member. The truss element has two nodal displacements, v_1 and v_2, one at each end. For any given set of displacements at the ends, a function is required to convert these into displacements along the length of the element. The obvious selection for the functions is the linear set shown in Figure 8.1. Notice how the given functions distribute the end displacements throughout the element. These distribution functions are called the *shape functions*. The shape functions can be put into matrix form along with the end displacements to form an equation that describes the displacement within the element. This displacement

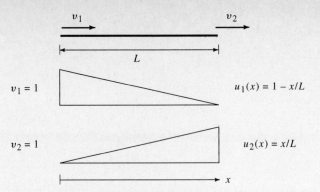

Figure 8.1 Unit displacement functions for truss member.

anywhere within the element is described by the following matrix equation:

$$\mathbf{u(x)} = \left\langle 1 - \frac{x}{L}, \frac{x}{L} \right\rangle \begin{bmatrix} v_1 \\ v_2 \end{bmatrix} \qquad (8.1)$$

Notice that the displacement is a function of x, the position in the element. Also note that the displacement anywhere in the element, $u(x)$, is the sum of the displacements caused by both end displacements distributed throughout the element. Equation (8.1) can be rewritten as

$$\mathbf{u(x)} = \mathbf{H(x)} \cdot \mathbf{v} \qquad (8.2)$$

The matrix $\mathbf{H(x)}$ is called the shape function matrix, and $\mathbf{v}$ represents the element nodal displacements. We now need a differential equation that converts the displacement into strain. For our one-dimensional example this is

$$\varepsilon_x = \frac{\delta u(x)}{\delta x} \qquad (8.3)$$

It is useful to rewrite equation (8.3) in the form of two matrices. This alternative form puts the differential operator into the form of an operator matrix, $\mathbf{D}$. Equation (8.3) is rewritten as

$$\varepsilon_x = \mathbf{D} \cdot \mathbf{u(x)} \qquad (8.4)$$

where the operator matrix has the form

$$\mathbf{D} = \left\langle \frac{\delta}{\delta x} \right\rangle \qquad (8.5)$$

If we apply this operator to the displacement, $u(x)$, given in equation (8.2), we can find the strain as a function of the element displacements, $\mathbf{v}$, and the shape function matrix,

$H(x)$. This gives the form for the strain in terms of the shape function matrix $H(x)$.

$$\varepsilon_x = D \cdot H(x) \cdot v \tag{8.6}$$

Note that the nodal displacements, v, are constants with respect to x and need not be operated on—differentiated. Therefore, only the derivatives of the shape functions need to be taken. For our case of the axial element, the strain can then be written by substituting the shape functions, $H(x)$, into equation (8.6) and applying the D operator, giving

$$\varepsilon_x = \left\langle -\frac{1}{L}, \frac{1}{L} \right\rangle \begin{bmatrix} v_1 \\ v_2 \end{bmatrix} \tag{8.7}$$

Typically, the differential operator times the shape function matrix is called B, the strain–displacement matrix. The strain is then commonly written in the shorter form

$$\varepsilon_x = B \cdot v \tag{8.8}$$

We also need the relationship of Hooke's law, which converts strain into stress:

$$\sigma = E \cdot \varepsilon \tag{8.9}$$

Therefore, if we calculate the internal strain energy, as was done in previous chapters to develop a stiffness matrix, virtual strain times stress, with substitutions we have

$$\delta W_i = \int_{\text{volume}} \bar{\varepsilon}^T \sigma = \int_{\text{volume}} (\bar{v}^T \cdot B^T \cdot E \cdot B \cdot v) \, dV \tag{8.10}$$

Equating internal to external virtual work and removing the arbitrary virtual displacements, $\bar{v}^T$, we get

$$S = \int_{\text{volume}} (B^T \cdot E \cdot B) \, dV \, v \tag{8.11}$$

Looking at equation (8.11), we see that this is the familiar stiffness form of the element relationship between forces and displacements. As a result, we can see that the integral is just the element stiffness. Taking that portion out of the equation, we have

$$K_e = \int_{\text{volume}} (B^T \cdot E \cdot B) \, dV \tag{8.12}$$

This is the classic form of the finite-element stiffness formulation. For our axial element example we substitute for B from equation (8.7) and E is just the familiar Young's modulus. In the general case, E is the constitutive matrix. For the linear case E is just the three-dimensional representation of Hooke's law. Integrating over the Y and Z coordinates for part of the volume integral, we get the area of the cross section. Multiplying the matrices after the partial integration for the area and removing the constants from the integral, we get

$$\mathbf{K}_e = AE \int_{\text{length}} \begin{bmatrix} \dfrac{1}{L^2} & -\dfrac{1}{L^2} \\ -\dfrac{1}{L^2} & \dfrac{1}{L^2} \end{bmatrix} dx \tag{8.13}$$

Integrating the matrix term by term over the length, we obtain the familiar form for a truss stiffness as

$$\mathbf{K}_e = \frac{AE}{L} \begin{bmatrix} 1 & -1 \\ -1 & 1 \end{bmatrix} \tag{8.14}$$

This is the final result we are looking for. Of course, this is identical to the standard truss stiffness matrix developed by traditional stiffness methods. However, the shape function process described can be extended to other types of elements where traditional stiffness by definition methods are not possible.

8.1.2 Available Elements

Energy derivations (here virtual work) are commonly used to form the stiffness for a variety of element types. The most common elements are the membrane (planar), plate, shell, and solid elements. Each of these elements has a given set of nodes and displacements associated with those nodes. The common forms of these elements are given in Figure 8.2. These elements have additional restrictions on their behavior that depend on their derivation. However, the result is always a stiffness matrix that can then be treated like any other stiffness matrix and may be rotated and transformed as desired. When combining

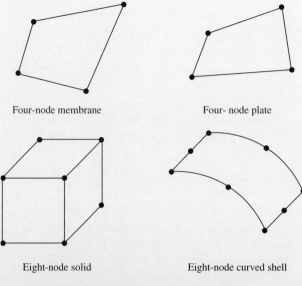

Four-node membrane   Four- node plate

Eight-node solid   Eight-node curved shell

Figure 8.2 Common finite elements.

these elements, the same concerns about boundary conditions and matching DOFs at the nodes must be accounted for. Additional concerns are also generated since the shape function assumption can affect the accuracy of the results. The standard beam element can be derived in a similar fashion using the cubic beam functions given in Section 7.6.1.

Element derivation has become increasingly complex. Techniques that include nonlinearities while reducing the number of unknowns in the element have become very theoretically demanding. However, the use of these sophisticated elements is identical to their simpler counterparts. The three elements most commonly used by structural engineers are the membrane, plate/shell, and solid elements. Each of these three is discussed briefly here and then in detail in subsequent chapters. The membrane element is a two-dimensional flat extensional element. The common versions are triangular and rectangular elements. The triangular elements vary from three to six nodes. The rectangular elements vary from four to nine nodes. There are two in-plane displacement DOFs at each node of the element (Figure 8.3). The elements can be used to model two-dimensional elasticity problems, plane strain and plane stress. It can reproduce the two normal and one shear stress in the plane of the element.

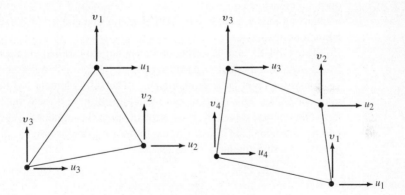

Figure 8.3 Local DOFs for planar elements.

The membrane element has no rotational stiffness or stiffness normal to the plane of the element. It can be situated arbitrarily in space, but the resultant forces must lie in the plane of the element. This is similar to the three-dimensional truss element. The three-node triangular element can model constant stress. The nine-node rectangular element can model linear variations of stress. Triangular elements are popular with automatic mesh generation and adaptive mesh generation schemes. Loads on membrane elements can only consist of in-plane loads.

The flat plate element is a two-dimensional element that acts like a flat plate. It is found in triangular and rectangular versions. There are two out-of-plane rotations and the normal displacement as DOF (Figure 8.4). These elements model plate bending behavior in two dimensions. The element can model the two normal moments and the cross moment in the plane of the element. Some versions will also give the transverse shear as a result. The three-node triangular version models constant moment. The higher node elements

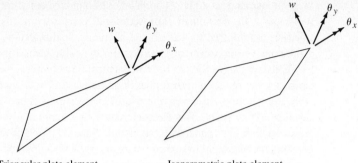

Triangular plate element   Isoparametric plate element

Figure 8.4 Local DOFs for plate bending element.

can model linear variation of moments across the element. This element has no rotational stiffness normal to the plane and no in-plane stiffness. Flat shell elements are commonly created by superimposing the membrane and plate elements on top of one another. Loading on plate elements can consist of any combination of forces normal to the plate and out-of-plane moments. Loading on shell elements can consist of the combination of plate and membrane loadings.

The solid element is a three-dimensional extensional element (Figure 8.5). Versions found vary from the four-node tetrahedron to the 27-node brick element. The most common version is the eight-node brick element. The solid element has three translations at each node for DOFs. This element can model a full three-dimensional stress state. The eight-node element has some linear variation of stress throughout the element. The solid element has no rotational stiffness. Tetrahedrons are popular for use in mesh generation and adaptive mesh refinement schemes.

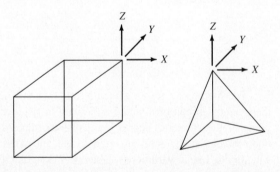

Eight-node isoparametric solid   Four-node tetrahedron solid

Figure 8.5 Local DOFs for solid elements.

8.2 Shape Functions

As shown in Section 8.1, one of the major variables in deriving a finite element is the choice of the shape function $\mathbf{H}(\mathbf{x})$. Remember that it is the shape function that describes how the nodal displacements are distributed throughout the element. In the truss member it was assumed that the end displacements are distributed linearly. This is an exact assumption for a uniform-cross-section member.

The choice of the shape function is directly related to the number of unknowns in the element. The number of unknowns defines the order of approximation that the shape function can have. Again for the truss member, there were two unknowns (one at each end). This gave the linear shape functions since two points give a straight line. As a result, to increase the accuracy for an analysis you need to increase the number of unknowns. This can be handled by two methods: (1) you can increase the number of unknowns and use piecewise continuous linear shape functions, or (2) you can increase the number of unknowns and increase the order of the shape function within an element. In the next section we show the effects of these assumptions on a simple example.

8.2.1   Tapered Extensional Example

For a constant-cross-section truss member, there is no need to increase the accuracy above linear shape functions. Instead, let's look at another simple problem. We want to see the effects of shape function selection on accuracy when trying to solve a tapered member subjected to axial load for different modeling conditions (Figure 8.6). This problem is simple enough that it can be solved exactly and the approximate finite-element results can be compared to this exact result. The problem has a unit thickness and is subjected to an axial load of 20. Note that no units are given. Any consistent units are acceptable. The exact solution for the displacement is

$$u(x) = -0.0074074 \ln(10 - 0.09x) + 0.0170561 \qquad (8.15)$$

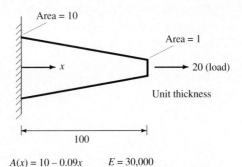

Figure 8.6 Axially loaded tapered problem.

The solution was achieved by integrating the strain, $\varepsilon = \sigma/E$. The exact solution for the stress, force over area, in the section is

$$\sigma(x) = \frac{20}{10 - 0.09x} \qquad (8.16)$$

We will use the finite-element method and solve the problem using three different assumed shape functions. We start by choosing a single unknown at the tip of the member. We also use two unknown displacements, one at the tip and one at the midpoint. Using the two unknowns, we have a choice: we can use a quadratic shape function or we can use

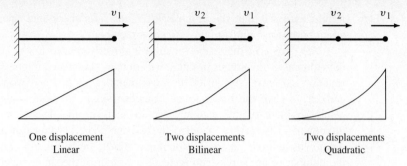

Figure 8.7 Three shape function models for tapered member.

two linear functions. We will perform the analysis using both. Figure 8.7 shows the finite-element models and assumed shape functions.

Using the appropriate shape functions, forming the stiffness matrices using equation (8.12), and then solving for the displacements and stresses for each problem, we get the results for displacements and stresses shown in Figures 8.8 and 8.9. There are some important things to notice about the results. First let's look at the displacement plots, Figure 8.8. The exact displacement is the natural log function given in equation (8.15). The assumed displacement functions try to approximate this function by a linear, bilinear, and a quadratic function, respectively. The finite-element method says that you can use the linear function, *but in the limit as the number of linear segments goes to infinity, the answer will be exact.* You can see that the bilinear is a better approximation than the linear. A trilinear would be even better; and so on. Also, the quadratic is even better than the other two at approximating the natural log function.

Next we will look at the stresses (Figure 8.9); here we see an even more dramatic result. The stress in each linear element is constant. This is obvious since the stress is

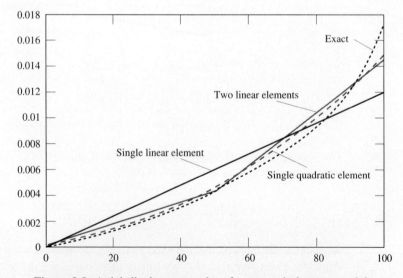

Figure 8.8 Axial displacement plots for tapered element models.

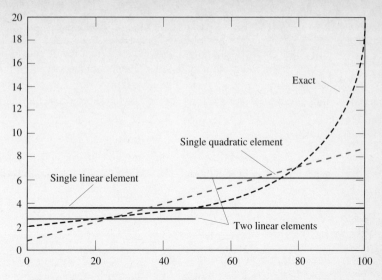

Figure 8.9 Axial stress for different tapered element models.

proportional to the strain and the strain is the derivative of the displacement. Another fact about finite elements is that it is a displacement-based method. This assumes that the displacements are continuous. However, it generally says nothing about the strain or stress. The displacement is continuous but the stresses are not continuous. This can be seen from the bilinear result. The stresses are constant but different between the two halves of the tapered element sections. The size of the stress difference indicates the accuracy of the model.

Finally, notice that the stress from the quadratic element is linear. Again, this is due to the fact that the stress is proportional to the derivative of the displacement. Notice how in all the cases, the finite-element stress underestimated the true stress.

8.2.2 Shape Function Accuracy

In summary, a finite element's accuracy is dependent on the choice of the shape function for that element and the number of elements used in the model. In the limit as the number of elements approach infinity, the model will give exact results. The displacements are assumed to be continuous between elements. However, stresses are not continuous between elements. The best accuracy is with many higher-order elements. Equal accuracy can be achieved by a larger number of lower-order elements. Clearly, the number of elements needed depends on the displacement field you are trying to match. A linear displacement field needs only linear elements.

8.3 Numerical Integration

In Section 8.2 we saw how the choice of the number of unknowns dictated the order of the finite-element shape function. The more nodes and unknowns, the higher the order. Once the shape function is chosen, the element can be formed by performing an integral over the element volume. In simple elements, this integration is easy to perform exactly. However, when we use irregular-shaped elements in two and three dimensions, it is very difficult to perform the integration exactly. As a result, most finite-element programs rely on numerical integration techniques. Most engineers are familiar with the trapezoidal rule and Simpson's rule for numerical integration. While these methods are easy to understand, they are computationally inefficient. A far superior method is Gauss quadrature.

8.3.1 Gauss Quadrature

Gauss quadrature is a very simple method to use and is exact for simple polynomials with very little computation. The basic formula for the method is

$$\int_{-1}^{1} F(x)\, dx = \sum_{i=1}^{\text{number of points}} A_i F(\mu_i) \tag{8.17}$$

This formula says that in order to get the integral for a function $F(x)$, you just need to sum the function at some given evaluation points, μ_i, times some weighting values, A_i, for the given number of evaluation points. Both the evaluation points and the weights are given numbers. In fact, the weights (A_i) and the evaluation points (μ_i) never change. As an example, the two-point Gauss quadrature points and weights are

$$\mu_1 = \frac{1}{\sqrt{3}}$$

$$\mu_2 = -\frac{1}{\sqrt{3}} \tag{8.18}$$

$$A_1 = A_2 = 1.0$$

These formulas say that if the function to be integrated is evaluated at the two Gauss points and the results are summed, you get the exact integral (from -1 to 1). For example, we will integrate the following equation:

$$\int_{-1}^{1} 5x^3 - 3.2x^2 - 6.7\, dx \tag{8.19}$$

Using the Gauss formulas, we get

$$\int_{-1}^{1} F(x)\, dx = 1.0F\left(\frac{1}{\sqrt{3}}\right) + 1.0F\left(-\frac{1}{\sqrt{3}}\right) \tag{8.20}$$

$$= (-6.8044) + (-8.7289) = -15.533 \tag{8.21}$$

The Gauss quadrature procedure is exact depending on the number of points used for the sum. In the case above for two-point integration, the method is exact for up to a cubic equation. The formula for exactness is

$$P = 2N - 1 \tag{8.22}$$

where P is the order of the polynomial that can be integrated exactly and N is the number of Gauss sampling points. Therefore, the procedure using three Gauss points ($N = 3$) is exact for up to a fifth-order polynomial. The Gauss points do not have to be rederived once they are found. As a matter of fact, they can be looked up in many sources for up to 20 points.

It is clear that this method requires a very small number of function evaluations in order to get exact integration results. Also notice that the method is defined as integrating a function from -1 to 1. This restriction can be lifted by using the mapping formula,

$$\int_A^B F(x) = \frac{B - A}{2} \sum_{i=1}^{\text{number of points}} F\left(\frac{B - A}{2}x + \frac{B + A}{2}\right) \tag{8.23}$$

This formula takes the points defined on (-1 to 1) and shifts and scales (or maps) them into the new limits, A to B. Using the mapping method also has some additional benefits. What most finite-element programs do is define their element on this -1-to-1 coordinate system. Then, whatever the actual shape of the element, it is mapped to the -1-to-1 system. This allows us to use nonrectangular shaped elements. The mapping of a nonrectangular element into a -1-to-1 coordinate system develops the isoparametric finite element.

8.3.2 Mapping Errors

There can also be problems when using the mapping to shift to a -1-to-1 coordinate system. In isoparametric elements the mapping procedure uses the displacement shape functions as mapping functions. Therefore, any real location X in the element can be found from its -1-to-1 coordinate by the formula

$$X = \sum_{i=1}^N H_i(\mu)X_i \tag{8.24}$$

where $H(\mu)_i$ is the shape function for nodes i and X_i is the coordinate for node i. This mapping will work for one-, two-, and three-dimensional elements. If we look at the one-dimensional axial problem again, we can develop the coordinate mapping for the quadratic (two-node) element. In doing so, we want the center node to be able to move (not be constrained to be at the midpoint). Figure 8.10 gives the coordinates in terms of the position of the center node. Substituting the quadratic shape functions and the coordinates

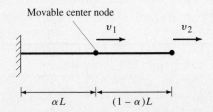

Figure 8.10 Axial model with movable center node.

given in Figure 8.10 into equation (8.24), we get the mapping function:

$$X = \tfrac{1}{2}(\mu + \mu^2)L + \alpha(1 - \mu^2)L \tag{8.25}$$

where α is the scale factor for where the center node lies. If $\alpha = 0$, the center node is all the way at the left support. If $\alpha = 1.0$, the center node is all the way at the right end. Figure 8.11 is a plot of this function, (8.25), for different values of α. μ is the normalized ordinate going from -1 to 1.

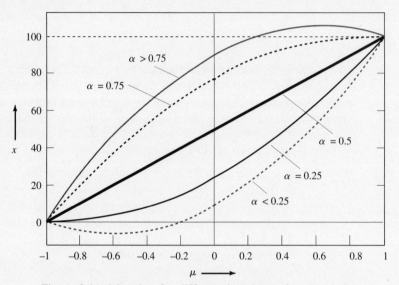

Figure 8.11 Mapping for different locations of center node.

Notice in Figure 8.11 that if α is either larger than $\tfrac{3}{4}$ or smaller than $\tfrac{1}{4}$, the mapping goes outside the actual length of the element. This has the effect of saying that more than one location on the element maps to the same point in the real X coordinate. The resulting element is shown in Figure 8.12. Clearly, this is not a valid element configuration. It also means that the element mapping is not valid because it is not invertible. The mapping from one coordinate to the other is handled by a transformation matrix called the *Jacobian*. The Jacobian matrix needs to be inverted to switch from one coordinate system to the other. The validity of a mapping can be determined by the invertibility of this mapping

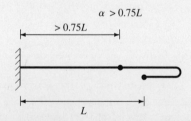

Figure 8.12 Bad mapping: element longer than true length.

matrix. Therefore, if the Jacobian matrix is singular or noninvertible, the mapping is not valid. This usually means that the nodes are not in the correct locations, ouside the $\frac{1}{4}$ points, or there are bad nodal coordinates.

8.3.3 Two-Dimensional Extension

These same mapping and integration procedures are used in two and three dimensions. When going to more dimensions, we need to extend the Gauss quadrature scheme. The usual method is to use the same points as in the one-dimensional case but in all coordinate directions. As an example, for a two-dimensional element, the two-point Gauss locations become a 2×2 pattern. Their locations are given in Figure 8.13. This same extension is done in three dimensions, leading to a $2 \times 2 \times 2$-point integration scheme. Similarly, any number of points can be extended to multiple dimensions. Although these are not necessarily optimal for multiple dimensions, they are the most common.

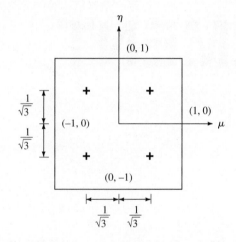

Figure 8.13 Two-dimensional locations for 2×2 Gauss quadrature.

8.4 Problems

8.1. What does the shape function do?

8.2. How can you improve a finite-element model's results?

8.3. What is the relationship between the shape functions and the stress in an element?

8.4. How do finite-element programs perform the integrations required to form an element stiffness, and what equations are used?

8.5. What rule of thumb can be used to avoid mapping errors?

8.6. What does the difference in stress at a common node between adjacent elements say about the accuracy of a finite-element model?

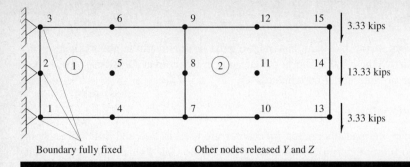

Boundary fully fixed · Other nodes released Y and Z

MEMBRANE ELEMENT

9.1 Introduction

In previous sections we have discussed bending and axial (beam and truss) elements. These elements are discrete members of a structure whose individual stiffness matrices can be developed by basic principles. A structure is composed of a number of these individual elements connected to one another. These elements have a direct counterpart in the real structure (e.g., beams and columns and bracing members).

General finite elements, on the other hand, are generally a small portion of a continuum. As an example, the analysis of a plate with a hole in it is shown with its subdivision of elements in Figure 9.1. Notice that while the structure shown is composed of smaller elements interconnected to make the whole, there is no direct structural counterpart for each element. Each element is only a small, irregular part of the whole structure. This is generally true for the finite elements we will discuss. They will be used to analyze continuum problems in two or three dimensions.

The first true finite element we will look at is the membrane element. The membrane element was the first finite element developed. The term *finite element* was coined in a paper by Ray Clough in 1965, which described the formation of a triangular membrane element. The membrane can be considered a two-dimensional extension of the truss member. It is an extension of the truss member because it is an in-plane force member only. This means that the element can resist forces only in the plane of the element. In addition, the only DOFs the element has are in-plane displacements.

Membrane elements can be used to describe any continuum problem that is two-dimensional in nature. Either the plane-stress or plane-strain condition can be modeled using these elements. The *plane-strain condition* says that the out-of-plane strain is zero. The *plane-stress condition* says that the out-of-plane stress is zero. These are shown in Figure 9.2. As an example, the problems shown in Figure 9.3 can be analyzed in two dimensions using membrane elements.

9.2 Membrane Theory

The membrane element is a flat element. It is generally assumed to have constant thickness. It can be triangular, rectangular, four-sided polygonal, or have curved sides. The element is generally found in configurations of three, four, six, eight, nine, and variable 3–9 nodes. Some examples are given in Figure 9.4. Whatever the shape or number of nodes, the element has two translational DOFs per node. These DOFs *must* lie in the plane of the

325

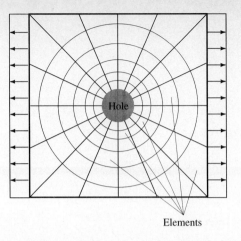

Figure 9.1 Two-dimensional mesh for plate with a hole in the center.

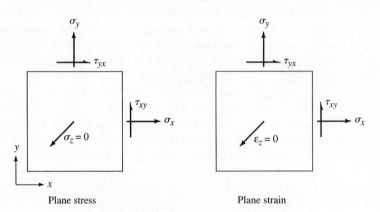

Figure 9.2 Stress states for membrane element.

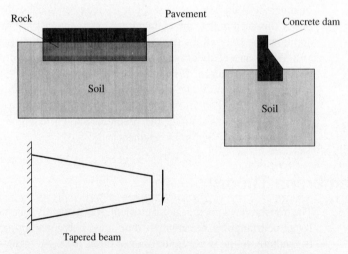

Figure 9.3 Two-dimensional model examples.

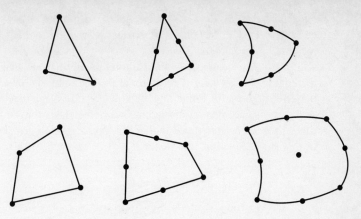

Figure 9.4 Various configurations of membrane elements.

element. The results from the element consist of two normal stresses and a shear stress in the plane of the element (Figure 9.5). The stress results are generally given at each node in the element. The difference in element behavior is dictated by the choice of the number of nodes and hence the number of DOFs for the element. The three-node triangle has linear shape functions and hence constant strain and stress. This element is referred to as the constant-strain triangle. The four-node element has slightly better response than the three-node element. The six-node triangle has quadratic shape functions and linear stress and strain. The eight- and nine-node elements have better response than the six-node element.

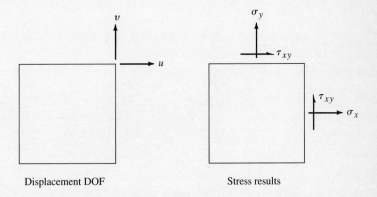

Figure 9.5 Membrane DOFs and stress results.

9.3 Two-Dimensional Shape Functions

Remember, the shape function describes how the nodal displacements are distributed throughout the element. In the membrane element, the X and Y displacements are assumed to be independent. As a result, the same shape functions are reused for the X and Y

displacements. For the four-node element, the plot of one of the shape functions is shown in Figure 9.6. On the plot, the vertical axis is the value of the shape function at that location in the element. This value is the percentage of the corner node displacement that will reach that location. As an example, notice that the shape functions are linear along the edge of the element. Therefore, at the midpoint on an edge, the displacement is half the value at the node. Remember that there is one shape function for each DOF. Therefore, the total displacement at any point in an element is the sum of the nodal displacements times their shape function value at that point. Remember that the stress is the derivative of this shape function, and hence it will be constant with some linear variation.

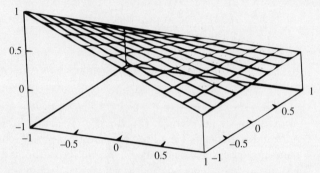

Figure 9.6 Corner node shape function for four-node element.

Now let's look at the nine-node element. This element is quadratic with some cubic components. We can also plot the shape functions for this element. Now we have three different shapes. One shape exists for each different type of node: the corner, the midside, and the center nodes. The plots shown in Figures 9.7 to 9.9 show the three different shape functions. In these figures for the nine-node element, you can see the quadratic variation. Again, remember that the stress is the derivative of the shape function and hence it will be linear with some quadratic parts.

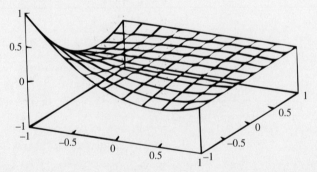

Figure 9.7 Corner node shape function for nine-node element.

The plots of the shape functions are shown to reinforce the capabilities of the particular elements to model displacements. The more nonlinear the displacement field in a problem, the more elements required (like the tapered axial problem). In addition, higher-order elements can approximate a more complex displacement field with fewer elements.

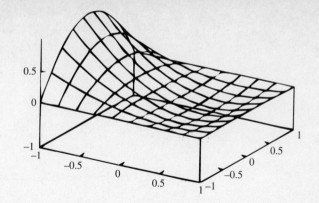

Figure 9.8 Midside node shape function for nine-node element.

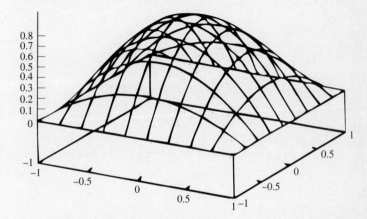

Figure 9.9 Center node shape function for nine-node element.

9.4 Cantilever Beam Example Using Membranes

Let's look at an example to see how to use these elements. The first example will be to model a cantilever beam and try to match flexural beam theory. The example we want to model is shown in Figure 9.10. The flexural beam theory solution for this problem is

$$\delta_{tip} = \frac{PL^3}{3EI} \tag{9.1}$$

Using the values for length, height, depth, and load, we get

$$\delta_{tip} = \frac{20(120^3)}{3(29,000)(864)} = 0.4598 \tag{9.2}$$

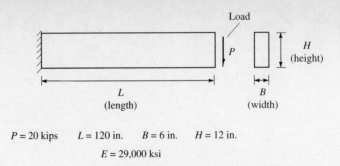

$$P = 20 \text{ kips} \quad L = 120 \text{ in.} \quad B = 6 \text{ in.} \quad H = 12 \text{ in.}$$
$$E = 29{,}000 \text{ ksi}$$

Figure 9.10 Cantilever beam subjected to tip load.

If we used a single-beam element, we would get the exact beam theory solution. However, by using membrane elements, we can study the stresses throughout the cross section as well as learn how to use two-dimensional finite elements.

9.4.1 Four-Node-Element Model

The first model we will use is with four four-node elements. The element configuration, nodal numbering, and loading are given in Figure 9.11. Modeling a structure using membrane elements requires similar information to all previous analyses we have done using beams and trusses. A general analysis program requires coordinates, boundary conditions, loads, and element information. For membrane elements, we need to provide the nodes, element thicknesses, and Poisson's ratio. Using this information, the element stiffness for each membrane element can be formed and assembled as previous elements.

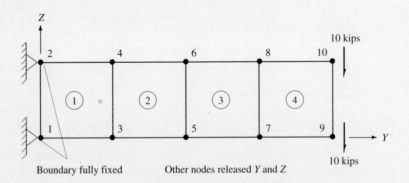

Figure 9.11 Model with four four-node elements for cantilever beam.

The following input file for SSTAN would be used for the given model:

```
Cantilever: Four Four-Node Elements
10,1,1  : 1 load case, 1 element type
:
Coordinates
1 x=0 y=0 z=0
9 y=120 g=1,9,2
```

```
2 y=0 z=12
10 y=120 g=2,10,2
:
Boundary
1,2 DOF=F,F,F,F,F,F
3,10 DOF=F,R,R,F,F,F
:
Plane
4,1
1 E=29000 U=0.0  : Poisson=0 to remove Poisson effect
1 q=1,3,2,4 m=1 h=6 g=4,1
:
Loads
9 l=1 f=0,0,-10
10 l=1 f=0,0,-10
:
```

Looking at the input file, we can see how to use membrane elements. The coordinates are just those of the nodes used to define each element. The choice of each node's location will be discussed later. The boundary conditions show that the nodes are allowed to move in the Y and Z planes, in-plane. The left end nodes, 1 and 3, are fully fixed. The membrane header defines the membrane elements. Only one material property is defined, steel with $E = 29,000$ ksi. Since we are analyzing the structure in two dimensions, we would normally include the true Poisson's ratio so that Poisson's effects would be modeled. Here, since we are trying to model beam theory which does not include Poisson's effect, we will set $\nu = 0$. Finally, we need to give the element connectivity. This is similar to beams and trusses in that we need to define the nodes connected to each element. We must also define which property to use and the element thickness. In this example, element generation is also used. Using this input file, we get the following output from SSTAN:

```
************************************************************
****  SSTAN  -  Simple Structural Analysis Program  ****
Copyright 1993  By  Dr. Marc Hoit, University of Florida
************************************************************
```

```
Cantilever: Four Four-Node Elements
NUMBER OF JOINTS                        =   10
NUMBER OF DIFFERENT ELEMENT TYPES   =    1
NUMBER OF LOAD CONDITIONS               =    1
NUMBER OF LOAD COMBINATIONS             =    0
NODE    BOUNDARY CONDITION CODES           NODAL POINT COORDINATES
NUMBER  X    Y    Z    XX   YY   ZZ        X            Y            Z
   1    F    F    F    F    F    F       .000         .000         .000
   2    F    F    F    F    F    F       .000         .000       12.000
   3    F    R    R    F    F    F       .000       30.000         .000
   4    F    R    R    F    F    F       .000       30.000       12.000
   5    F    R    R    F    F    F       .000       60.000         .000
   6    F    R    R    F    F    F       .000       60.000       12.000
   7    F    R    R    F    F    F       .000       90.000         .000
   8    F    R    R    F    F    F       .000       90.000       12.000
```

```
   9   F   R   R   F   F   F        .000     120.000          .000
  10   F   R   R   F   F   F        .000     120.000        12.000
```

EQUATION NUMBERS
```
   N   X    Y    Z   XX   YY   ZZ
   1   0    0    0    0    0    0
   2   0    0    0    0    0    0
   3   0    1    2    0    0    0
   4   0    3    4    0    0    0
   5   0    5    6    0    0    0
   6   0    7    8    0    0    0
   7   0    9   10    0    0    0
   8   0   11   12    0    0    0
   9   0   13   14    0    0    0
  10   0   15   16    0    0    0
```

--------------- SHELL/PLANE/PLATE ELEMENTS ----------

- MATERIAL PROPERTIES -
MATERIAL I.D. NUMBER   =   1
 MODULUS OF ELASTICITY =   .2900E+05
 POISSON'S RATIO   = .0000
 SHEAR MODULUS   =   .1450E+05
 SELF WEIGHT   = .0000

- ELEMENT DEFINITIONS -
```
EL #  MAT   N1   N2   N3   N4   N5   N6   N7   N8   N9    THICK
   1    1    1    0    3    0    0    0    2    0    4     6.00
UNIFORM LOAD =      .000
   2    1    3    0    5    0    0    0    4    0    6     6.00
UNIFORM LOAD =      .000
   3    1    5    0    7    0    0    0    6    0    8     6.00
UNIFORM LOAD =      .000
   4    1    7    0    9    0    0    0    8    0   10     6.00
UNIFORM LOAD =      .000
```

THE NODE NUMBERING USED PRODUCED A HALF BANDWIDTH OF   5

TOTAL STORAGE REQUIRED   =   142
TOTAL STORAGE AVAILABLE =   35000

*** CONCENTRATED NODAL LOAD ***
```
NODE   LOAD       X          Y          Z         XX         YY         ZZ
   9      1    .00E+00    .00E+00   -.10E+02    .00E+00    .00E+00    .00E+00
  10      1    .00E+00    .00E+00   -.10E+02    .00E+00    .00E+00    .00E+00
```

SOLUTION CONVERGED IN   1 ITERATION(S)

*** PRINT OF FINAL DISPLACEMENTS ***

DISPLACEMENTS FOR LOAD CONDITION 1
```
NODE        X           Y            Z           XX            YY            ZZ
   1   .00000E+00   .00000E+00   .00000E+00   .00000E+00    .00000E+00    .00000E+00
   2   .00000E+00   .00000E+00   .00000E+00   .00000E+00    .00000E+00    .00000E+00
   3   .00000E+00  -.36573E-02  -.97179E-02   .00000E+00    .00000E+00    .00000E+00
   4   .00000E+00   .36573E-02  -.97179E-02   .00000E+00    .00000E+00    .00000E+00
   5   .00000E+00  -.62696E-02  -.35110E-01   .00000E+00    .00000E+00    .00000E+00
   6   .00000E+00   .62696E-02  -.35110E-01   .00000E+00    .00000E+00    .00000E+00
```

```
 7  .00000E+00 -.78370E-02 -.70951E-01  .00000E+00  .00000E+00  .00000E+00
 8  .00000E+00  .78370E-02 -.70951E-01  .00000E+00  .00000E+00  .00000E+00
 9  .00000E+00 -.83595E-02 -.11202E+00  .00000E+00  .00000E+00  .00000E+00
10  .00000E+00  .83595E-02 -.11202E+00  .00000E+00  .00000E+00  .00000E+00
```

```
-------- PLANE/PLATE/SHELL STRESS RESULTS --------

------- ELEMENTS ARE IN THE Y-Z PLANE ------
 EL#  LD#  PNT#     SX           SY           SXY
-------------------------------------------------------------------------
  1    1    1   -.3535E+01   -.2525E-15   -.4697E+01
            3   -.3535E+01   -.4210E-15    .4141E+01
            4    .3535E+01   -.4210E-15    .4141E+01
            2    .3535E+01   -.2525E-15   -.4697E+01
  2    1    3   -.2525E+01   -.2238E-14   -.3434E+01
            5   -.2525E+01    .6893E-14    .2879E+01
            6    .2525E+01    .6893E-14    .2879E+01
            4    .2525E+01   -.2238E-14   -.3434E+01
  3    1    5   -.1515E+01   -.3141E-14   -.2172E+01
            7   -.1515E+01    .1401E-14    .1616E+01
            8    .1515E+01    .1401E-14    .1616E+01
            6    .1515E+01   -.3141E-14   -.2172E+01
  4    1    7   -.5051E+00   -.4341E-14   -.9091E+00
            9   -.5051E+00   -.3150E-15    .3535E+00
           10    .5051E+00   -.3150E-15    .3535E+00
            8    .5051E+00   -.4341E-14   -.9091E+00
```

Looking at the output, we can see the types of results given by membrane elements. First, by looking at the output displacements, we can see that the tip displacement at node 9 or 10 is −0.11202. The error is 75 percent compared to the theoretical result in equation (9.2). This is clearly not very close to the real solution. The reason is that the four-node elements are too stiff for modeling bending effects. The output file also gives the stress in each element at its corner nodes. We can see the same jump between the elements as was shown in the tapered axial element example in Chapter 8. For instance, look at the jump in the stress for node 4 in elements 1 and 2. From these results we can see that the solution is not acceptably accurate.

9.4.2   Nine-Node-Element Model

Let's try the same problem using nine-node elements. Remember, nine-node elements have quadratic displacement modeling capabilities. Using these elements should improve our results. To compare with the preceding example, we will use the same number of nodes along the length and the same spacing. The main difference between this and the preceding model is the use of nine-node membrane elements. Also notice that the load distribution has changed. This is due to the difference in elements and will be explained later. The model we will use is given in Figure 9.12.

The input file for the model with two nine-node elements is almost identical to the preceding one. The main differences are in the membrane connectivity line. Here the full nine nodes are given. The following input file is used:

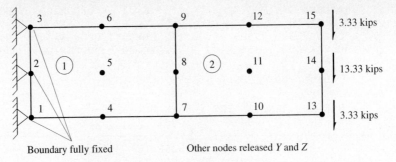

Figure 9.12 Model with two nine-node elements for cantilever beam.

```
Cantilever: Two Nine-Node Elements
15,1,1  : 1 element type, 1 load case
:
Coordinate
1 x=0 y=0 z=0
13 y=120 g=1,13,3
3 y=0 z=12 g=1,3,1
14 y=120 z=6 g=2,14,3
15 z=12 g=3,15,3
:
Boundary
1,3,1 DOF=F,F,F,F,F,F
4,15  DOF=F,R,R,F,F,F
:
Plane
2,1,0
1 E=29000 U=0.0
1 N=1,4,7,2,5,8,3,6,9 g=2,1 m=1 h=6
:
Loads
13 1=1 f=0,0,-3.333333
14 1=1 f=0,0,-13.333333
15 1=1 f=0,0,-3.333333
:
```

After running SSTAN, the output generated is as follows:

```
**********************************************************
****  SSTAN  -  Simple Structural Analysis Program  ****
Copyright 1993  By  Dr. Marc Hoit, University of Florida
**********************************************************

Cantilever: Two Nine-Node Elements
NUMBER OF JOINTS                      =  15
NUMBER OF DIFFERENT ELEMENT TYPES  =  1
NUMBER OF LOAD CONDITIONS          =  1
NUMBER OF LOAD COMBINATIONS        =  0
```

```
NODE    BOUNDARY CONDITION CODES            NODAL POINT COORDINATES
NUMBER  X     Y     Z     XX    YY    ZZ         X            Y           Z
    1   F     F     F     F     F     F        .000         .000        .000
    2   F     F     F     F     F     F        .000         .000       6.000
    3   F     F     F     F     F     F        .000         .000      12.000
    4   F     R     R     F     F     F        .000       30.000        .000
    5   F     R     R     F     F     F        .000       30.000       6.000
    6   F     R     R     F     F     F        .000       30.000      12.000
    7   F     R     R     F     F     F        .000       60.000        .000
    8   F     R     R     F     F     F        .000       60.000       6.000
    9   F     R     R     F     F     F        .000       60.000      12.000
   10   F     R     R     F     F     F        .000       90.000        .000
   11   F     R     R     F     F     F        .000       90.000       6.000
   12   F     R     R     F     F     F        .000       90.000      12.000
   13   F     R     R     F     F     F        .000      120.000        .000
   14   F     R     R     F     F     F        .000      120.000       6.000
   15   F     R     R     F     F     F        .000      120.000      12.000

EQUATION NUMBERS
    N   X     Y     Z     XX    YY    ZZ
    1   0     0     0     0     0     0
    2   0     0     0     0     0     0
    3   0     0     0     0     0     0
    4   0     1     2     0     0     0
    5   0     3     4     0     0     0
    6   0     5     6     0     0     0
    7   0     7     8     0     0     0
    8   0     9    10     0     0     0
    9   0    11    12     0     0     0
   10   0    13    14     0     0     0
   11   0    15    16     0     0     0
   12   0    17    18     0     0     0
   13   0    19    20     0     0     0
   14   0    21    22     0     0     0
   15   0    23    24     0     0     0

--------------- SHELL/PLANE/PLATE ELEMENTS ----------
- MATERIAL PROPERTIES -
MATERIAL I.D. NUMBER    =    1
 MODULUS OF ELASTICITY =   .2900E+05
 POISSON'S RATIO        = .0000
 SHEAR MODULUS          =   .1450E+05
 SELF WEIGHT            = .0000

- ELEMENT DEFINITIONS -
EL #  MAT   N1    N2    N3    N4    N5    N6    N7    N8    N9    THICK
   1    1    1     4     7     2     5     8     3     6     9    6.00
UNIFORM LOAD =      .000
   2    1    7    10    13     8    11    14     9    12    15    6.00
UNIFORM LOAD =      .000

THE NODE NUMBERING USED PRODUCED A HALF BANDWIDTH OF     9

TOTAL STORAGE REQUIRED  =      309
TOTAL STORAGE AVAILABLE =    35000
```

```
*** CONCENTRATED NODAL LOADS ***
NODE   LOAD      X          Y          Z          XX         YY         ZZ
  13    1     .00E+00    .00E+00   -.33E+01    .00E+00    .00E+00    .00E+00
  14    1     .00E+00    .00E+00   -.13E+02    .00E+00    .00E+00    .00E+00
  15    1     .00E+00    .00E+00   -.33E+01    .00E+00    .00E+00    .00E+00


SOLUTION CONVERTED IN   1 ITERATION(S)
*** PRINT OF FINAL DISPLACEMENTS ***
DISPLACEMENTS FOR LOAD CONDITION  1
NODE      X           Y            Z           XX          YY          ZZ
  1   .00000E+00  .00000E+00   .00000E+00   .00000E+00  .00000E+00  .00000E+00
  2   .00000E+00  .00000E+00   .00000E+00   .00000E+00  .00000E+00  .00000E+00
  3   .00000E+00  .00000E+00   .00000E+00   .00000E+00  .00000E+00  .00000E+00
  4   .00000E+00 -.14751E-01  -.38970E-01   .00000E+00  .00000E+00  .00000E+00
  5   .00000E+00  .47417E-16  -.38970E-01   .00000E+00  .00000E+00  .00000E+00
  6   .00000E+00  .14751E-01  -.38970E-01   .00000E+00  .00000E+00  .00000E+00
  7   .00000E+00 -.25862E-01  -.14259E+00   .00000E+00  .00000E+00  .00000E+00
  8   .00000E+00  .10431E-15  -.14259E+00   .00000E+00  .00000E+00  .00000E+00
  9   .00000E+00  .25862E-01  -.14259E+00   .00000E+00  .00000E+00  .00000E+00
 10   .00000E+00 -.31993E-01  -.28932E+00   .00000E+00  .00000E+00  .00000E+00
 11   .00000E+00  .15134E-15  -.28932E+00   .00000E+00  .00000E+00  .00000E+00
 12   .00000E+00  .31993E-01  -.28932E+00   .00000E+00  .00000E+00  .00000E+00
 13   .00000E+00 -.34483E-01  -.45760E+00   .00000E+00  .00000E+00  .00000E+00
 14   .00000E+00  .14538E-15  -.45760E+00   .00000E+00  .00000E+00  .00000E+00
 15   .00000E+00  .34483E-01  -.45760E+00   .00000E+00  .00000E+00  .00000E+00


-------- PLANE/PLATE/SHELL STRESS RESULTS --------
------- ELEMENTS ARE IN THE Y-Z PLANE ------


 EL#  LD#  PNT#      SX          SY          SXY
-------------------------------------------------------------------
  1    1    1    -.1602E+02    .2925E-08   -.3667E+00
            7    -.8981E+01   -.3201E-08   -.3667E+00
            9     .8981E+01    .3201E-08   -.3667E+00
            3     .1602E+02   -.2925E-08   -.3667E+00
  2    1    7    -.7686E+01    .3208E-08   -.3667E+00
           13    -.6475E+00   -.4090E-08   -.3667E+00
           15     .6475E+00    .4091E-08   -.3667E+00
            9     .7686E+01   -.3209E-08   -.3667E+00
```

Looking at the displacements from the model with two nine-node elements at the tip, nodes 13, 14, and 15, we get −0.4576. This is almost exact. The error is only 0.47 percent compared to the beam theory result in equation (9.2). Although the nine-node model had more nodes through the depth than the four-node model, this is not why the four-node model is so bad. As a matter of fact, if 40 four-node elements are used, the tip displacement is −0.4481. The error is still 2.5 percent compared to the theory.

We can also look at the stress results. Recall that the top fiber stress in a beam can be calculated by the equation

$$\sigma_y = \frac{Mz}{I} \tag{9.3}$$

This gives the beam theory top fiber stress as 16.67 ksi at the support. Looking at the outputs: From the four-node model we get 3.54 ksi; from the nine-node model we get 16.02 ksi; from the 40-element four-node model we get 15.96 ksi. Again, the nine-node-element model gives the best results. The first example with four-node elements gave very poor results.

9.5 Shear Locking

The problem with the four-node element lies in how the element models the bending (moment) effect. Figure 9.13 gives the displaced shapes that each element uses to approximate beam bending. The four-node element can model bending only by linear displacements. The nine-node element has quadratic capabilities. The linear displaced shape causes shear to occur in the element, whereas in true bending none exists (Figure 9.14). The nine-node element does not generate this fictitious shear. This phenomenon is called *shear locking*. Four-node elements do exist that do not exhibit this problem. By changing the formulation, you can create an element that gives better behavior. The two most common methods are by creating a nonconforming element or by using reduced integration. A nonconforming element adds additional shape functions that contain the bending shape (Figure 9.15). However, this additional shape function is not continuous across element boundaries. Therefore, the effect is to allow gaps to open between elements. Reduced integration uses a different number of Gauss points during the integration. The sampling points are chosen so that this shear is not included. For the current example, a single point in the center of the element is used for integrating the shear terms. As can be seen, this will neglect the shear developed in the element as a result of the linear displaced shape.

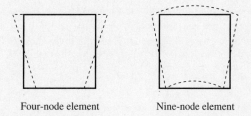

Four-node element   Nine-node element

Figure 9.13 Displaced shapes to approximate bending.

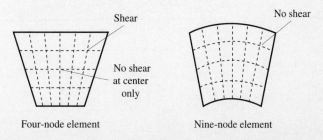

Shear

No shear

No shear at center only

Four-node element   Nine-node element

Figure 9.14 Shear at Gauss points due to displaced shape.

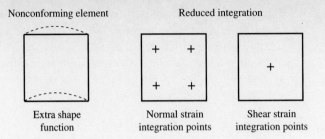

Figure 9.15 Methods for removing shear locking.

9.6 Mesh Correctness and Convergence

As discussed in Chapter 8, the accuracy of the solution depends on the number of elements and the order of the shape functions. As the number of elements increase, the piecewise displacement approximation approaches the true displacement field. Recall that two linear elements provided a better response than that of a single linear element. Also, a single quadratic element performed even better.

9.6.1 Stress Difference

The stress results also follow the same pattern. More elements provide better stress results. However, since we only guarantee the continuity of the displacements, the stresses are discontinuous. This means that at a node where two elements meet, the stresses do not match. However, as the number of elements increase, the stresses between elements get closer. As an example, Figure 9.16 is a plot of the stress along the top of the cantilever beam. The results are plotted for the four four-node membranes, the two nine-node membranes, and the 40 four-node membranes. Notice that for the four four-node elements, the difference between the elements is 28 percent. This large percentage error indicates a poor mesh (or not enough elements). Looking at the model with two nine-node elements, we see a closer difference. Here the error is 14.0 percent. This indicates that the mesh is

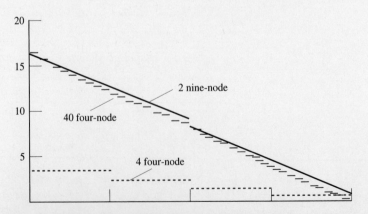

Figure 9.16 Stress plot for cantilever beam.

marginal but probably sufficient. Finally, we look at the 40-element model. Here the error is much better and only 3 percent. The 40-element model is very good.

The difference in element stresses at a node is an important measure of model correctness. In general, we do not have the exact displacements to check our model. Hence the stress check is necessary to verify convergence of our model. *If the difference in stresses between elements is small, the finite-element mesh is good.*

9.6.2 Element Meshing

Defining a mesh is critical to finding a correct solution with a minimum number of elements. More elements are needed where the displacement field is highly nonlinear (or in a high-stress-gradient area). Fewer elements are needed as the response becomes linear, and only a single element is required in a constant-stress field. As an example, the following stress function could be modeled by the mesh shown in Figure 9.17. This change in number of elements is handled by mesh changes from a single to multiple or higher-order elements. This is where variable-node elements are useful. Examples of some mesh changes are given in Figure 9.18.

Another important principle is that stress concentrations are localized phenomena and do not affect the solution at a reasonable distance from the concentration. In other

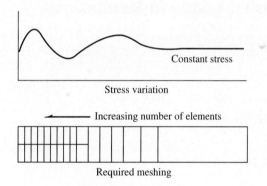

Figure 9.17 Mesh variation with stress gradient.

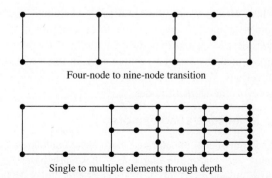

Figure 9.18 Mesh changes for increased accuracy.

words, bad stress differences at one portion of a model do not affect the results at a well-modeled portion. For example, if the results are only needed at one of the notches of a double-notched beam, a poor mesh is acceptable at the other notch. Figure 9.19 shows a mesh that could be used if results are necessary at only one notch location.

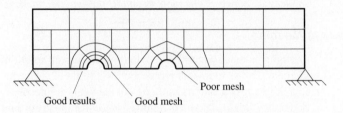

Figure 9.19 Notched beam with localized stress errors.

In summary, *a small stress difference between elements means a good mesh. In addition, large stress differences in localized areas will not affect the result a reasonable distance away.*

9.7 Distributed Loads in Membranes

As you recall from frame structures, matrix structural analysis does not handle distributed loads directly. The matrix equations correspond to concentrated loads applied at the active DOF for the structure. To handle distributed loads, they first need to be converted into equivalent concentrated loads. For beam elements, these turned out to be the fixed-end forces. For finite elements, we also need to convert distributed loads into concentrated equivalent loads.

The method is actually the same as for beam elements, we find the set of concentrated loads that create the same amount of virtual work as the distributed loads. For finite elements, this turns out to be of the form

$$R_{\text{concentrated}} = \int_{\text{volume}} \mathbf{H}(\mathbf{x}, \mathbf{y}, \mathbf{z}) \cdot \mathbf{b}(\mathbf{x}, \mathbf{y}, \mathbf{z}) \, dV \qquad (9.4)$$

where $\mathbf{H}(\mathbf{x}, \mathbf{y}, \mathbf{z})$ are the familiar shape functions and $\mathbf{b}(\mathbf{x}, \mathbf{y}, \mathbf{z})$ is a vector of the body forces applied to each of the three component directions. The $\mathbf{b}$ vector can also be considered as a distributed load with the appropriate change of integration. In two dimensions, $\mathbf{b}(\mathbf{x})$ has X and Y components. In three dimensions it has components for all three directions. When this equation is integrated, you get the equivalent concentrated loads to apply to the element nodes.

9.7.1 Four- and Nine-Node Load Factors

The equivalent concentrated loads were calculated for a uniform load on the edge of both four- and nine-node elements. These are given in Figure 9.20. Notice that the load distributions refer only to the number of nodes on the edge to which the uniform load is applied. Hence the results are given for distribution to two nodes and three nodes. The

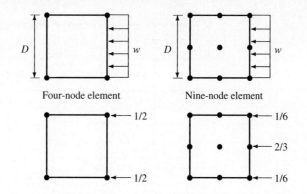

Figure 9.20 Equivalent concentrated loads for uniform edge load.

total amount of uniform distributed load ($D \times W \times$ thickness) is distributed half to each node for the four-node element. For the three-node edge the values are $\frac{2}{3}$ to the center and $\frac{1}{6}$ to the corners for the nine-node element.

Condensed Example: Notched Beam—Membrane Elements

Given the following notched beam, use SSTAN and membrane elements to find the maximum stress near the notch. The following structure is to be analyzed:

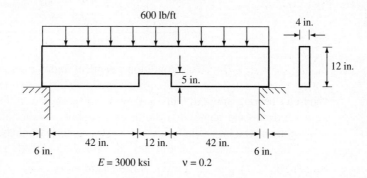

Notice that there is a 6-in. bearing support area at each end. In addition, the beam and loading are symmetric. We will use symmetry to reduce the number of nodes and elements in the analysis. As a first attempt at the mesh, we will use the nodes shown in the figure at the top of p. 342. Notice that there are more and smaller elements near the notch. This is required to attempt to capture the stress concentration in that area. In addition, note how the smaller elements are merged into the larger using the serendipity eight-node elements on the smaller elements. The element numbering, equivalent loads, and boundary conditions are given in the second figure on p. 342. The loads come from using the equivalent load factors from Figure 9.20. Notice how the symmetry boundary conditions put rollers at the line center of the beam. Also notice that a pin support is used for the

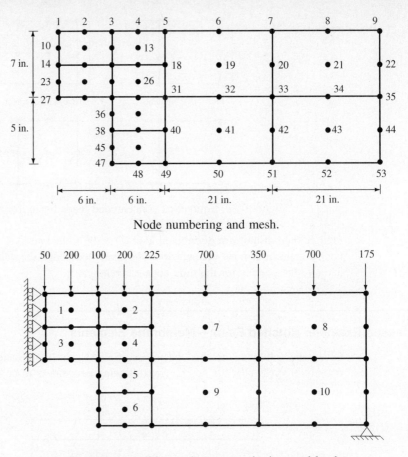

Node numbering and mesh.

Boundary conditions, element numbering, and loads.

right-end bearing support. This is done in the example to simplify the model. If very accurate results are required at the bearing support, smaller elements with nodes spaced at a minimum of 6 in. would be used. Then the supports would be applied over the true 6 in.

Using the preceding information, the following SSTAN input file is created:

```
Example:  Notched Beam
53,1,1            : 53 nodes, 1 type (plane), 1 load
:
Coordinates
1  x=0  y=12  z=0
5  x=12  g=1,5
10  x=0  y=12-1.75
13  x=9  g=10,13
14  x=0  y=12-3.5
18  x=12  g=14,18
23  x=0  y=12-5.25
26  x=9  g=23,26
```

```
27 x=0 y=5
31 x=12 g=27,31
36 x=6 y=5-1.25
37 x=9
38 x=6 y=5-2.5
40 x=12 g=38,40
45 x=6 y=5-3.75
46 x=9
47 x=6 y=0
49 x=12 g=47,49
9 x=54 y=12 g=5,9
22 y=12-3.5 g=18,22
35 y=5 g=31,35
44 y=5-2.5 g=40,44
53 y=0 g=49,53
:
Boundary
1,53 dof=r,r,f,f,f,f : Set to 2 dimen. (X-Y displacements)
1    dof=f,r,f,f,f,f : Symmetry boundary conditions
10   dof=f,r,f,f,f,f
14   dof=f,r,f,f,f,f
23   dof=f,r,f,f,f,f
27   dof=f,r,f,f,f,f
53   dof=f,f,f,f,f,f : Simple support (pinned)
:
Plane
10,1
1 e=3000000 u=0.2    : E (units) lb/in^2
1 n=14,15,16,10,11,12,1,2,3 h=4 m=1
2 n=16,17,18,12,13,0,3,4,5
3 n=27,28,29,23,24,25,14,15,16
4 n=29,30,31,25,26,0,16,17,18
5 n=38,39,40,36,37,0,29,30,31
6 n=47,48,49,45,46,0,38,39,40
7 n=31,32,33,18,19,20,5,6,7
8 n=33,34,35,20,21,22,7,8,9
9 n=49,50,51,40,41,42,31,32,33
10 n=51,52,53,42,43,44,33,34,35
:
Loads
1 l=1 f=0 -50        : Loads in pounds
2     f=0 -200
3     f=0 -100
4     f=0 -200
5     f=0 -225
```

```
6       f=0  -700
7       f=0  -350
8       f=0  -700
9       f=0  -175
:
```

After running SSTAN, we can either plot the results or look at the OUTPUT file. The output from the three elements at the notch (3, 4, and 5) is given below.

```
------- ELEMENTS ARE IN THE X-Y PLANE ------
```

EL#	LD#	PNT#	SX	SY	SXY
3	1	27	.2428E+03	-.1237E+02	.3693E+02
		29	.2830E+03	.3705E+02	-.5184E+02
		16	-.2910E+03	.3846E+02	.4482E+02
		14	-.2137E+03	-.1975E+02	-.2974E+02
4	1	29	.1889E+03	.1040E+03	-.6625E+02
		31	-.1682E+03	-.6148E+02	.5293E+02
		18	-.2379E+03	-.4358E+02	.6498E+02
		16	-.2869E+03	.2853E+02	.9149E+02
5	1	38	-.8147E+02	-.1956E+02	.4368E+02
		40	.2926E+03	.7750E+02	-.1920E+03
		31	-.2096E+03	.8302E+00	.4773E+01
		29	.2284E+03	.9895E+02	-.1246E+03

The common node between these elements is 29. Looking at the X stress at 29 from each element we have 283, 189, and 228. The difference between these values is too great and the mesh needs to be refined and the problem reanalyzed. This will be left as a homework exercise.

9.8 STANPLOT: Stress and Displacement Contours

STANPLOT has the additional capability of plotting stress, moment, and displacement contours for membrane, plate, shell, axisymmetric, and solid elements. This option allows a visual inspection of stresses, moments, and displacement concentrations. The *contours* option on the menu gives the choice of stress (average or local) or displacement. Then an appropriate submenu is given to choose which result to plot. The stresses are plotted by two choices, averaging the results from different elements at common nodes or local element values. Averaging gives a slightly false impression of the "goodness" of the result. The stress (or moment) difference still needs to be checked to validate the mesh. This can be done using the local stress option. Figure 9.21 is a local stress plot of the earlier condensed example of a notched beam for stress in the X direction. Notice the discontinuity at element edges, where there is a stress jump. The plot of local stress shows where elements need to be refined.

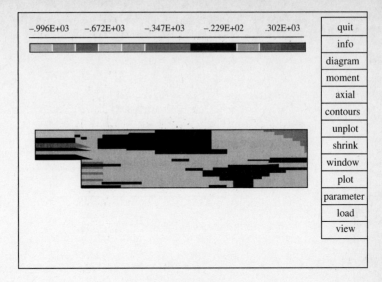

−.996E+03	−.672E+03	−.347E+03	−.229E+02	.302E+03	quit

Figure 9.21

9.9 Problems

9.1. Use the membrane element to solve the following problems.

   (a) Find the maximum positive and negative stress near the hole for the plate with hole subjected to uniform stress (model using symmetry, only $\frac{1}{4}$ of the plate).

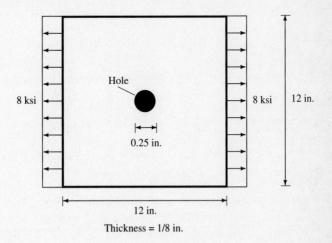

Thickness = 1/8 in.

(b) Compare the stress distribution of this simple beam with a bearing pad against a simple beam with pin supports at the neutral axis of the beam. (*Hint*: Use symmetry.)

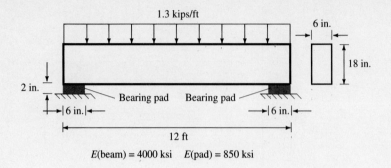

E(beam) = 4000 ksi E(pad) = 850 ksi

(c) Compare the tip displacement and the stress along the top of the beam for each model. Plot the stress along the top of the beam (all on one plot) for all models.

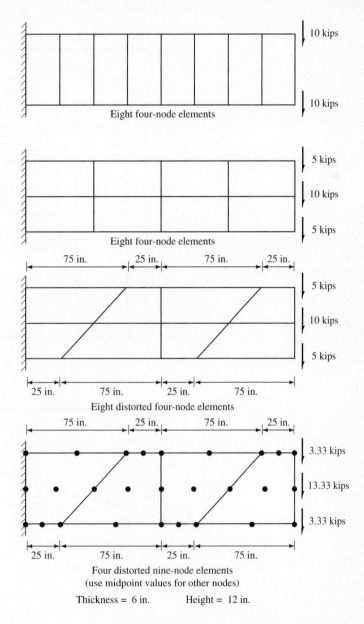

Eight four-node elements

Eight four-node elements

Eight distorted four-node elements

Four distorted nine-node elements
(use midpoint values for other nodes)

Thickness = 6 in.   Height = 12 in.

(d) Find the shear stress and moments applied to the columns between the two windows for both floors in this shear wall model.

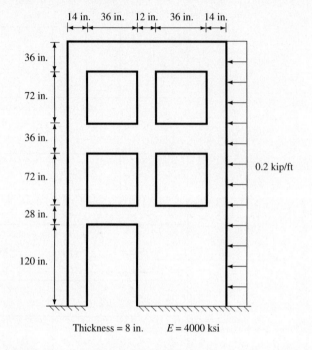

Thickness = 8 in. $E = 4000$ ksi

(e) Find the tip deflection and the maximum stresses at the fixed end.

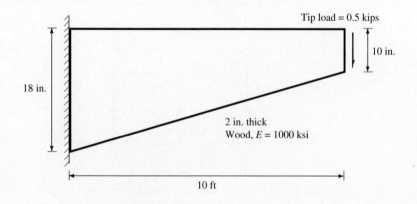

(f) Find the tip deflection and the maximum stresses at the fixed end.

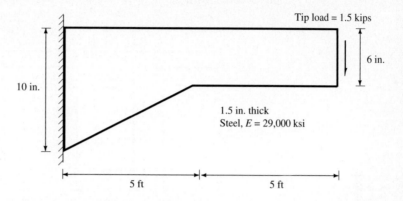

Tip load = 1.5 kips

6 in.

10 in.

1.5 in. thick
Steel, $E = 29,000$ ksi

5 ft   5 ft

(g) Find the deflection at the center of the beam and the maximum stress at the fixed ends and the center of the beam.

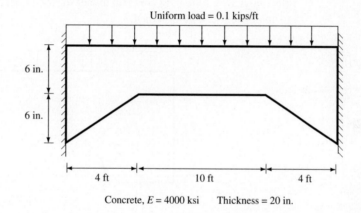

Uniform load = 0.1 kips/ft

6 in.

6 in.

4 ft   10 ft   4 ft

Concrete, $E = 4000$ ksi   Thickness = 20 in.

(h) Find the maximum stress in the weld material due to the stress concentration.

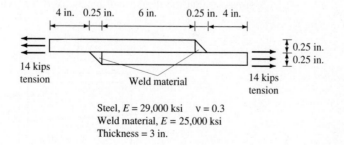

4 in.   0.25 in.   6 in.   0.25 in.   4 in.

0.25 in.
0.25 in.

14 kips
tension

Weld material

14 kips
tension

Steel, $E = 29,000$ ksi $\nu = 0.3$
Weld material, $E = 25,000$ ksi
Thickness = 3 in.

(i) Analyze the cross section of the concrete T-beam for the stress concentrations due to the nonsymmetric loading.

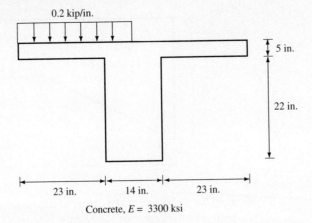

Concrete, $E = 3300$ ksi

(j) Analyze the following concrete tunnel for the maximum stress so that reinforcement steel can be designed.

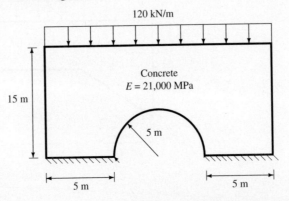

(k) The following concrete box is to support the given load. Find the maximum stress so that the design can be checked for failure.

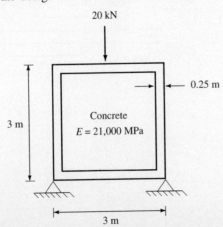

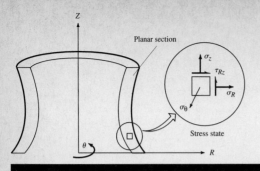

Planar section

σ_z

τ_{Rz}

σ_R

σ_θ

Stress state

Axisymmetric Solids

10.1 Introduction

The next type of element we will look at is the axisymmetric solid or solid of revolution. This element is used to represent solids of revolution. However, this representation is accomplished using a planar element and accounting for the circumferential strain through a change in the membrane element theory. Some solid-of-revolution structures are shown in Figure 10.1. If we slice one of these structures with a plane containing the axis of revolution, we can look at only one side of the slice for our analysis. This has the effect of modeling the structure as a planar structure. For the coordinate system shown in Figure 10.2, we see that there exist radial, circumferential, and vertical stresses and hence strain. By modeling the system as a planar structure, we have reduced the effort of creating a model to creating a planar mesh for a cross section of the structure. An example of a mesh of constant-size elements that could be used to get a complete solution to the three-dimensional structure is shown in Figure 10.3. Notice how the mesh is treated like a planar section of one side of the radial slice. This simplifies modeling and mesh development.

10.2 Axisymmetric Theory

Let's take out a single element and look at the difference between this and a membrane element. Figure 10.4 shows one element from the mesh given in Figure 10.3 and a portion

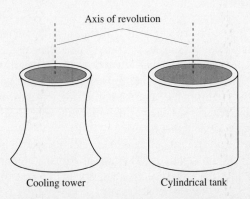

Figure 10.1 Axisymmetric examples.

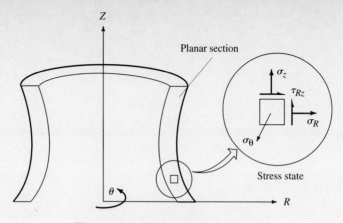

Figure 10.2 Sliced solid of rotation.

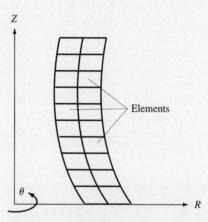

Figure 10.3 Mesh for planar section.

of the complete ring that it represents. This happens to be a four-node element. This single element can clearly be stretched vertically, hence stretching the entire ring, causing strain and stress in the Z direction. It can clearly be stretched in the R direction, hence expanding the entire ring, causing radial strain and stress. Notice that movement in the R direction implies that the entire ring expands or contracts. This means that the radial movement is restrained by the rest of the ring. Finally, the circumferential strain is directly related to the radial strain.

Figure 10.5 shows a radial stretching of the entire ring, which also expands the entire ring circumferentially. This means that the circumference of the element increases as a result of the radial change. The amount of circumferential change is exactly

$$\delta\theta = 2\pi r_{\text{original}} - 2\pi r_{\text{stretched}} \qquad (10.1)$$

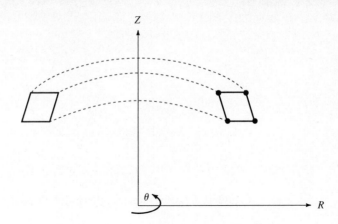

Figure 10.4 Single axisymmetric element.

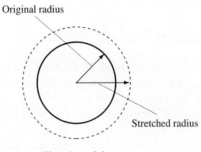

Figure 10.5 Circumferential change due to radius change.

As a result of this relationship, the only unknowns required for this type of analysis are displacements in the R and Z directions. The circumferential strain and stress can be recovered by a simple formula from the radial (R) displacements as given in equation (10.1). This means that the axisymmetric element only needs two planar DOFs which are exactly the same as the membrane element. The addition of the circumferential strain is handled by changing the **B**, strain displacement matrix. As a result, the final axisymmetric element we will work with has the same external configuration capabilities as the membrane element. It can be found in triangular, rectangular, and three- to nine-node versions. The possible DOFs for an axisymmetric element consist of only the two in-plane displacements (Figure 10.6). In addition, because of its axisymmetric representation, the R displacement has an additional constraint, due to the ring, on its movement. Due to this constraint from the ring, no boundary conditions are required in the R direction. Stiffness exists in the radial direction because movement means expansion of the ring section and hence circumferential strain. Additional boundary conditions may be needed for modeling a particular structure. As a result of this minor change to the membrane element, many

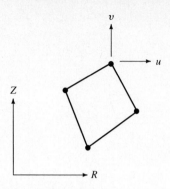

Figure 10.6 Axisymmetric element with possible displacements.

analysis programs just use the same element as the membrane, with an adjustment to the stiffness to account for the axisymmetric nature.

10.3 SSTAN Axisymmetric Element

The axisymmetric element in SSTAN is a modified version of the membrane element. To use the axisymmetric option, the structure must be oriented in the $Y-Z$ plane. In addition, the Z axis must be the axis of revolution. These types of restrictions are common on axisymmetric elements. Just like the membrane element, the only DOFs allowed for this element are the in-plane displacements. In SSTAN, this means only the Y and Z displacements. Clearly, Y corresponds to the radial direction.

10.4 Axisymmetric Circular Plate Example

As an example of using axisymmetric elements, let's analyze a circular plate subjected to a point load in the center. We assume that the plate is simply supported around the entire perimeter. The structure and the finite-element model are shown in Figure 10.7. It is important to realize that with an axisymmetric element, boundary conditions act completely around the structure. Therefore, the pin support shown goes completely around the structure. As a result, we need to look further at the required boundary conditions and loads for these structures.

Since the Z axis is the center of the structure, we need to examine the required boundary conditions at the center. Just like the ring described for a single element, the left edge of the plate model is the inside radius of the ring. Here the radius happens to be zero. As a result, this implies a "pinhole" in the center of the plate. If no boundary conditions are specified, the "hole" is allowed to "get bigger" as a result of loading. To restrain this "opening" we need to fix the displacements at this edge in the Y direction. Figure 10.8 shows the two possible boundary conditions on the inner edge of a model and its resulting implications.

Remember, the axisymmetric element provides restraint against movement in the radial direction through circumferential stiffness. Therefore, no radial boundary conditions

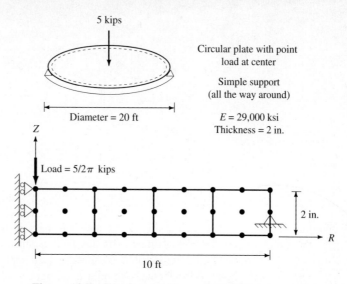

Figure 10.7 Circular simply supported plate example.

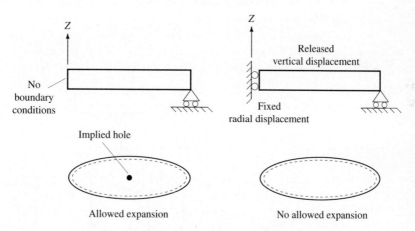

Figure 10.8 Effect of boundary conditions at Z axis.

are required. A movement in the radial direction implies an expansion or contraction of the entire structure.

10.5 Loads on Axisymmetric Models

Now let's look at the load requirements for axisymmetric models. Since the element we are using is axisymmetric, it is implied that we are using a 1-rad slice. Like the boundary conditions, any load is assumed to act around the entire circumference of the structure. Therefore, any load applied to the planar element representation needs to be loaded with a 1-rad portion of the load. If a point load is applied to the center, at the axis of rotation, it needs to be divided by 2π to account for its 1-rad portion. A line load acting around

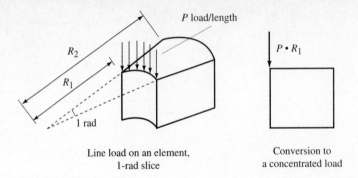

Figure 10.9 Conversion of line load to concentrated load.

the circumference at some radius R also needs to have its 1-rad portion calculated. Here the 1-rad portion is the load times its radius. This is shown in Figure 10.9.

As another example, let's look at the uniform load applied over the surface of a plate. Figure 10.10 shows a distributed load on a 1-rad portion of an element. This distributed loading needs to be adjusted for the fact that it acts completely around the structure. Since we are working with a 1-rad slice wedge, we need to collapse the surface load on this element to an equivalent line load on the planar element. This causes a uniform load to be converted into a trapezoidal line load on the edge of the planar model (Figure 10.10). Notice that the values at each edge of the element are the uniform load times the radius. This is because the arc over which the uniform load acts changes with the radius as $R\theta$, with θ equal to 1 rad.

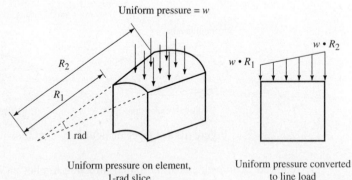

Figure 10.10 Conversion of pressure to line load.

10.5.1 Nodal Loads for Trapezoidal Line Loads

The concentrated loads to be used for a trapezoidal loading are shown in Figure 10.11.

10.6 SSTAN Axisymmetric Plate Example

We will now use SSTAN to analyze the axisymmetric plate example given in Section 10.4. The element mesh is given in Figure 10.12. The required SSTAN input file follows.

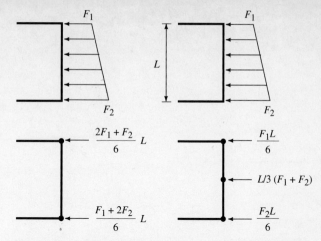

Figure 10.11 Concentrated loads for trapezoidal loading.

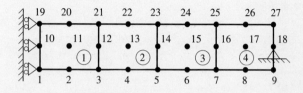

Figure 10.12 Axisymmetric node and element numbering.

```
Circular Plate
27,1,1 : 27 nodes, axisymmetric, 1 point load
:
Coordinates
1 X=0 Y=0 Z=0
9 y=10*12 g=1,9
10 y=0 z=1
18 y=10*12 g=10,18
19 y=0 z=2
27 y=10*12 g=19,27
:
Boundary
1,27 dof=f,r,r,f,f,f
18   dof=f,f,f,f,f,f
1,19,9 dof=f,f,r,f,f,f
:
Axisymmetric
4,1  : 4 elements, 1 material
1 e=29000 u=.3
1 n=1,2,3,10,11,12,19,20,21 m=1 g=4,1 : Generate 4 by 1
:
```

```
          Loads
          19 1=1 f=0,0,-5/2/3.1415  : Divide by 2 pi - 1 radian slice
          :
```

Notice that we are analyzing the center point load example. As a result, the center load is divided by 2π. Also note that the boundary conditions for nodes 1, 10, and 15, the left edge, are fixed in the Y (radial) direction so as not to allow a "hole" to exist. Selected portions of the output file follow.

```
*** AXISYMMETRIC ELEMENT *****
    ELEMENTS ARE AXISYMMETRIC
*********************************

-ELEMENT LOAD MULTIPLIERS -   Y-GRAVITY  Z-GRAVITY
    1        .000         .000
MATERIAL I.D. NUMBER  =        1
WEIGHT (GRAVITY LOAD) =      .0000E+00
E (YOUNG'S MODULUS)   =      .2900E+05
NU (POISSON'S RATIO)  =      .3000
G (SHEAR MODULUS)     =      .1115E+05

EL #  MAT   N1    N2    N3    N4    N5    N6    N7    N8    N9    THICK
   1    1    1     2     3    10    11    12    19    20    21     .00
   2    1    3     4     5    12    13    14    21    22    23     .00
   3    1    5     6     7    14    15    16    23    24    25     .00
   4    1    7     8     9    16    17    18    25    26    27     .00

THE NODE NUMBERING USED PRODUCED A HALF BANDWIDTH OF   20
TOTAL STORAGE REQUIRED  =     1150
TOTAL STORAGE AVAILABLE =     35000

*** CONCENTRATED NODAL LOADS ***
 NODE  LOAD       X          Y          Z        XX         YY         ZZ
   19    1    .00E+00    .00E+00   -.80E+00    .00E+00    .00E+00    .00E+00

SOLUTION CONVERGED IN   1 ITERATION(S)
*** PRINT OF FINAL DISPLACEMENTS ***
DISPLACEMENTS FOR LOAD CONDITION  1
NODE        X            Y            Z           XX           YY           ZZ
   1   .00000E+00   .00000E+00  -.16874E+00   .00000E+00   .00000E+00   .00000E+00
   2   .00000E+00   .72019E-03  -.16296E+00   .00000E+00   .00000E+00   .00000E+00
   3   .00000E+00   .12220E-02  -.14811E+00   .00000E+00   .00000E+00   .00000E+00
   4   .00000E+00   .14570E-02  -.12795E+00   .00000E+00   .00000E+00   .00000E+00
   5   .00000E+00   .16474E-02  -.10460E+00   .00000E+00   .00000E+00   .00000E+00
   6   .00000E+00   .17286E-02  -.79243E-01   .00000E+00   .00000E+00   .00000E+00
   7   .00000E+00   .17829E-02  -.52869E-01   .00000E+00   .00000E+00   .00000E+00
   8   .00000E+00   .17652E-02  -.26230E-01   .00000E+00   .00000E+00   .00000E+00
   9   .00000E+00   .17287E-02   .30977E-05   .00000E+00   .00000E+00   .00000E+00
  10   .00000E+00   .00000E+00  -.16877E+00   .00000E+00   .00000E+00   .00000E+00
```

```
11  .00000E+00   .51770E-06  -.16298E+00   .00000E+00   .00000E+00   .00000E+00
12  .00000E+00   .21245E-07  -.14812E+00   .00000E+00   .00000E+00   .00000E+00
13  .00000E+00   .14915E-06  -.12796E+00   .00000E+00   .00000E+00   .00000E+00
14  .00000E+00   .17190E-07  -.10461E+00   .00000E+00   .00000E+00   .00000E+00
15  .00000E+00   .68003E-07  -.79249E-01   .00000E+00   .00000E+00   .00000E+00
16  .00000E+00   .93041E-08  -.52873E-01   .00000E+00   .00000E+00   .00000E+00
17  .00000E+00   .28646E-07  -.26233E-01   .00000E+00   .00000E+00   .00000E+00
18  .00000E+00   .00000E+00   .00000E+00   .00000E+00   .00000E+00   .00000E+00
19  .00000E+00   .00000E+00  -.16875E+00   .00000E+00   .00000E+00   .00000E+00
20  .00000E+00  -.71914E-03  -.16296E+00   .00000E+00   .00000E+00   .00000E+00
21  .00000E+00  -.12220E-02  -.14811E+00   .00000E+00   .00000E+00   .00000E+00
22  .00000E+00  -.14567E-02  -.12795E+00   .00000E+00   .00000E+00   .00000E+00
23  .00000E+00  -,16474E-02  -.10460E+00   .00000E+00   .00000E+00   .00000E+00
24  .00000E+00  -.17284E-02  -.79243E-01   .00000E+00   .00000E+00   .00000E+00
25  .00000E+00  -.17829E-02  -.52869E-01   .00000E+00   .00000E+00   .00000E+00
26  .00000E+00  -.17652E-02  -.26230E-01   .00000E+00   .00000E+00   .00000E+00
27  .00000E+00  -.17287E-02   .29724E-05   .00000E+00   .00000E+00   .00000E+00
```

```
EL# LD# NODE     Srr       Szz       Stt       Srz     S(MAX)    S(MIN)   ANGLE
 1  1----------------------------------------------------------------------------
            3    1.57       .01      1.79       .05      1.57       .01    1.81
           21    1.40       .01      1.67       .03      1.40       .01    1.04
            2     .00       .00       .00       .05       .05      -.05   45.09
           12     .00       .00       .00       .02       .03      -.02   46.50
           10   -1.57      -.01     -1.79       .05      -.01     -1.58   88.19
           11   -1.40      -.01     -1.67       .02       .00     -1.40   88.99
 2  1----------------------------------------------------------------------------
            3     .97      -.01      1.48       .02       .97      -.01    1.19
            5     .75       .00      1.16       .01       .75       .00    1.06
           23     .64       .01      1.00       .01       .64       .01     .95
           21     .00       .00       .00       .02       .02      -.02   43.60
            4     .00       .00       .00       .01       .01      -.01   44.90
           14     .00       .00       .00       .01       .01      -.01   46.21
           22    -.97       .01     -1.48       .02       .01      -.97   88.77
           12    -.75       .00     -1.16       .01       .00      -.75   88.94
           13    -.64      -.01     -1.00       .01      -.01      -.64   89.08
 3  1----------------------------------------------------------------------------
            5     .47      -.01       .94       .01       .47      -.01    1.22
            7     .36       .00       .77       .01       .36       .00    1.32
           25     .29       .01       .66       .01       .29       .01    1.43
           23     .00       .00       .00       .01       .01      -.01   44.07
            6     .00       .00       .00       .01       .01      -.01   44.88
           16     .00       .00       .00       .01       .01      -.01   45.68
           24    -.47       .01      -.94       .01       .01      -.47   88.75
           14    -.36       .00      -.77       .01       .00      -.36   88.68
           15    -.29      -.01      -.66       .01      -.01      -.29   88.60
```

```
4    1------------------------------------------------------------------
     7     .17      -.01       .63       .01       .17      -.01    2.15
     9     .10       .00       .52       .01       .10       .00    3.28
    27     .04       .00       .43       .01       .04       .00    6.71
    25     .00       .00       .00       .01       .01      -.01   44.58
     8     .00       .00       .00       .01       .01      -.01   44.86
    18     .00       .00       .00       .01       .01      -.01   45.12
    26    -.17       .01      -.63       .01       .01      -.17   87.81
    16    -.10       .00      -.52       .01       .00      -.10   86.69
    17    -.04       .00      -.43       .01       .00      -.04   83.30
```

Looking at the results, we see that the center vertical deflection (node 1) is 0.1607, as compared to the result from plate theory, which is 0.1711. This is an error of only 6.1 percent. The error is caused by a combination of things, including boundary conditions and the inclusion of Poisson's ratio, which are modeled differently in plate theory. This is similar to the errors in trying to match beam theory results using membrane elements. Setting Poisson's ratio, v, equal to zero would better match plate theory. However, using a true value of v more closely models the real structure.

10.7 Axisymmetric Results Evaluation

We also need to look at the stress results from an axisymmetric analysis (Figure 10.13). The stress results are: normal stress in the R, Z, and θ directions and shear stress in the R–Z plane. The circumferential stress is calculated from the radial displacements and printed. In the SSTAN output, the radial stress is S_{rr}, the vertical stress is S_{zz}, and the circumferential stress is S_{tt}. The shear stress is S_{rz}. SSTAN also gives the principal stresses in the R–Z plane. These consist of the maximum and minimum normal stresses, and the angle at which they occur. The shear stress is zero at the given angle. Notice that the stresses are given at the nodes and are stress per unit length along the edge.

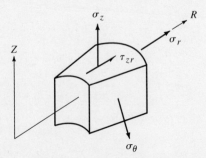

Figure 10.13 Stress results for axisymmetric elements.

Condensed Example: Steel Collar—Axisymmetric Elements

The steel collar shown here is to be analyzed for the bearing that loads in will undergo. The collar is cylindrical in shape (a solid of revolution), and the maximum stress is desired. The following figure gives the dimensions, loads, and boundary conditions applied to the collar.

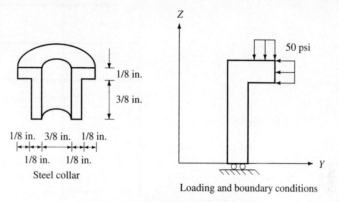

Steel collar

Loading and boundary conditions

Loading and boundary conditions.

The following mesh will be used and modeling will be done with axisymmetric elements:

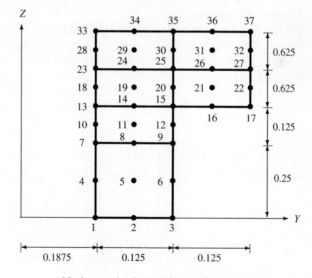

Node numbering and coordinate system.

The element numbering and nodal boundary conditions to be used are given below.

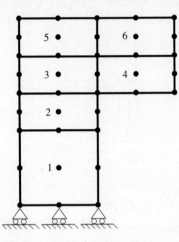

Boundary conditions and element numbering.

To apply the distributed loads, they must be converted to equivalent concentrated loads. For the axisymmetric structures, the magnitude depends on the radius. For the vertical portion of the load, the radius changes along the element. This causes the load to act as a trapezoidal loading on the element. The load factors are shown in Figure 10.11. The final required values are

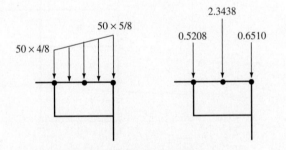

Load conversion for uniform pressure.

The horizontal portion of the load is at a constant radius and continues to remain constant. The equivalent loads for this portion become

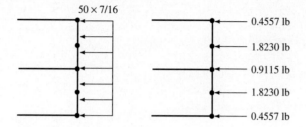

50 × 7/16

0.4557 lb

1.8230 lb

0.9115 lb

1.8230 lb

0.4557 lb

Conversion of side pressure to concentrated loads.

Using the preceding figures, we can create the required SSTAN input file:

```
Example: Steel Collar - Axisymmetric
37,1,1       : 37 nodes, 1 type (axisymmetric), 1 load
:
Coordinates : MUST BE IN Y-Z PLANE!
1 y=0.1875 z=0
3 y=0.3125              g=1,3
4 y=0.1875 z=0.125
6 y=0.3125              g=4,6
7 y=0.1875 z=0.25
9 y=0.3125              g=7,9
10 y=0.1875 z=0.3125
12 y=0.3125             g=10,12
13 y=0.1875 z=0.375
17 y=0.4375             g=13,17
18 y=0.1875 z=0.40625
22 y=0.4375             g=18,22
23 y=0.1875 z=0.4375
27 y=0.4375             g=23,27
28 y=0.1875 z=0.46875
32 y=0.4375             g=28,32
33 y=0.1875 z=0.5
37 y=0.4375             g=33,37
:
Boundary
1,37  DOF=F,R,R,F,F,F  : Set all to 2 dimen. (Y-Z plane)
1,3   DOF=F,R,F,F,F,F
:
Axisymmetric
6,1
1 e=29000 u=.2
1 n=1,2,3,4,5,6,7,8,9 g=1,2
```

```
3 n=13,14,15,18,19,20,23,24,25 g=2,2 : Generate elements
:
Loads
35   F=0,0,-0.5208
36   F=0,0,-2.3438
37   F=0,-0.4557,-0.6510
32   F=0,-1.8230
27   F=0,-0.9115
22   F=0,-1.8230
17   F=0,-0.4557
:
```

After running SSTAN, we can look at a plot using STANPLOT to determine where the maximum stress occurs. From the plot, the maximum stress occurs near node 23. The output for elements 2 and 3 is as follows:

EL#	LD#	NODE	Srr	Szz	Stt	Srz	S(MAX)	S(MIN)	ANGLE
2	1								
		7	36.14	-127.23	-129.80	-83.83	71.50	-162.59	-22.87
		9	63.30	-105.83	-67.86	-15.68	64.74	-107.28	-5.25
		15	79.60	-93.00	-30.70	25.20	83.20	-96.60	8.14
		13	-14.11	57.41	-130.94	-55.99	88.08	-44.79	-61.28
		8	-22.07	-113.29	-125.77	-49.72	-.21	-135.16	-23.73
		12	-26.84	-215.72	-122.66	-45.96	-16.25	-226.31	-12.98
		14	-64.36	242.04	-132.09	-28.15	244.61	-66.93	-84.79
		10	-107.44	-120.76	-183.67	-83.76	-30.07	-198.13	-42.73
		11	-133.28	-338.44	-214.62	-117.13	-80.16	-391.55	-24.39
3	1								
		13	69.37	251.00	-87.65	28.08	255.24	65.13	81.41
		15	-175.09	-104.28	-195.86	-43.43	-83.65	-195.72	-64.59
		25	-321.76	-317.45	-260.78	-86.34	-233.24	-405.97	-45.72
		23	15.96	196.97	-91.47	59.06	214.54	-1.60	73.44
		14	-93.57	-62.02	-143.95	-10.30	-58.95	-96.63	-73.43
		20	-159.29	-217.41	-175.44	-51.91	-128.86	-247.85	-30.38
		24	-37.45	142.95	-95.28	90.04	180.20	-74.70	67.52
		18	-12.05	-19.76	-92.04	22.84	7.25	-39.06	40.21
		19	3.18	-117.38	-90.10	-17.49	5.67	-119.86	-8.09

Again, looking at the nodes in common between the two elements (nodes 13, 14, and 15), we can see that the stress discontinuity is too large and the model needs to be refined.

10.8 Problems

10.1 Given the following simply supported circular plate with a circular hole and a ring load along the edge, find the maximum stress and displacement. The plate has a 6-ft radius and the hole has a 1-ft radius. The thickness of the plate is 2 in.

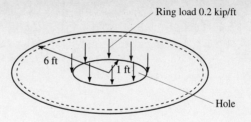

Steel, $E = 29{,}000$, thickness = 2 in.

10.2 Given the following cylindrical water tank, find the maximum stress in the tank. The tank is partially filled with water at a depth H. The pressure exerted by the water can be calculated using γZ, where Z is the depth and γ is the specific gravity of water. γ for water is 64.4 lb/ft^3. The tank is resting on a concrete pad. Do not be overly concerned with the stress concentration at the connection between the floor and the sidewalls. The thickness of the tank is $\frac{3}{4}$ in.

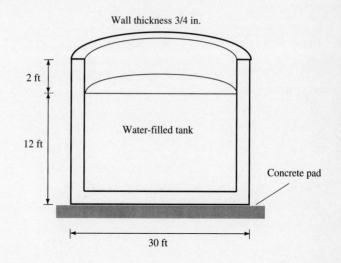

10.3. Given the following tank, which holds sand, find the maximum stress in the tank walls. Treat the sand as a fluid in order to consider the pressure exerted by it on the walls. Use a specific gravity of 110 lb/ft³. The wall thickness is 1 in.

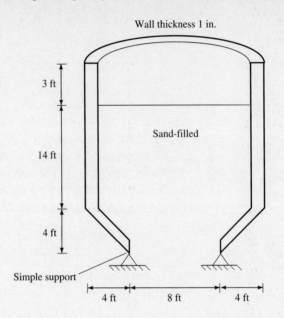

Wall thickness 1 in.

3 ft

Sand-filled

14 ft

4 ft

Simple support

4 ft   8 ft   4 ft

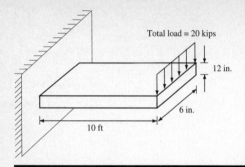

Total load = 20 kips

12 in.

6 in.

10 ft

FLAT PLATE AND SHELL ELEMENTS

11.1   Introduction

The next finite element we examine is the flat plate bending element. This element can be thought of as a two-dimensional extension of a beam element. Beam elements provide both shear and bending resistance. Plate elements provide this same resistance but in two directions. Figure 11.1 shows the difference between beams and plates. Plate bending elements can be used to model very common structural elements such as floor slabs, floor diaphragms, bridge decks, and even I-beams. Whenever out-of-plane bending effects need to be considered, plate elements can be used. They are also useful for thin-walled structures such as pipes and tanks. Most modeling situations require that both the out-of-plane (plate bending) and in-plane (membrane) effects be modeled. Some example structures that use plate and membrane elements are shown in Figure 11.2.

True plate elements do not include in-plane effects. In-plane effects are handled by membrane elements. Similarly, in a beam element the bending and axial effects are uncoupled. This is the same in two dimensions. These two elements are commonly merged to get a complete in- and out-of-plane element, referred to as a flat shell element. We will discuss a true plate element before discussing flat shell elements. To do this we must cover a small amount of theory.

11.2   Plate Theory

There are two common versions of plate theory used in finite elements: Kirchhoff and Mindlin. Kirchhoff plate bending theory is derived in a similar fashion to beam bending but includes bending in both directions. The derivation assumes that the normal displacement, w, controls. In Kirchhoff theory the rotation, θ, in the plate is the derivative of w. This is the same as beam theory. In Mindlin theory shear deformation is included and the rotation is the sum of the derivative of w and the shear angle.

11.2.1   Kirchhoff Theory

In Kirchhoff theory, the normal to the surface remains normal. Hence this theory ignores shear deformations (just like beam theory). To derive this type of finite element, a shape function that describes the distribution of the normal displacement $w(x, y)$ throughout the element is needed. This shape function has the property that its derivative is equal to the slope of the surface. An important implication of this is that the slope is continuous across elements. This is called a C^1 *element*, meaning that it has continuous first derivatives

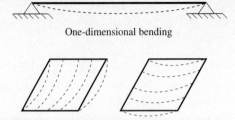

One-dimensional bending

Two-dimensional bending

Figure 11.1

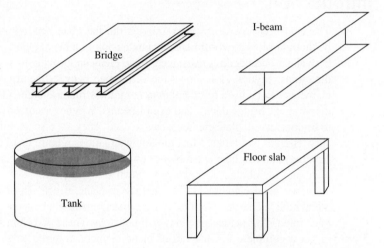

Figure 11.2 Example structures using plate and shell elements.

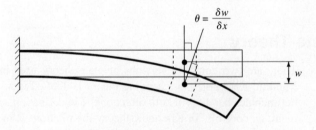

$$\theta = \frac{\delta w}{\delta x}$$

No shear–normal remains normal

Figure 11.3 Kirchhoff plate theory.

between elements. Figure 11.3 shows the relationship between w and θ for Kirchhoff theory.

11.2.2 Mindlin Theory

The second theory, Mindlin theory, includes shear deformations. As a result, the normal to the surface does not remain normal. Similarly, the derivative of the shape function for the normal displacement $w(x, y)$ is not equal to the rotation. In Mindlin theory the rotation of the surface is the sum of the derivative of $w(x, y)$ and the shear angle change. Figure 11.4 shows the relationship between the displacement $w(x, y)$, shear angle γ, and the derivative of the displacement. This sum of angles to get the total rotation implies that independent shape functions can be used for the displacement w and the rotations (θ_x, θ_y). This is the most common formulation found in flat plate and shell elements used in current computer programs. This means that there will not be rotational continuity across element boundaries (since shear exists). There is, of course, still rotational continuity at the nodes. Hence the elements are considered to be C^0 elements. Figure 11.5 shows this lack of continuity across the edge between two elements. In both the Kirchhoff and Mindlin formulations, the pure plate bending element has three DOFs per node; the normal displacement w and the out-of-plane rotations (θ_x, θ_y). These are shown in Figure 11.6.

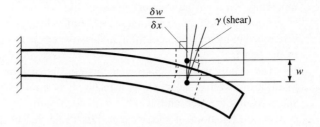

Shear included–sum of derivative and shear

Figure 11.4 Mindlin plate theory.

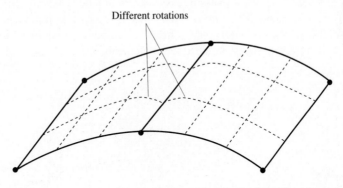

Figure 11.5 Lack of rotational continuity for Mindlin plate.

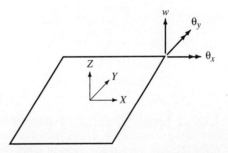

Figure 11.6 Plate degrees of freedom.

11.2.3 Generalized Stress

In plate theory, most derivations refer to the equations for generalized stress and strain. This is because the equations for plate behavior can be converted to the form

$$M(x, y) = \mathbf{E}^* \psi(\text{curvature}) \qquad (11.1)$$

where $\mathbf{E}^*$ is a modified constitutive matrix. Note that this is just like the equation for stress and strain except that we have moments replacing stresses and curvature replacing strain. In plates, the displacement unknowns are the normal displacement and the two rotations. Following the analogy of generalized stress, moments are equivalent to stress and curvature is equivalent to strain. This means that when using these elements in modeling, we treat the moment gradient like we would stress to determine the level of shape function and number of elements required for an accurate analysis. In addition, the difference in moment at a common node between two elements indicates the adequateness of the mesh.

The results from all plate elements consist of moments. Some plate elements also give the transverse shear, Q, as a result. It is important to note that the moments and shear results are per unit length of plate. Figure 11.7 gives the most common sign convention for moment and shear results. Note that positive M_x and M_y moments cause tension in the top ($+Z$) face of the element. Flat plate elements can be found in three- to nine-node versions, just like the membrane elements (Figure 11.8). The same concepts of shape function order are true for plates as they were for the membranes. Three-node triangular plates model constant moments exactly. Nine-node elements model linear moments with some second-order effects. It is important to remember that in plates, moments are equivalent to stress and curvature is equivalent to strain, in terms of modeling. In other words, we need more elements in a high moment gradient area for plates.

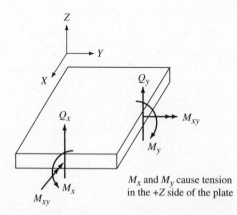

Figure 11.7 Definition of positive plate results.

11.3 Surface Load Distribution Factors

Very often the load on a plate element takes the form of a uniform pressure. The effect of the pressure needs to be distributed to the nodes using formula (9.4), which depends

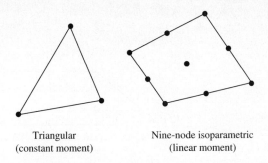

Triangular   Nine-node isoparametric
(constant moment)   (linear moment)

Figure 11.8 Common flat plate configurations.

on the shape functions and the variation of the load. The nodal distribution values for a uniform surface load on the face of nine- and eight-node plates are given in Figure 11.9. Notice for the eight-node element that the corner node loads are in the direction opposite to that of the actual applied loads (and the midsize node loads). The total surface load is: surface area $\times$ load. The midsize nodes have $\frac{1}{3}$ of the total surface load applied in the direction of the applied load. The four corner nodes have $\frac{1}{12}$ of the total surface load applied in the direction opposite to that of the applied surface load. For the nine-node element, all the loads are applied in the direction of the real load.

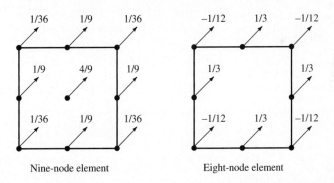

Nine-node element   Eight-node element

Figure 11.9 Equivalent concentrated loads for uniform surface load.

11.4 Cantilever Plate Example

Now let's look at a simple plate example with SSTAN. The plate element in SSTAN is a variable three- to nine-node Mindlin element. It may be oriented in any direction in three-dimensional space. As an example we will model a cantilever beam using the flat plate element. The example structure given in Figure 11.10 is simulating a 10-ft cantilever beam 6 in. wide with a tip load of 20 kips. To model this structure, we will use a single eight-node element. Although this is probably not as accurate as might be needed, it will demonstrate the use of flat plate elements. The single-element model is shown in Figure 11.11.

Notice that the tip load has been converted to concentrated loads at the nodes in the proportions given for a uniform load on a three-node edge. These factors were given in Section 9.7.1. The required input file for this example is as follows:

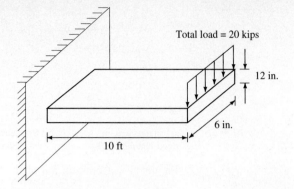

Figure 11.10 Cantilever beam using plate element.

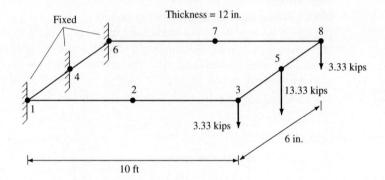

Figure 11.11 Single plate cantilever beam model.

```
Single Plate: Cantilever Beam Example
8,1,1 : 8 nodes, plate element only
:
C--- NOTE: IN Y-Z PLATE - X is normal displacement
Coordinates
1 x=0 y=0 z=0
3 y=120 g=1,3,1
4 y=0 z=3
5 y=120
6 y=0 z=6
8 y=120 g=6,8,1
:
Boundary
1,8   DOF=R,F,F,F,R,R : X is vertical displacement
1,4,3 DOF=F,F,F,F,F,F
6     DOF=F,F,F,F,F,F
:
Plate
1,1
```

```
1 E=29000 u=0.0          : U=0 to match beam theory
1 N=1,2,3,4,0,5,6,7,8 M=1 H=12
:

Loads
3,8,5 L=1 F=-3.3333
5     L=1 F=-13.3333
:
```

By looking at the input file we can see that the only DOFs allowed are the normal displacement, X, and the two rotations, θ_y and θ_z. Also notice how Poisson's ratio is set to zero to remove these effects and model beam theory more closely. Finally, see how an eight-node element is specified by leaving the middle node equal to zero. After running SSTAN, we get the following output:

```
************************************************************
****  SSTAN  -  Simple Structural Analysis Program  ****
Copyright 1993  By  Dr. Marc Hoit, University of Florida
************************************************************
```

```
Single Plate: Cantilever Beam Example
NUMBER OF JOINTS                   =   8
NUMBER OF DIFFERENT ELEMENT TYPES  =   1
NUMBER OF LOAD CONDITIONS          =   1
NUMBER OF LOAD COMBINATIONS        =   0
```

NODE	BOUNDARY CONDITION CODES						NODAL POINT COORDINATES		
NUMBER	X	Y	Z	XX	YY	ZZ	X	Y	Z
1	F	F	F	F	F	F	.000	.000	.000
2	R	F	F	F	R	R	.000	60.000	.000
3	R	F	F	F	R	R	.000	120.000	.000
4	F	F	F	F	F	F	.000	.000	3.000
5	R	F	F	F	R	R	.000	120.000	3.000
6	F	F	F	F	F	F	.000	.000	6.000
7	R	F	F	F	R	R	.000	60.000	6.000
8	R	F	F	F	R	R	.000	120.000	6.000

```
EQUATION NUMBERS
```

N	X	Y	Z	XX	YY	ZZ
1	0	0	0	0	0	0
2	1	0	0	0	2	3
3	4	0	0	0	5	6
4	0	0	0	0	0	0
5	7	0	0	0	8	9
6	0	0	0	0	0	0
7	10	0	0	0	11	12
8	13	0	0	0	14	15

```
--------------- SHELL/PLANE/PLATE ELEMENTS ----------

- MATERIAL PROPERTIES -
MATERIAL I.D. NUMBER     =    1
 MODULUS OF ELASTICITY =    .2900E+05
 POISSON'S RATIO       =  .0000
 SHEAR MODULUS         =    .1450E+05
 SELF WEIGHT           =  .0000
```

```
- ELEMENT DEFINITIONS -
EL #  MAT   N1   N2   N3   N4   N5   N6   N7   N8   N9    THICK
   1    1    1    2    3    4    0    5    6    7    8    12.00
UNIFORM LOAD =      .000

THE NODE NUMBERING USED PRODUCED A HALF BANDWIDTH OF      8

TOTAL STORAGE REQUIRED  =        167
TOTAL STORAGE AVAILABLE =      35000

*** CONCENTRATED NODAL LOADS ***
  NODE   LOAD        X           Y           Z          XX          YY          ZZ
    3     1    -.33E+01    .00E+00     .00E+00     .00E+00     .00E+00     .00E+00
    8     1    -.33E+01    .00E+00     .00E+00     .00E+00     .00E+00     .00E+00
    5     1    -.13E+02    .00E+00     .00E+00     .00E+00     .00E+00     .00E+00

SOLUTION CONVERGED IN   1 ITERATION(S)

*** PRINT OF FINAL DISPLACEMENTS ***

DISPLACEMENTS FOR LOAD CONDITION   1
NODE          X             Y             Z            XX            YY            ZZ
  1   .00000E+00    .00000E+00    .00000E+00    .00000E+00    .00000E+00    .00000E+00
  2  -.14506E+00    .00000E+00    .00000E+00    .00000E+00   -.22544E-12    .43103E-02
  3  -.46253E+00    .00000E+00    .00000E+00    .00000E+00   -.45316E-12    .57471E-02
  4   .00000E+00    .00000E+00    .00000E+00    .00000E+00    .00000E+00    .00000E+00
  5  -.46253E+00    .00000E+00    .00000E+00    .00000E+00   -.40111E-14    .57471E-02
  6   .00000E+00    .00000E+00    .00000E+00    .00000E+00    .00000E+00    .00000E+00
  7  -.14506E+00    .00000E+00    .00000E+00    .00000E+00    .22317E-12    .43103E-02
  8  -.46253E+00    .00000E+00    .00000E+00    .00000E+00    .44514E-12    .57471E-02

-------- PLANE/PLATE/SHELL STRESS RESULTS --------
-------ELEMENTS ARE IN THE Y-Z PLANE ------


  EL#  LD#  PNT#     MX          MY          MYX          SXZ         SYZ
-----------------------------------------------------------------------------
   1    1    1    .4000E+03   .2427E-09   .7801E-08   .3333E+01   .2347E-10
             3   -.7792E-10  -.6250E-06  -.3856E-08   .3333E+01   .6249E-06
             8   -.8341E-10  -.6250E-06   .3943E-08   .3333E+01  -.6249E-06
             6    .4000E+03   .2427E-09  -.7790E-08   .3333E+01  -.2603E-10
```

We can see from the output that the tip displacement at nodes 3, 5, and 8 is 0.46253. The beam theory result is 0.45977. The plate answer is greater than beam theory since it includes shear deformations. It still has only a 0.6 percent error from the beam theory solution. In addition, the moment results at the fixed end are 400 kip-in./in. at all three fixed nodes. Multiplying this times the true width, we get the support moment of 2400 kip-in., which is the exact moment.

11.5 Flat Slab Example

As another example using flat plate elements, we will solve a rectangular simply supported plate subjected to a uniform load. The problem being solved is shown in Figure 11.12.

For the finite-element model, we will use symmetry and analyze only one-fourth of the plate. For the load we will use the factors for the equivalent point loads on an eight-node face. The finite-element model used is given in Figure 11.13. Following the element and

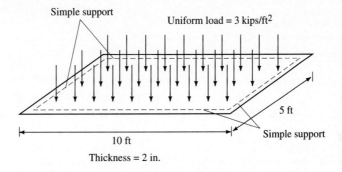

Figure 11.12 Rectangular plate with uniform load.

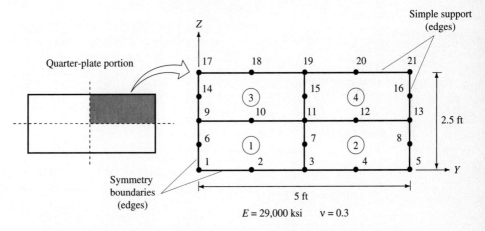

Figure 11.13 Analysis model using one-fourth of the plate.

nodal numbering in Figure 11.13, we can create the required input file. Notice the boundary conditions used for the model. On the simply supported edges only the rotation is released. On the symmetry boundaries, the normal displacement and the in-plane rotation are released. Using SSTAN, we find that the required input file is

```
Rectangular Simply Supported Plate: Uniform Load
21,1,1
:
Coordinates
1 x=0  y=0  z=0
5 y=60  g=1,5
```

```
6 y=0  z=7.5
8 y=60 g=6,8
9 y=0  z=15
13 y=60 g=9,13
14 y=0  z=22.5
16 y=60 g=14,16
17 y=0  z=30
21 y=60 g=17,21
:
Boundary
1,21   DOF=R,F,F,F,R,R : Set all to allowed plate
17,21  DOF=F,F,F,F,R,F : Simple support edge
5,21,8 DOF=F,F,F,F,F,R : Simple support edge
8,16,8 DOF=F,F,F,F,F,R : Simple support edge
21     DOF=F,F,F,F,F,F : Corner has both supports
1,5    DOF=R,F,F,F,F,R : Sym. bound. cond. - X-Y plane
1,17,8 DOF=R,F,F,F,R,F : Sym. bound. cond. - X-Z plane
6,14,8 DOF=R,F,F,F,R,F : Sym. bound. cond. - X-Z plane
1      DOF=R,F,F,F,F,F : Ctr. bound. cond. both planes
:
Plate
4,1
1 E=29000 U=.3
1 N=1,2,3,6,0,7,9,10,11 M=1 H=2
2 N=3,4,5,7,0,8,11,12,13
3 N=9,10,11,14,0,15,17,18,19
4 N=11,12,13,15,0,16,19,20,21
:
Loads
1 1=1 f=.78125
2     f=-3.125
3     f=.78125*2
4     f=-3.125
6     f=-3.125
7     f=-3.125*2
9     f=.78125*2
10    f=-3.125*2
11    f=.78125*4
12    f=-3.125*2
14    f=-3.125
15    f=-3.125*2
```

Looking closely at the input file, we can see how the model is specified. First, note that the structure lies in the Y–Z plane. Next, under Boundary, all nodes are first released in the X translation and Y and Z rotations. These are the only possible DOFs for a flat plate. Next, the simply supported edges are specified. Finally, the symmetry boundary conditions are given. Nodes 1 through 5 are along an X–Y symmetry plane. As a result, the X translation and Z rotation need to be released, and the Y rotation needs to be fixed. Nodes 1, 6, 9, 14, and 17 are along an X–Z symmetry plane. As a result, the X translation and the Y rotation are released and the Z rotation is fixed. Finally, node 1 is in both symmetry planes; therefore, it must adhere to both sets of conditions. As a result, only the X translation is released.

Next, let's look at the loads. The total load on a single element is the pressure times its face area, which comes to 9.375 kips. This total load needs to be distributed to the element nodes according to the ratio $\frac{1}{3}$ for midsides, $-\frac{1}{12}$ to the corner nodes. If we combine the effects for all elements, we get the given loads. If we look at node 11, we get the corner value from four elements $[4(-\frac{1}{12})]$. Node 10 gets the midside value from two elements $[2(\frac{1}{3})]$. Some selected results of running SSTAN are as follows:

```
- ELEMENT DEFINITIONS -
EL #   MAT   N1    N2    N3    N4    N5    N6    N7    N8    N9    THICK
   1    1     1     2     3     6     0     7     9    10    11    2.00
UNIFORM LOAD =      .000
   2    1     3     4     5     7     0     8    11    12    13    2.00
UNIFORM LOAD =      .000
   3    1     9    10    11    14     0    15    17    18    19    2.00
UNIFORM LOAD =      .000
   4    1    11    12    13    15     0    16    19    20    21    2.00
UNIFORM LOAD =      .000

THE NODE NUMBERING USED PRODUCED A HALF BANDWIDTH OF    14
TOTAL STORAGE REQUIRED   =        659
TOTAL STORAGE AVAILABLE =     38000

*** CONCENTRATED NODAL LOADS ***
 NODE  LOAD       X            Y            Z           XX           YY           ZZ
    1    1      .78E+00     .00E+00      .00E+00      .00E+00      .00E+00      .00E+00
    2    1    -.31E+01     .00E+00      .00E+00      .00E+00      .00E+00      .00E+00
    3    1      .16E+01     .00E+00      .00E+00      .00E+00      .00E+00      .00E+00
    4    1    -.31E+01     .00E+00      .00E+00      .00E+00      .00E+00      .00E+00
    6    1    -.31E+01     .00E+00      .00E+00      .00E+00      .00E+00      .00E+00
    7    1    -.63E+01     .00E+00      .00E+00      .00E+00      .00E+00      .00E+00
    9    1      .16E+01     .00E+00      .00E+00      .00E+00      .00E+00      .00E+00
   10    1    -.63E+01     .00E+00      .00E+00      .00E+00      .00E+00      .00E+00
   11    1      .31E+01     .00E+00      .00E+00      .00E+00      .00E+00      .00E+00
   12    1    -.63E+01     .00E+00      .00E+00      .00E+00      .00E+00      .00E+00
   14    1    -.31E+01     .00E+00      .00E+00      .00E+00      .00E+00      .00E+00
   15    1    -.63E+01     .00E+00      .00E+00      .00E+00      .00E+00      .00E+00

SOLUTION CONVERGED IN    1 ITERATION(S)
*** PRINT OF FINAL DISPLACEMENTS ***

DISPLACEMENTS FOR LOAD CONDITION   1
NODE        X              Y              Z             XX             YY             ZZ
    1  -.12924E+00   .00000E+00    .00000E+00    .00000E+00    .00000E+00    .00000E+00
    2  -.12258E+00   .00000E+00    .00000E+00    .00000E+00    .00000E+00   -.95794E-03
    3  -.99312E-01   .00000E+00    .00000E+00    .00000E+00    .00000E+00   -.20920E-02
    4  -.57893E-01   .00000E+00    .00000E+00    .00000E+00    .00000E+00   -.34010E-02
```

```
    5   .54843E-03   .00000E+00   .00000E+00   .00000E+00   .00000E+00  -.42567E-02
    6  -.11998E+00   .00000E+00   .00000E+00   .00000E+00   .25163E-02   .00000E+00
    7  -.92469E-01   .00000E+00   .00000E+00   .00000E+00   .19174E-02  -.19359E-02
    8   .00000E+00   .00000E+00   .00000E+00   .00000E+00   .00000E+00  -.38447E-02
    9  -.92252E-01   .00000E+00   .00000E+00   .00000E+00   .47227E-02   .00000E+00
   10  -.87633E-01   .00000E+00   .00000E+00   .00000E+00   .44566E-02  -.67096E-03
   11  -.71270E-01   .00000E+00   .00000E+00   .00000E+00   .35973E-02  -.14600E-02
   12  -.41291E-01   .00000E+00   .00000E+00   .00000E+00   .21030E-02  -.24743E-02
   13   .00000E+00   .00000E+00   .00000E+00   .00000E+00   .00000E+00  -.28893E-02
   14  -.50578E-01   .00000E+00   .00000E+00   .00000E+00   .62775E-02   .00000E+00
   15  -.39190E-01   .00000E+00   .00000E+00   .00000E+00   .48831E-02  -.80392E-03
   16   .00000E+00   .00000E+00   .00000E+00   .00000E+00   .00000E+00  -.15864E-02
   17  -.58238E-03   .00000E+00   .00000E+00   .00000E+00   .68156E-02   .00000E+00
   18   .00000E+00   .00000E+00   .00000E+00   .00000E+00   .65566E-02   .00000E+00
   19   .00000E+00   .00000E+00   .00000E+00   .00000E+00   .53664E-02   .00000E+00
   20   .00000E+00   .00000E+00   .00000E+00   .00000E+00   .30922E-02   .00000E+00
   21   .00000E+00   .00000E+00   .00000E+00   .00000E+00   .00000E+00   .00000E+00
```

```
-------- PLANE/PLATE/SHELL STRESS RESULTS --------
------- ELEMENTS ARE IN THE Y-Z PLANE ------
```

EL#	LD#	PNT#	MX	MY	MXY	SXZ	SYZ
1	1	1	-.3657E+01	-.8230E+01	-.3508E-02	-.1028E+00	.1410E+00
		3	-.3616E+01	-.6581E+01	-.7318E-02	.2027E+00	-.2889E+00
		11	-.2682E+01	-.5076E+01	.1200E+01	.1478E-01	.3269E+00
		9	-.2765E+01	-.6365E+01	-.1865E-01	.3029E-01	.2475E+00
2	1	3	-.4040E+01	-.6715E+01	-.8178E-02	-.4362E-01	.5758E-01
		5	-.1026E+01	-.5613E+00	.1347E+00	.5197E+00	.3191E+00
		13	-.2981E+00	-.3430E+00	.2406E+01	.2431E+00	-.7338E+00
		11	-.3322E+01	-.5274E+01	.1216E+01	.1575E+00	.5780E+00
3	1	9	-.2700E+01	-.6269E+01	-.1691E-01	-.5698E+00	.3024E+00
		11	-.2715E+01	-.5304E+01	.1148E+01	.4805E+00	.2735E+00
		19	-.1500E+00	-.3974E+00	.1689E+01	-.1027E+01	.5380E+00
		17	-.1057E+00	-.2495E+00	-.1041E+00	.1333E+01	.5574E+00
4	1	11	-.3355E+01	-.5442E+01	.1327E+01	-.1600E+00	-.2402E+00
		13	-.2232E+00	-.1542E+00	.2505E+01	.4703E+00	.9019E+00
		21	-.5983E-01	-.1052E+00	.3609E+01	-.2886E+00	-.3957E+00
		19	-.1296E+00	-.3377E+00	.1955E+01	.5525E+00	.8202E+00

We should first check if our solution has converged. To do this, we need to look at the moments from adjoining elements. When we look at node 7, we see that it is attached between elements 1 and 2. If we look at the moments from both those elements at node 7, we can check the "goodness" of our mesh. From this check we see that the error between elements is still 15 percent. This is probably acceptable if we use an average of these values. However, at the center (node 1), because we used symmetry we cannot average the moments and the result may have too much error. We should consider refining the mesh before completing the design.

Let's compare the results with the theoretical solution. Looking at the center displacement (at node 1 in the X direction), we get 0.1293 in. The theoretical solution is 0.1287 in., a 0.5 percent error. Also, the moment at the center is 8.21 kip-in./in. The theoretical solution is 7.63 kip-in./in., a 7.6 percent error. The displacement is very close; the moment is a little further away.

11.5.1   STANPLOT Stress and Displacement Contours

STANPLOT has the additional capability of plotting stress, moment, and displacement contours for membrane, plate, shell, axisymmetric, and solid elements. This option allows a visual inspection of stresses, moments, and displacement concentrations. The *contours* option on the menu gives the choice of stress (average and local) or displacement. Then an appropriate submenu is given to choose which result to plot. For example, in plate elements the M_{xx} moment can be plotted. The stresses are plotted by two choices, averaging the results from different elements at common nodes or local element values. Averaging gives a slightly false impression of the ''goodness'' of the result. The stress (or moment) difference still needs to be checked to validate the mesh. This can be done using the local stress option. Figure 11.14 is an average stress plot of the *X*-translation contours for the simply supported plate example above.

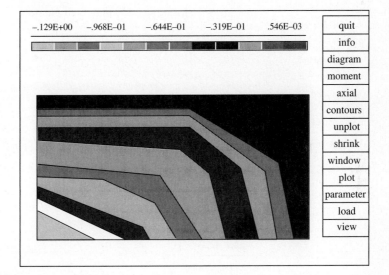

Figure 11.14

11.6   Flat Shell Elements

Shell elements combine the effects of plate bending and in-plane (membrane) effects. There exist formulations for both flat and curved shell elements. The curved element formulation is a much more complicated derivation. The flat shell, however, can be considered to be merely the addition of the membrane and flat plate elements (Figure 11.15). This is the most common form of shell element. The flat shell element can be used to model structures where both bending and stretching effects need to be considered. Many small flat shell elements can be used to form curved surfaces. The modeling of bridge decks, wide-flange beams, and curved shell structures are three such structures where flat shell elements are commonly used (Figure 11.16). In SSTAN we have an element that can act as a shell, plate, or membrane element. Thus SSTAN will automatically

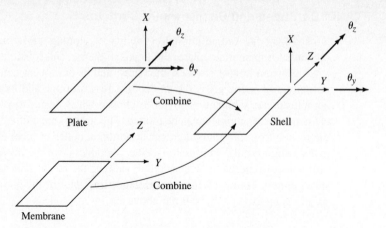

Figure 11.15 Flat shell element.

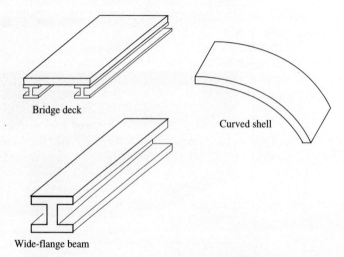

Figure 11.16 Common applications of flat shell elements.

put the effects of both elements at the same locations when the shell is used. The shell element in SSTAN is a variable three- to nine-node element.

11.6.1   Bridge Example

As an example of the use of flat shell elements, let's analyze a bridge structure. This structure simulates a one-lane steel girder bridge with a complete deck. The bridge span is 50 ft. The deck width is one lane but the portion between the girders is only 8 ft. Note that there will be deck overhang past the girders (which will not be included in this simple model). In addition, the load applied will be the equivalent of one HS-20 truck. However, the load will be applied as a uniform load over one of the bridge deck elements (one-fourth of the deck). This is a simplified model used to demonstrate the various modeling techniques needed. In a more accurate model, there would be more deck elements, the

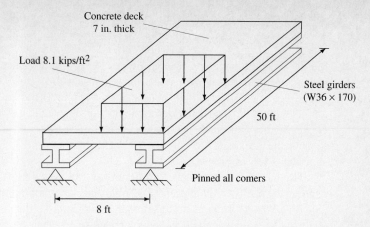

Figure 11.17 Single-lane bridge model.

overhang would be included, and the loading would more accurately represent a true HS-20 truck. The structure we wish to model is shown in Figure 11.17.

The load on the structure is a uniform load on one-fourth of the deck. Because of the nonsymmetric load, we cannot use symmetry. The finite-element discretization we will be using is given in Figure 11.18. Notice that rigid links are used to connect the shells to the beams. This is to account for the difference in the neutral axis of the deck and the girders. SSTAN has rigid-end offsets for beams, so these could have been used. Instead, we will again use small rigid beams to simulate the rigid-end offsets. Notice how the shell element is used to include the bending and in-plane effects of the deck. The in-plane effects are critical when the shift of the neutral axis is included. This is because the deck has an Ad^2 effect on the bending stiffness (moment of inertia), due to its shift from the neutral axis. Recall that the calculation of the moment of inertia of a section consisting of smaller parts includes Ad^2 of a part when that part's neutral axis is shifted a distance d from the total-section neutral axis.

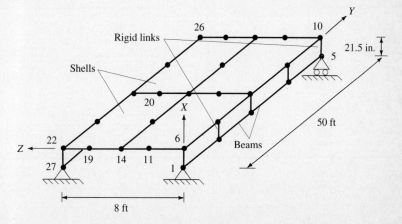

Figure 11.18 Bridge finite-element model.

The same model could be analyzed using rigid-end offsets for the girder elements instead of the rigid links. To do this, we would not have the extra nodes below the deck to which the girders are connected. Instead, the girders would be connected directly to the deck nodes, including an end offset. Using this model can have slightly different results since offsets on the girders mean that supports can be applied only at the deck nodes, not at girder nodes. If the system is not free to expand (e.g., has abutments), the girders cannot develop axial deformations and there will be a difference in the models. Generally, bridge systems are free to expand (have rollers) and they develop axial forces in both the deck and the girders. Comparing the results using rigid ends and pinned supports is left as an exercise.

The model we use is given in Figure 11.18. Notice how the rigid links are used to connect the shells to the girders. The rigid link lengths are the distance from the neutral axis of the deck to the neutral axis of the girders. Also notice the roller supports at one end to allow expansion and the development of in-plane forces in the deck.

The required input file for analysis using SSTAN is as follows:

```
Bridge Model: Using Shells and Rigid Links
33,2,1
:
Coordinates
1 x=-21.5 y=0 z=0      : Girder nodes
5 y=50*12 g=1,5        : Girder nodes
6 x=0 y=0
10 y=50*12 g=6,10
11 y=0 z=2*12
13 y=50*12 g=11,13
14 y=0 z=4*12
18 y=50*12 g=14,18
19 y=0 z=6*12
21 y=50*12 g=19,21
22 y=0 z=8*12
26 y=50*12 g=22,26
27 x=-21.5 y=0              : Girder nodes
31 y=50*12 g=27,31         : Girder nodes
32 y=-1 x=0 z=0            : K node for right girder
33 z=8*12                 : K node for left girder
:
Boundary
1,10 dof=r,r,r,r,r,r      : Girder and deck connection nodes
11,21 dof=r,r,r,f,r,r     : Deck (no girder) - fix X rotation
22,31 dof=r,r,r,r,r,r     : Girder and deck connection nodes
1,27,26 dof=f,f,f,r,r,r   : Pin support
5,31,26 dof=f,r,r,r,r,r   : Roller supports
:
Shell
4,1
1 e=3000 u=.2
```

```
1 n=6,7,8,11,0,12,14,15,16 h=7 m=1
2 n=8,9,10,12,0,13,16,17,18
3 n=14,15,16,19,0,20,22,23,24
4 n=16,17,18,20,0,21,24,25,26
:

Beam
18,2
1 a=50 i=10500,320 e=29000 j=4        : W36x170
2 a=50 i=10500,320 e=29000000 j=4     : Links - rigid
1 1,2,6 m=1 g=3,1,1,1,0               : Girders
5 1,6,32 m=2 g=4,1,1,1,0              : Links
10 27,28,22 m=1 g=3,1,1,1,0           : Girders
14 27,22,33 m=2 g=4,1,1,1,0           : Links
:

Loads
6,8,2 l=1 f=6.75
7          f=-27
11,12      f=-27
14,16,2    f=6.75
15         f=-27
:

CLOADS    : Symmetry load case - modeling check
16 l=1 f=-50
:
```

Looking at the input file, note that three types of elements are used. The rigid link beam elements have properties 1000 times that of the girder to which they connect. The load must be distributed to the nodes according to uniform load distribution on an eight-node face. The boundary conditions for the nodes to which the girders are connected have all six DOFs released. Also note how at the deck nodes where the rigid links connect, all six DOFs are also released. This is required so that the girder can bend outward. Figure 11.19 shows the deformed plot of the structure.

The other deck nodes, those with no rigid links connected, have the X rotation fixed. This is because neither the plate nor the membrane has normal rotational stiffness. Finally, note that we have commented out an additional load case, a single point load at the center of the deck. This was used as a check of the symmetry and modeling. This load case was run and the results checked for symmetry of displacements, stresses, and moments. If all results are symmetric, we have more confidence in our model. This model showed very good symmetry. Next, we use the true loading to create the results, of which selected parts are given below.

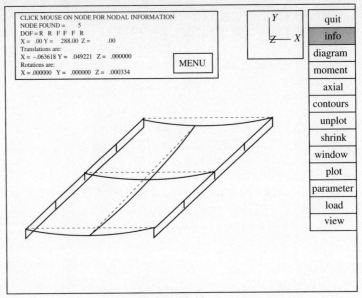

Figure 11.19

```
************************************************************
****  SSTAN  -  Simple Structural Analysis Program  ****
Copyright 1993  By  Dr. Marc Hoit, University of Florida
************************************************************
```

```
Bridge Model: Using Shells and Rigid Links
NUMBER OF JOINTS                    =  33
NUMBER OF DIFFERENT ELEMENT TYPES =  2
NUMBER OF LOAD CONDITIONS          =  1
NUMBER OF LOAD COMBINATIONS        =  0

***** FRAME MEMBERS *****
NUMBER OF MEMBER PROPERTIES =    2
MEMBER PROPERTY NUMBER --------- =   1
AXIAL AREA, A ------------------ =        50.0000
SHEAR AREA, 3 Axis ------------- =          .0000
SHEAR AREA, 2 Axis ------------- =          .0000
TORSIONAL MOMENT OF INERTIA, J  =          4.0000
MOMENT OF INERTIA, I(2) -------- =     10500.0000
MOMENT OF INERTIA, I(3) -------- =       320.0000
MODULUS OF ELASTICITY, E ------- =  .290000E+05
SHEAR MODULUS, G --------------- =  .111538E+05(USED FOR JG/L CAL.)

MEMBER PROPERTY NUMBER --------- =  2
AXIAL AREA, A ------------------ =        50.0000
SHEAR AREA, 3 Axis ------------- =          .0000
SHEAR AREA, 2 Axis ------------- =          .0000
TORSIONAL MOMENT OF INERTIA, J  =          4.0000
MOMENT OF INERTIA, I(2) -------- =     10500.0000
MOMENT OF INERTIA, I(3) -------- =       320.0000
MODULUS OF ELASTICITY, E ------- =  .290000E+08
SHEAR MODULUS, G --------------- =  .111538E+05(USED FOR JG/L CAL.)
```

```
MEMB  CONNECTIVITY MATERIAL                END ECCENTRICITIES
NUM. K=0 FOR Z-AXIS  SET  *****   I-END  ****   ****  J-END   *****  PIN
       I    J    K       X-VALUE Y-VALUE Z-VALUE X-VALUE Y-VALUE Z-VALUE OPT
   1    1    2    6    1    .00     .00     .00     .00     .00     .00   0
UNIFORM LOAD =      .000
   2    2    3    6    1    .00     .00     .00     .00     .00     .00   0
UNIFORM LOAD =      .000
   3    3    4    6    1    .00     .00     .00     .00     .00     .00   0
UNIFORM LOAD =      .000
   4    4    5    6    1    .00     .00     .00     .00     .00     .00   0
UNIFORM LOAD =      .000
   5    1    6   32    2    .00     .00     .00     .00     .00     .00   0
UNIFORM LOAD =      .000
   6    2    7   32    2    .00     .00     .00     .00     .00     .00   0
UNIFORM LOAD =      .000
   7    3    8   32    2    .00     .00     .00     .00     .00     .00   0
UNIFORM LOAD =      .000
   8    4    9   32    2    .00     .00     .00     .00     .00     .00   0
UNIFORM LOAD =      .000
   9    5   10   32    2    .00     .00     .00     .00     .00     .00   0
UNIFORM LOAD =      .000
  10   27   28   22    1    .00     .00     .00     .00     .00     .00   0
UNIFORM LOAD =      .000
  11   28   29   22    1    .00     .00     .00     .00     .00     .00   0
UNIFORM LOAD =      .000
  12   29   30   22    1    .00     .00     .00     .00     .00     .00   0
UNIFORM LOAD =      .000
  13   30   31   22    1    .00     .00     .00     .00     .00     .00   0
UNIFORM LOAD =      .000
  14   27   22   33    2    .00     .00     .00     .00     .00     .00   0
UNIFORM LOAD =      .000
  15   28   23   33    2    .00     .00     .00     .00     .00     .00   0
UNIFORM LOAD =      .000
  16   29   24   33    2    .00     .00     .00     .00     .00     .00   0
UNIFORM LOAD =      .000
  17   30   25   33    2    .00     .00     .00     .00     .00     .00   0
UNIFORM LOAD =      .000
  18   31   26   33    2    .00     .00     .00     .00     .00     .00   0
UNIFORM LOAD =      .000
--------------- SHELL/PLANE/PLATE ELEMENTS ----------
- MATERIAL PROPERTIES -
MATERIAL I.D. NUMBER   =    1
 MODULUS OF ELASTICITY =   .3000E+04
 POISSON'S RATIO       = .2000
 SHEAR MODULUS         =   .1250E+04
 SELF WEIGHT           = .0000
- ELEMENT DEFINITIONS -
EL # MAT  N1   N2   N3   N4   N5   N6   N7   N8   N9    THICK
   1    1    6    7    8   11    0   12   14   15   16    7.00
UNIFORM LOAD =      .000
   2    1    8    9   10   12    0   13   16   17   18    7.00
UNIFORM LOAD =      .000
   3    1   14   15   16   19    0   20   22   23   24    7.00
UNIFORM LOAD =      .000
   4    1   16   17   18   20    0   21   24   25   26    7.00
UNIFORM LOAD =      .000
```

```
THE NODE NUMBERING USED PRODUCED A HALF BANDWIDTH OF    34

TOTAL STORAGE REQUIRED  =    6154
TOTAL STORAGE AVAILABLE =    38000
```

*** CONCENTRATED NODAL LOADS ***

NODE	LOAD	X	Y	Z	XX	YY	ZZ
6	1	.68E+01	.00E+00	.00E+00	.00E+00	.00E+00	.00E+00
8	1	.68E+01	.00E+00	.00E+00	.00E+00	.00E+00	.00E+00
7	1	-.27E+02	.00E+00	.00E+00	.00E+00	.00E+00	.00E+00
11	1	-.27E+02	.00E+00	.00E+00	.00E+00	.00E+00	.00E+00
12	1	-.27E+02	.00E+00	.00E+00	.00E+00	.00E+00	.00E+00
14	1	.68E+01	.00E+00	.00E+00	.00E+00	.00E+00	.00E+00
16	1	.68E+01	.00E+00	.00E+00	.00E+00	.00E+00	.00E+00
15	1	-.27E+02	.00E+00	.00E+00	.00E+00	.00E+00	.00E+00

```
SOLUTION CONVERGED IN   1 ITERATION(S)
```

*** PRINT OF FINAL DISPLACEMENTS ***

DISPLACEMENTS FOR LOAD CONDITION 1

NODE	X	Y	Z	XX	YY	ZZ
1	.00000E+00	.00000E+00	.00000E+00	-.23654E-03	-.34067E-03	.37439E-03
2	-.17675E-01	.34680E-03	-.82204E-02	.16385E-04	-.12239E-02	.22669E-03
3	-.22485E-01	.10835E-02	-.11370E-02	.10783E-03	-.55176E-03	-.40316E-04
4	-.14714E-01	.15987E-02	.85091E-03	.56260E-04	-.37575E-03	-.22743E-03
5	.00000E+00	.16880E-02	.78457E-02	.16617E-03	.48526E-03	-.29461E-03
6	-.82881E-07	.37439E-02	.34069E-02	-.23654E-03	-.34074E-03	.37439E-03
7	-.17675E-01	.26137E-02	.40189E-02	.16385E-04	-.12240E-02	.22669E-03
8	-.22485E-01	.68031E-03	.43806E-02	.10783E-03	-.55174E-03	-.40316E-04
9	-.14714E-01	-.67563E-03	.46085E-02	.56260E-04	-.37577E-03	-.22743E-03
10	-.56003E-08	-.12580E-02	.29931E-02	.16617E-03	.48526E-03	-.29461E-03
11	-.17417E-01	.27783E-02	.29502E-02	.00000E+00	-.48478E-03	.28244E-03
12	-.35915E-01	.72979E-03	.46469E-02	.00000E+00	-.16516E-03	-.23336E-03
13	.10097E-02	-.47841E-03	.32884E-02	.00000E+00	.23446E-03	-.23178E-03
14	-.22371E-01	.20971E-02	.25235E-02	.00000E+00	.34576E-03	.12379E-03
15	-.36881E-01	.17695E-02	.40555E-02	.00000E+00	.32406E-03	.26742E-03
16	-.37403E-01	.68426E-03	.48274E-02	.00000E+00	.40299E-03	-.31874E-03
17	-.15509E-01	.10496E-02	.44345E-02	.00000E+00	.13163E-03	-.41481E-03
18	-.14076E-02	-.46445E-04	.36298E-02	.00000E+00	.23298E-04	-.32253E-04
19	-.12029E-01	.16481E-02	.19665E-02	.00000E+00	.58462E-03	.17003E-03
20	-.25894E-01	.62914E-03	.48919E-02	.00000E+00	.68869E-03	-.18595E-03
21	-.10365E-03	.11160E-03	.38599E-02	.00000E+00	-.13713E-03	-.10385E-03
22	-.30201E-07	.11704E-02	.13603E-02	.38340E-03	-.13599E-03	.11704E-03
23	-.55694E-02	.94473E-03	.41330E-02	.10268E-03	.11518E-02	.71981E-04
24	-.71137E-02	.56025E-03	.49837E-02	-.79930E-04	.78716E-03	-.10885E-04
25	-.47409E-02	.24885E-03	.43094E-02	-.95340E-04	.57184E-03	-.72344E-04
26	-.15460E-08	.61828E-04	.40502E-02	-.20885E-03	-.32574E-03	-.95521E-04
27	.00000E+00	.00000E+00	.00000E+00	.38340E-03	-.13605E-03	.11704E-03
28	-.55693E-02	.22493E-03	.15651E-01	.10268E-03	.11517E-02	.71980E-04
29	-.71137E-02	.66910E-03	.12855E-01	-.79931E-04	.78718E-03	-.10886E-04
30	-.47409E-02	.97228E-03	.10028E-01	-.95340E-04	.57183E-03	-.72343E-04
31	.00000E+00	.10170E-02	.79288E-03	-.20885E-03	-.32573E-03	-.95521E-04
32	.00000E+00	.00000E+00	.00000E+00	.00000E+00	.00000E+00	.00000E+00
33	.00000E+00	.00000E+00	.00000E+00	.00000E+00	.00000E+00	.00000E+00

```
---------------------- FRAME MEMBER RESULTS ----------------
MEM LOAD NODE        1-2 PLANE              1-3 PLANE      AXIAL FORCE
 #    #    #      MOMENT      SHEAR      MOMENT      SHEAR
 1    1    1   .49673E-05  .16098E+01 -.72903E+02  .33546E+02 -.93121E+01
           2   .86931E+02 -.16098E+01 -.17386E+04 -.33546E+02  .93121E+01
                              AXIAL TORQUE =  -.72975E+00
 2    1    2  -.86931E+02 -.26376E+01  .16449E+04 -.51583E+01 -.19781E+02
           3  -.55501E+02  .26376E+01 -.13664E+04  .51583E+01  .19781E+02
                              AXIAL TORQUE =   .55535E+00
 3    1    3   .55501E+02  .17274E+01  .13942E+04 -.12560E+02 -.13834E+02
           4   .37776E+02 -.17274E+01 -.71599E+03  .12560E+02  .13834E+02
                              AXIAL TORQUE =   .14542E+00
 4    1    4  -.37776E+02 -.69956E+00  .76691E+03 -.14374E+02 -.23995E+01
           5  -.34895E-05  .69956E+00  .93060E+01  .14374E+02  .23995E+01
                              AXIAL TORQUE =   .71138E+00
 5    1    1   .72975E+00  .12032E+02  .72903E+02 -.93121E+01  .12018E+02
           6   .11959E+03 -.12032E+02  .20218E+02  .93121E+01 -.12018E+02
                              AXIAL TORQUE =   .49673E-05
 6    1    2  -.12851E+01  .42474E+01  .93661E+02 -.10469E+02  .38704E+02
           7   .43760E+02 -.42474E+01  .11026E+02  .10469E+02 -.38704E+02
                              AXIAL TORQUE =  -.34409E-06
 7    1    3   .40993E+00 -.43650E+01 -.27873E+02  .59467E+01  .74017E+01
           8  -.44060E+02  .43650E+01 -.31594E+02 -.59467E+01 -.74017E+01
                              AXIAL TORQUE =  -.45288E-05
 8    1    4  -.56596E+00  .24269E+01 -.50913E+02  .11435E+02  .18143E+01
           9   .24835E+02 -.24269E+01 -.63433E+02 -.11435E+02 -.18143E+01
                              AXIAL TORQUE =  -.11815E-05
 9    1    5   .71138E+00 -.69956E+00 -.93060E+01  .23995E+01  .81205E+00
          10  -.77070E+01  .69956E+00 -.14689E+02 -.23995E+01 -.81205E+00
                              AXIAL TORQUE =  -.34895E-05
10    1   27  -.80513E-05 -.17867E+01 -.37690E+02  .10807E+02 -.60398E+01
          28  -.96483E+02  .17867E+01 -.54590E+03 -.10807E+02  .60398E+01
                              AXIAL TORQUE =   .10640E+01
11    1   28   .96483E+02  .24111E+01  .53317E+03 -.24409E+01 -.11927E+02
          29   .33718E+02 -.24111E+01 -.40137E+03  .24409E+01  .11927E+02
                              AXIAL TORQUE =  -.30121E+00
12    1   29  -.33718E+02 -.13469E+01  .42529E+03 -.29163E+01 -.81410E+01
          30  -.39014E+02  .13469E+01 -.26781E+03  .29163E+01  .81410E+01
                              AXIAL TORQUE =  -.17793E+00
13    1   30   .39014E+02  .72249E+00  .26137E+03 -.48395E+01 -.12018E+01
          31   .43859E-05 -.72249E+00 -.34809E-01  .48395E+01  .12018E+01
                              AXIAL TORQUE =  -.74157E+00
14    1   27  -.10640E+01 -.11855E+02  .37690E+02 -.60398E+01  .43792E+01
          22  -.11749E+03  .11855E+02  .22708E+02  .60398E+01 -.43792E+01
                              AXIAL TORQUE =  -.80513E-05
15    1   28   .13652E+01 -.41979E+01  .12723E+02 -.58870E+01  .13248E+02
          23  -.43344E+02  .41979E+01  .46147E+02  .58870E+01 -.13248E+02
                              AXIAL TORQUE =  -.21563E-05
16    1   29  -.12328E+00  .37580E+01 -.23923E+02  .37858E+01  .47543E+00
          24   .37704E+02 -.37580E+01 -.13935E+02 -.37858E+01 -.47543E+00
                              AXIAL TORQUE =   .33571E-05
17    1   30   .56365E+00 -.20694E+01  .64437E+01  .69392E+01  .19232E+01
          25  -.21258E+02  .20694E+01 -.75836E+02 -.69392E+01 -.19232E+01
                              AXIAL TORQUE =   .20021E-05
```

```
 18     1    31 -.74157E+00   .72249E+00   .34809E-01   .12018E+01   .22417E+00
            26  .79665E+01  -.72249E+00  -.12053E+02  -.12018E+01  -.22417E+00
                                                    AXIAL TORQUE =    .43859E-05

-------- PLANE/PLATE/SHELL STRESS RESULTS --------
------- ELEMENTS ARE IN THE Y-Z PLANE ------
   EL#  LD#  PNT#      SX            SY           SXY
                       MX            MY           MXY            SXZ           SYZ
----------------------------------------------------------------------------------
    1    1     6  -.5213E-01    -.7533E-01   -.5994E-01
               8  -.1180E+00     .4167E-01    .1213E-01
              16  -.7354E-01     .2196E-01    .2527E-02
              14  -.4923E-02    -.5585E-01    .6752E-02
               6   .2010E+00     .1928E+01    .1094E+01   -.6150E-01   -.1022E+01
               8  -.1114E+01    -.2752E+01   -.1338E+01   -.3475E-01    .3102E+00
              16  -.2507E+01    -.4761E+01   -.3712E-01    .7355E-01   -.5127E+00
              14  -.7467E+00    -.7547E+01   -.2224E+00   -.5327E-01    .5338E+00
    2    1     8  -.9578E-01    -.6181E-02    .3673E-01
              10  -.7541E-02     .5296E-02    .1087E-01
              18   .1061E-01     .2432E-01   -.2083E-02
              16  -.4680E-01    -.2498E-01   -.7403E-02
               8  -.7938E+00    -.1970E+01   -.5101E+00    .7915E-01   -.7346E+00
              10   .1652E+00     .9203E+00   -.9376E+00    .2963E-01    .3211E+00
              18   .1127E+01     .7351E+00    .4193E+00   -.6785E-02   -.3558E+00
              16  -.1300E+01    -.3801E+01    .4059E-01    .7912E-01    .7097E+00
    3    1    14  -.9148E-02    -.5928E-01    .7064E-02
              16  -.7657E-01     .2448E-01   -.1289E-02
              24  -.1681E-01     .4550E-01   -.1482E-01
              22  -.2398E-01    -.7862E-01    .4526E-01
              14  -.1910E+00    -.4811E+01    .9849E-01   -.1116E+00   -.8824E-01
              16  -.2269E+01    -.3610E+01   -.3272E-02    .9510E-01    .7134E+00
              24  -.4922E+00    -.1471E+01    .1038E+01   -.1526E+00   -.4581E+00
              22   .8290E+00     .4534E+01   -.1742E+01    .3703E-01    .5306E+00
    4    1    16  -.4696E-01    -.2317E-01   -.4020E-02
              18   .9473E-02     .2128E-01   -.1546E-02
              26  -.4133E-02     .5299E-02   -.8208E-02
              24  -.2079E-01    -.8872E-02   -.2202E-01
              16  -.1077E+01    -.2673E+01    .6134E+00    .8283E-01   -.4934E+00
              18   .1028E+01     .2520E+00   -.4545E-01   -.2402E-01    .2432E+00
              26   .4885E-01     .3246E+00    .1074E+01    .6839E-02   -.2387E+00
              24  -.2838E+00    -.7307E+00    .4645E+00    .7602E-01    .5412E+00
```

We need to check the accuracy of our mesh and chosen link elements. To check the link element, we need to have the rotations the same at both ends of each link. Looking at nodes 3 and 8, we see that they rotate the same amount. To check the accuracy of the mesh, we need to look at the stress and the moment from two adjoining elements at a common node. For example, we can look at node number 16 from elements 1 and 2. Element 1 gives an M_{xx} value of −2.507; element 2's value is −1.503. These are not in very good agreement. Therefore, the deck mesh needs to be refined and the analysis rerun. However, this will not be done for this example since the modeling concepts for this type of structure have been explained sufficiently.

11.7 Problems

11.1. Given the following plate structure, find the maximum moment in the plate and the maximum deflection. The plate is 8 in. thick and is made of concrete. The plate is fixed on one edge, simply supported on another, and two edges are free.
  (a) Use a loading of 25 lb/ft².
  (b) Use a line load along both free edges of 75 lb/ft.

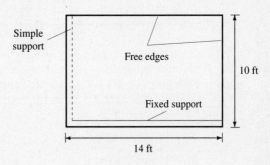

11.2. Given the following plate structure, find the maximum moment in the plate. The plate is 2.5 in. thick and made of steel. The plate is simply supported on three sides.
  (a) Use a line load on the free edge of 35 lb/ft.
  (b) Use a uniformly distributed load of 13 lb/ft².

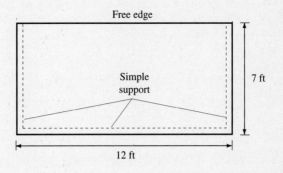

11.3. Given the following plate structure, find the maximum moment in the plate. The plate has a hole in the middle and is loaded with a uniform load on the shaded area. The plate is made of steel and is 4 in. thick.

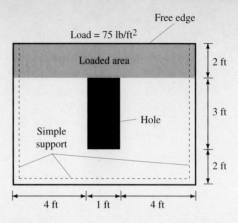

11.4. Given the following plate structure, find the maximum moment in the plate. The plate is made of steel and is 2.5 in. thick. Use a uniform load of 45 lb/ft².

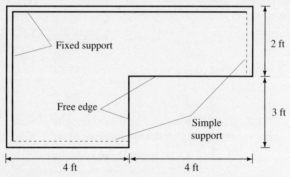

11.5. The following bracket needs to be checked to see if it will withstand the surface load shown. The surface load acts only on the 5-in. arm. The bracket is welded only on the two sides. Find the maximum stress in the bracket due to the load.

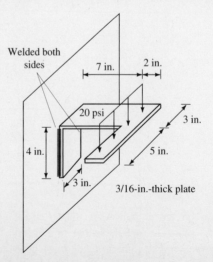

11.6. The following cantilever I-beam is subjected to a triangular tip load. The beam needs to be checked for the maximum stress due to the combined effects of beam action and torsion. Model the flanges and web as shell elements and find the maximum stress.

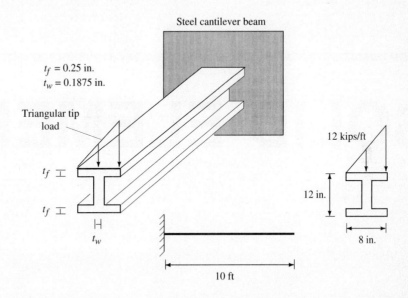

11.7. Reanalyze the bridge structure from the example in this chapter using rigid-end offsets instead of rigid links. Compare the results with those in the example.

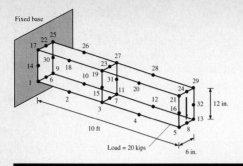

SOLID ELEMENTS

12.1   Introduction

The final element we look at is the three-dimensional solid element. This element is the most general of the finite elements. It fully represents a three-dimensional stress and strain state. It is a fundamental building block that can be used to create a structure of any shape. Clearly, as a structure becomes more complex, the effort required to define the geometry and mesh becomes extremely time consuming. As a result of the required sophistication of many analyses, many people are moving to computer-based solid geometry modeling of structures. This especially includes mechanical components. Solid elements are a natural choice for this type of modeling since any solid object can be meshed by solid elements.

12.2   Solid Element Behavior and DOFs

Solid elements are found in varied configurations analogous to membrane elements. They can be found in 4-node tetrahedrons, 8-node bricks, 20-node bricks, and 28-node bricks. Of course, variable-node versions exist that allow from 8 to 28 nodes. Figure 12.1 shows some of these configurations.

  The solid element is a three-dimensional analog of the membrane element. It has three DOFs per node. It does not have any rotational stiffness at the nodes. The possible DOFs for a typical node of a solid element are shown in Figure 12.2. The solid element is capable of representing a full three-dimensional stress state. The stress results from the element are shown in Figure 12.3. The properties required for the solid element consist of Young's modulus, E, and Poisson's ratio, v. Note that no thickness is required as for the membrane element since these are three-dimensional elements. Solid elements model the full three-dimensional effects of Hooke's law.

12.3   Modeling Capabilities and Results

The displacement and stress modeling capabilities of solid elements are similar to those of the membrane elements. The four-node tetrahedron is a linear displacement constant-stress element. The eight-node solid is slightly better than linear and constant. The 20- and 28-node elements have a little better than quadratic displacements and linear stress. Of course, the 28-node element is a little better than the 20-node element. Serendipity

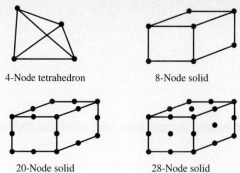

4-Node tetrahedron   8-Node solid

20-Node solid   28-Node solid

Figure 12.1 Solid element configurations.

Figure 12.2 Displacement DOFs for solid elements.

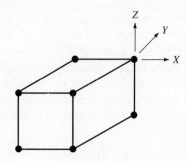

Figure 12.3 Solid element stress results.

elements, elements with between 8 and 28 nodes, are somewhere in between in terms of accuracy. SSTAN has a variable 8- to 20-node element.

12.4 Solid Element Loads

Solid elements can, of course, have concentrated loads applied to any of their nodes. However, more commonly a surface load needs to be applied. Like other distributed loads, this needs to be converted into an equivalent set of concentrated loads. For this conversion we again use finite-element equation (9.4). The nodal distribution values for a uniform

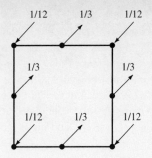

Figure 12.4 Equivalent concentrated loads for uniform surface loads on face of 20-node brick.

surface load on the face of a solid brick element are given on Figure 12.4. Note that these are the same as for a uniform load on an eight-node plate element, and that the corner node loads for the brick element are in the direction opposite that of the actual applied loads (and center node loads). The total surface load is: surface area × load. The midside nodes have $\frac{1}{3}$ of the total surface load applied in the direction of the applied load. The four corner nodes have $\frac{1}{12}$ of the total surface load applied in the direction opposite that of the applied surface load.

12.5 Solid Element Cantilever Example

As an example of how to use solid elements, we will model a cantilever beam, the same example as that used for the membrane element. For this model we will use two 20-node solid elements. The original structure is given in Figure 9.10. The model, including node and element numbering, is given in Figure 12.5. Notice that the support is fixed as in a

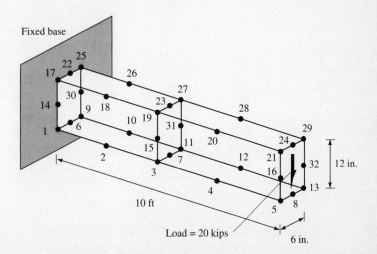

Figure 12.5 Solid element cantilever beam example.

true cantilever beam. However, as was the case in the membrane, the solid element includes Poisson's effects. Therefore, a fully fixed base will cause an additional constraint on the beam and reduce the displacements. To avoid this and try to match better with beam theory, we will use a Poisson's ratio of zero. In a true structure, the Poisson's effect would exist and the true Poisson's ratio should be used. We will use the 20-node solid element to get better accuracy. The tip load needs to be distributed to the nodes to model beam theory more closely. The distribution factors given for the eight-node face of the 20-node element are used. The input file required is as follows:

```
Cantilever: Solid Elements
32,1,1
:
Coordinates
1 x=0 y=0 z=0
5 x=10*12 g=1,5
6 x=0 y=3
8 x=10*12 g=6,8
9 x=0 y=6
13 x=10*12 g=9,13
14 x=0 y=0 z=6
16 x=10*12 g=14,16
17 x=0 z=12
21 x=10*12 g=17,21
22 x=0 y=3
24 x=10*12 g=22,24
25 x=0 y=6
29 x=10*12 g=25,29
30 x=0 z=6
32 x=10*12 g=30,32
:
Boundary
1,32 DOF=R,R,R,F,F,F
1 DOF=F,F,F,F,F,F
6 DOF=F,F,F,F,F,F
9 DOF=F,F,F,F,F,F
14 DOF=F,F,F,F,F,F
17 DOF=F,F,F,F,F,F
22 DOF=F,F,F,F,F,F
25 DOF=F,F,F,F,F,F
30 DOF=F,F,F,F,F,F
:
Brick
2,1
1 E=29000 U=0  : Poisson's=0 for beam theory
1 N=1,3,11,9,17,19,27,25,2,7,10,6,14,15,31,30,18,23,26,22  M=1
2 N=3,5,13,11,19,21,29,27,4,8,12,7,15,16,32,31,20,24,28,23 M=1
:
```

```
Loads
5,13,8    L=1 F=0,0,20/12   :-1/12 to corners
21,29,8   L=1 F=0,0,20/12   :-1/12 to corners
8,32,8    L=1 F=0,0,-20/3   : 1/3 to midsides
:
```

Note that the signs of the tip loads match the eight-node distribution pattern. Also note how for the boundary conditions only the *X*, *Y*, and *Z* translational DOFs are released and all rotational DOFs are fixed. For the brick element, if all 20 nodes must be specified for each element, this can be very time consuming and error prone. Plotting is very important for model verification. Also, improper connectivity or bad nodal coordinates can be caught by the Jacobian mapping operator. A negative on the diagonal or singular Jacobian indicates these types of problems. Selected output from the OUTPUT file created from running SSTAN follows.

```
NUMBER OF MEMBER PROPERTIES =   1

* * * * * INFORMATION FOR BRICK ELEMENT #   1 * * * * *

MAT #   1 E =  .290000E+05 POISSON = .0000
 N1  N2  N3  N4  N5  N6  N7  N8  N9 N10 N11 N12 N13 N14 N15 N16 N17 N18 N19 N20
  1   3  11   9  17  19  27  25   2   7  10   6  14  15  31  30  18  23  26  22

* * * * * INFORMATION FOR BRICK ELEMENT #   2 * * * * *

MAT #   1 E =  .290000E+05 POISSON = .0000
 N1  N2  N3  N4  N5  N6  N7  N8  N9 N10 N11 N12 N13 N14 N15 N16 N17 N18 N19 N20
  3   5  13  11  19  21  29  27   4   8  12   7  15  16  32  31  20  24  28  23

THE NODE NUMBERING USED PRODUCED A HALF BANDWIDTH OF   35

TOTAL STORAGE REQUIRED   =    2724
TOTAL STORAGE AVAILABLE  =   35000

*** CONCENTRATED NODAL LOADS ***
 NODE  LOAD        X            Y            Z           XX          YY          ZZ
    5    1    .00E+00      .00E+00      .17E+01     .00E+00     .00E+00     .00E+00
   13    1    .00E+00      .00E+00      .17E+01     .00E+00     .00E+00     .00E+00
   21    1    .00E+00      .00E+00      .17E+01     .00E+00     .00E+00     .00E+00
   29    1    .00E+00      .00E+00      .17E+01     .00E+00     .00E+00     .00E+00
    8    1    .00E+00      .00E+00     -.67E+01     .00E+00     .00E+00     .00E+00
   16    1    .00E+00      .00E+00     -.67E+01     .00E+00     .00E+00     .00E+00
   24    1    .00E+00      .00E+00     -.67E+01     .00E+00     .00E+00     .00E+00
   32    1    .00E+00      .00E+00     -.67E+01     .00E+00     .00E+00     .00E+00

SOLUTION CONVERGED IN   1 ITERATION(S)

*** PRINT OF FINAL DISPLACEMENTS ***

DISPLACEMENTS FOR LOAD CONDITION   1
NODE        X            Y            Z           XX          YY          ZZ
   1   .00000E+00   .00000E+00   .00000E+00   .00000E+00   .00000E+00   .00000E+00
   2  -.15340E-01  -.20619E-10  -.40725E-01   .00000E+00   .00000E+00   .00000E+00
```

```
 3 -.26384E-01 -.24379E-10 -.14741E+00  .00000E+00  .00000E+00  .00000E+00
 4 -.32929E-01  .42066E-10 -.29807E+00  .00000E+00  .00000E+00  .00000E+00
 5 -.35179E-01  .14842E-09 -.47072E+00  .00000E+00  .00000E+00  .00000E+00
 6  .00000E+00  .00000E+00  .00000E+00  .00000E+00  .00000E+00  .00000E+00
 7 -.26384E-01 -.58031E-12 -.14741E+00  .00000E+00  .00000E+00  .00000E+00
 8 -.35179E-01 -.13115E-11 -.47072E+00  .00000E+00  .00000E+00  .00000E+00
 9  .00000E+00  .00000E+00  .00000E+00  .00000E+00  .00000E+00  .00000E+00
10 -.15340E-01  .20232E-10 -.40725E-01  .00000E+00  .00000E+00  .00000E+00
11 -.26384E-01  .23218E-10 -.14741E+00  .00000E+00  .00000E+00  .00000E+00
12 -.32929E-01 -.43990E-10 -.29807E+00  .00000E+00  .00000E+00  .00000E+00
13 -.35179E-01 -.15105E-09 -.47072E+00  .00000E+00  .00000E+00  .00000E+00
14  .00000E+00  .00000E+00  .00000E+00  .00000E+00  .00000E+00  .00000E+00
15 -.38264E-13 -.51442E-12 -.14741E+00  .00000E+00  .00000E+00  .00000E+00
16 -.32725E-13 -.12269E-11 -.47072E+00  .00000E+00  .00000E+00  .00000E+00
17  .00000E+00  .00000E+00  .00000E+00  .00000E+00  .00000E+00  .00000E+00
18  .15340E-01  .20292E-10 -.40725E-01  .00000E+00  .00000E+00  .00000E+00
19  .26384E-01  .23339E-10 -.14741E+00  .00000E+00  .00000E+00  .00000E+00
20  .32929E-01 -.43849E-10 -.29807E+00  .00000E+00  .00000E+00  .00000E+00
21  .35179E-01 -.15090E-09 -.47072E+00  .00000E+00  .00000E+00  .00000E+00
22  .00000E+00  .00000E+00  .00000E+00  .00000E+00  .00000E+00  .00000E+00
23  .26384E-01 -.45204E-12 -.14741E+00  .00000E+00  .00000E+00  .00000E+00
24  .35179E-01 -.11499E-11 -.47072E+00  .00000E+00  .00000E+00  .00000E+00
25  .00000E+00  .00000E+00  .00000E+00  .00000E+00  .00000E+00  .00000E+00
26  .15340E-01 -.20552E-10 -.40725E-01  .00000E+00  .00000E+00  .00000E+00
27  .26384E-01 -.24243E-10 -.14741E+00  .00000E+00  .00000E+00  .00000E+00
28  .32929E-01  .42229E-10 -.29807E+00  .00000E+00  .00000E+00  .00000E+00
29  .35179E-01  .14860E-09 -.47072E+00  .00000E+00  .00000E+00  .00000E+00
30  .00000E+00  .00000E+00  .00000E+00  .00000E+00  .00000E+00  .00000E+00
31  .38779E-13 -.51804E-12 -.14741E+00  .00000E+00  .00000E+00  .00000E+00
32  .34261E-13 -.12342E-11 -.47072E+00  .00000E+00  .00000E+00  .00000E+00
```

```
* * * * BRICK ELEMENT STRESSES * * * *

* * * * STRESSES FOR BRICK ELEMENT #    1
LC NODE    Sxx         Syy         Szz         Sxy         Syz         Szx
 1    1 -.1690E+02   .5481E-07  -.2844E-12  -.3717E-06   .1370E-07  -.2923E+00
 1    3 -.8601E+01   .2848E-06  -.1268E-05  -.3717E-06  -.6460E-06  -.2923E+00
 1   11 -.8601E+01   .2848E-06  -.1268E-05   .3717E-06   .6460E-06  -.2923E+00
 1    9 -.1690E+02   .5481E-07   .1138E-11   .3717E-06  -.1370E-07  -.2923E+00
 1   17  .1690E+02  -.5481E-07   .6002E-12   .3717E-06   .1370E-07  -.2923E+00
 1   19  .8601E+01  -.2848E-06   .1268E-05   .3717E-06  -.6460E-06  -.2923E+00
 1   27  .8601E+01  -.2848E-06   .1268E-05  -.3717E-06   .6460E-06  -.2923E+00
 1   25  .1690E+02  -.5481E-07   .5150E-12  -.3717E-06  -.1370E-07  -.2923E+00

* * * * STRESSES FOR BRICK ELEMENT #    2
LC NODE    Sxx         Syy         Szz         Sxy         Syz         Szx
 1    3 -.8403E+01   .3582E-06  -.1268E-05  -.6005E-06  -.6277E-06  -.2923E+00
```

```
1     5 -.9913E-01  -.1319E-05   .5073E-05  -.6005E-06   .2533E-05  -.2923E+00
1    13 -.9913E-01  -.1319E-05   .5073E-05   .6005E-06  -.2533E-05  -.2923E+00
1    11 -.8403E+01   .3582E-06  -.1268E-05   .6005E-06   .6277E-06  -.2923E+00
1    19  .8403E+01  -.3582E-06   .1268E-05   .6005E-06  -.6277E-06  -.2923E+00
1    21  .9913E-01   .1319E-05  -.5073E-05   .6005E-06   .2533E-05  -.2923E+00
1    29  .9913E-01   .1319E-05  -.5073E-05  -.6005E-06  -.2533E-05  -.2923E+00
1    27  .8403E+01  -.3582E-06   .1268E-05  -.6005E-06   .6277E-06  -.2923E+00
```

Looking at the output file, we see that the tip displacement (at node 32) is −0.4707. This is an error of 2.4 percent from the exact answer of 0.4598. Note that the displacement is larger than the exact. Usually, the 20-node solid element uses $3 \times 3 \times 3$ integration, which creates an element that is a little too stiff. The 20-node solid in SSTAN uses a 14-point reduced integration that softens the element response. The 14-point element has some error when the aspect ratio (length to width or height) becomes too great. The nine-node element is using a reduced eight-point scheme for integration that softens the element and produces better behavior.

If we look at the stresses we see that the X stress at the fixed support is 16.9, compared to 16.67 from beam theory. This is an error of only 3.5 percent. In addition, we need to check the quality of the mesh. If we look at the X stress (at node 27) from adjoining elements, we see that element 1 has 8.6 and element 2 has 8.4. This is a difference of a little over 2 percent. Therefore, the mesh is clearly acceptable. The beam theory stress at this point should be 8.33. The average of these two values is 8.5, which is still very good.

12.6   Problems

12.1. Use a solid element to model the following cantilever beam subjected to an edge load. Find the maximum stress at the fixed end. Give an approximation of the true answer and how it was achieved.

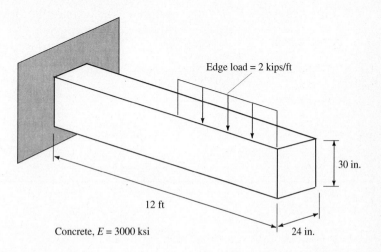

Edge load = 2 kips/ft

30 in.

12 ft

24 in.

Concrete, $E = 3000$ ksi

12.2. Use solid elements to model the following structures. These are the same problems as given in Chapter 9 for modeling with membrane elements. Convert any line loads to surface loads. Compare the results with using solid and membrane elements or theory. Check the effect of putting all the load on one edge of the beam.

(a) Compare the stress distribution of this simple beam with a bearing pad against a simple beam with pin supports at the neutral axis of the beam. (*Hint*: Use symmetry.)

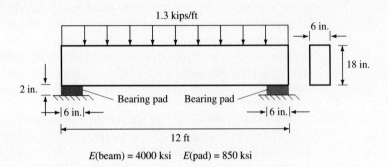

1.3 kips/ft

6 in.

18 in.

2 in.

Bearing pad   Bearing pad

6 in.

6 in.

12 ft

E(beam) = 4000 ksi E(pad) = 850 ksi

(b) Compare the tip deflection for each of the models given below. Also plot the stress along the top of the beams for all four models (on one plot). Compare this to beam theory.

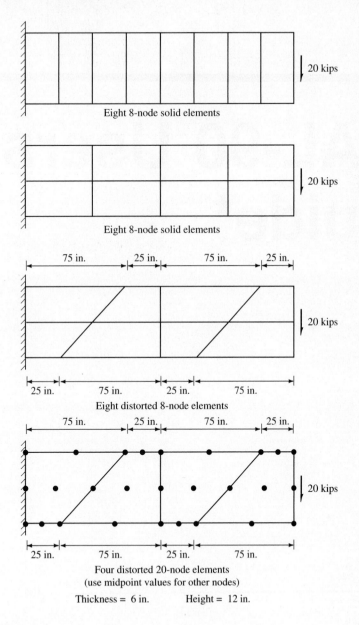

$$\begin{bmatrix} \cos(\theta) & -\sin(\theta) & 0 & 0 & 0 & 0 \\ \sin(\theta) & \cos(\theta) & 0 & 0 & 0 & 0 \\ 0 & 0 & 1 & 0 & 0 & 0 \\ 0 & 0 & 0 & \cos(\theta) & -\sin(\theta) & 0 \\ 0 & 0 & 0 & \sin(\theta) & \cos(\theta) & 0 \\ 0 & 0 & 0 & 0 & 0 & 1 \end{bmatrix} \begin{bmatrix} \dfrac{AE}{L} & 0 & 0 & -\dfrac{AE}{L} & 0 & 0 \\ 0 & \dfrac{12EI}{L^3} & \dfrac{6EI}{L^2} & 0 & -\dfrac{12EI}{L^3} & \dfrac{6EI}{L^2} \\ 0 & \dfrac{6EI}{L^2} & \dfrac{4EI}{L} & 0 & -\dfrac{6EI}{L^2} & \dfrac{2EI}{L} \\ -\dfrac{AE}{L} & 0 & 0 & \dfrac{AE}{L} & 0 & 0 \\ 0 & -\dfrac{12EI}{L^3} & -\dfrac{6EI}{L^2} & 0 & \dfrac{12EI}{L^3} & -\dfrac{6EI}{L^2} \\ 0 & \dfrac{6EI}{L^2} & \dfrac{2EI}{L} & 0 & -\dfrac{6EI}{L^2} & \dfrac{4EI}{L} \end{bmatrix} \begin{bmatrix} \cos(\theta) & \sin(\theta) & 0 & 0 & 0 & 0 \\ -\sin(\theta) & \cos(\theta) & 0 & 0 & 0 & 0 \\ 0 & 0 & 1 & 0 & 0 & 0 \\ 0 & 0 & 0 & \cos(\theta) & \sin(\theta) & 0 \\ 0 & 0 & 0 & -\sin(\theta) & \cos(\theta) & 0 \\ 0 & 0 & 0 & 0 & 0 & 1 \end{bmatrix}$$

CAL-90 Users' Guide*

CAL-90 Introduction

CAL-90 is a matrix manipulation and structural analysis language. The structure of CAL-90 is designed so that the program will operate on most computers. This includes microcomputers with 64K of memory as well as the CRAY supercomputer. CAL-90 is intended to operate as an interactive program. It is intended to be used as an education and development language. For these reasons, the interactive format was chosen. CAL-90 also has the ability to operate in a batch or macro type mode. This consists of taking commands from an input file rather than the terminal.

The enclosed version of CAL-90 is a single program consisting of a series of overlaid program segments that run using a common database. A 286 or better and a numeric coprocessor are suggested. A hard disk is recommended. The plot command uses PC-based graphics and requires VGA capabilities. Since the program is overlaid, the executable file must be located in the current directory or its location must be in the path statement. The PLOT.FON file must be located in the root directory in order to display plots.

There are a few general rules that must be followed for execution of any CAL-90 command. All commands are of the form

OP M1+ M2− M3 N=N1,N2,...

where   OP   is the operation to be executed. This can be one to six characters.
           Mi   are the names of the array to be used for that operation. The name can be from one to six characters, including numbers.
           Ni   is a set of i additional parameters. This can be any LETTER= (e.g., R=).

Some of the SYSTEM operations have single-character abbreviations. They are designated by the single character enclosed in parentheses. A command may require none or up to seven matrices. In this guide we describe the matrices required for each command. The "+" or "−" after the Mi designates the condition of the matrix after the command is completed. A "+" means that the array will be created by the command and therefore another array with the same name must not exist or it will be deleted. A "−" after the Mi indicates that the matrix is changed as a result of the command. A command may or may not require additional name lists. Again, each command and its required parameters are described.

Input Conventions

CAL-90 uses a free-format interpreter for all input. The free-format system allows a letter= designation for data. It also allows arithmetic data for real numbers. Following are the conventions used for input:

A "C" in column 1 of any line will cause the line to be echoed as a comment on the console.

Upper- or lowercase letters may be used interchangeably. The system is *not* case sensitive.

A backslash (\) at any location on a line indicates that from that point on, the next line will be interpreted as a continuation of the current line. A total of 4000 characters or 50 full lines may be continued.

A colon (:) indicates the end of information on a line. Information entered to the right of the colon is ignored by the program. Therefore, the colon can be used to provide additional comments within the input file.

Most data are specified by a letter= and then the data. For example, I= 53, 67.5 is one possible input. The letter and = *must not* have space between them. All other data are separated by spaces and/or commas.

If a blank identifier is specified, no letter=, the data strings are assumed to be the first data on the line. All letter= data can be put in any order on the line after the blank identifier data.

If fewer data exist than are specified, the values returned will be either a zero or a blank, according to the routine used.

Real numbers do not require decimal points.

E formats with + or − exponents are accepted.

Simple arithmetic statements may be used within the input for real numbers. The functions that can be used are +, −, *, /. The order of evaluation is sequential, not hierarchical as in the FORTRAN language. Parentheses are *not* acceptable.

System Commands

The SYSTEM commands are the basic operational commands for CAL-90. These commands perform the operating system commands within the language. The single letter in parentheses is the shorthand command, if one exists. Their form and description are as follows:

LIST (L)

This command lists the directory of all matrices known to the database.

HELP (H)

This command lists the commands that are available for the current segment. Specifying the topic for which help is needed will give the command description. Giving the command HELP and the topic on the same line will give the command description directly.

PRINT (P) M1

This command prints the matrix named M1 to the screen and the output file.

DELETE (D) M1−

This command deletes the matrix named M1.

STOP (S)

This command stops execution of the current segment and saves the database. When the program is executed another time, all previously existing arrays will be reloaded into memory.

START

This command reinitializes the database by deleting all matrices from memory.

LOAD M1+ R=R1 C=C1

This command loads the real matrix M1 with R1 rows and C1 columns. The matrix is entered one row per line.

MODIFY M1−

This command modifies any individual term of the matrix M1. This command can only be used interactively.

ZERO M1+ R=R1 C=C1

This command forms the matrix M1. The matrix is R1 rows by C1 columns. All the terms of M1 have a value of zero.

INPUT

This command is used to change the input file used for batch operation from the default file INPUT to any other file. It can only be used from the interactive mode. It prompts for the input file name.

SUBMIT SEPAR

This command starts the batch execution. It executes the commands following the SEPAR separator in the input file. The input file name is defaulted to the name ''INPUT.'' SEPAR can be a one- to six-character string, including numbers.

RETURN

This is the last command in a separator group of a submit or macro command. This command returns control to the interactive mode after executing the commands following a separator.

Basic Matrix Commands

The BASIC MATRIX commands perform the standard linear algebra operations required for most matrix operations.

ADD M1– M2

This command adds the matrices M1 and M2 and stores the result in M1.

SUB M1– M2

This command subtracts the matrix M2 from M1 and stores the results in M1.

MULT M1 M2 M3+

This command multiplies matrix M1 times M2 and stores the result in M3.

TRAN M1 M2+

This command forms the matrix M2 which is the transpose of matrix M1.

DUP M1 M2+

This command forms the matrix M2 which is the duplicate of M1.

SCALE M1– M2

This command scales all the terms of matrix M1 by the (1, 1) term of the matrix M2. Note that M2 can be a 1×1 matrix.

SOLVE M1– M2– S=S1

This command solves the set of equations $Ax = B$, when the matrix M1 is A and M2 is B. The result is written back into M1 and/or M2. M1 can be nonsymmetric. M2 can have any number of columns. The matrix is factored by LU decomposition:

where $S1 = 0$ Complete solution of $Ax = B$. x is returned in M2.
         $S1 = 1$ Triangularization of M1 only. The result is stored in M1.
         $S1 = 2$ Forward and back substitution of M2 only. The result is stored in M2.
                     Note: The M1 matrix must already be a triangularized matrix.

INVERT M1–

This command inverts the symmetric matrix M1. The inverse is stored in M1.

COND M1– M2– R=R1 S=S1

The command COND performs a static condensation on the set of equations $Ax = B$, where the matrix M1 is A and M2 is B. The first R1 equations will be condensed. The result is written back on top of A and/or B. The command will solve nonsymmetric as well as symmetric sets of equations. It uses the Gauss elimination procedure to triangularize the matrix A and then forward and back substitutes for B. S1 dictates the procedure to use:

where $S1 = 0$ only condenses the A matrix (DEFAULT).
         $S1 = 1$ only reduces the load matrix B.
         $S1 = 2$ only back substitutes B.

The condensed matrix and load are found in the lower portion of the M1 and M2 matrices, below the R1 row and after the R1 column. When S1 = 2, it assumes that the retained equation solutions have been stored in the reduced load vector, possibly using the STOSM command. *Note*: The command must be called multiple times for a complete process.

LEAST M1 M2 M3+ O=O1

The command LEAST performs a least squares fit of a polynomial of the order O1. It uses the M1 vector as the X values for the data and the M2 vector as the function values. The command can only fit polynomials.

Norm Operations

The NORM operations perform basic matrix norm operations, matrix blocking, and separation functions.

DUPSM M1 M2+ R=R1 C=C1 L=L1,L2

This command forms the matrix M2, which is a submatrix of M1. The submatrix is R1 rows by C1 columns. The submatrix starts at the location L1,L2 of the matrix M1.

STOSM M1− M2 L=L1,L2

This command stores the matrix M2 as a submatrix in matrix M1. The submatrix starts at the location L1,L2 of matrix M1.

DUPDG M1 M2+

This command forms the row matrix M2, which consists of the diagonal values of the matrix M1.

STODG M1− M2

This command stores the row or column matrix M2 on the diagonal of matrix M1.

MAX M1 M2+

This command forms the column matrix M2, in which each row contains the maximum absolute value of the corresponding row in matrix M1. The maximum and its column number are printed for each row.

NORM M1 M2+ T=T1

This command forms the row matrix M2, which contains the row norm of matrix M1:

where   T1 = 1   forms the sum of the absolute values of each row.
        T1 = 2   forms the square root of the sum of the squares of each row.

INVEL M1−

This command replaces each term of the matrix M1 with the inverse of the term.

SQREL M1−

This command replaces each term of the matrix M1 with the square root of the term.

LOG M1−

This command replaces each term of the matrix M1 with the natural log of the term.

PROD M1 M2+

This command forms a 1×2 matrix named M2 which contains the product of all the terms of the matrix M1. The product X is stored as two numbers in the form

$$X = P \times 10^E$$

where   M2(1) = P.
          M2(2) = E.

Direct Stiffness Operations

These commands perform the direct stiffness operations of stiffness matrix formation and assembly.

LOADI M1+ R=R1 C=C1

This command loads the integer matrix M1 with R1 rows and C1 columns. R1 is the number of local element degrees of freedom and C1 corresponds to the number of elements used. Each column in M1 identifies the connection between the local degrees of freedom for an element and the arbitrarily selected global degrees of freedom of the structure. The matrix is entered one row per line.

FRAME M1+ M2+ X=XI,XJ Y=YI,YJ E=E1 I=I1 A=A1

where   M1   is the element stiffness matrix in global directions.
          M2   is the force transformation matrix.
          XI   is the X coordinate of the i node.
          XJ   is the X coordinate of the j node.
          YI   is the Y coordinate of the i node.
          YJ   is the Y coordinate of the j node.
          E1   is the modulus of elasticity, E.
          I1   is the moment of inertia, I.
          A1   is the axial area, A.

This operation forms the 6×6 stiffness matrix M1 and the 6×6 force transformation matrix M2 for a two-dimensional beam element. The stiffness is formed for a two-dimensional beam element in global directions. The degrees of freedom are shown below. DOFs 1, 2, and 3 are always at node I. DOFs 3, 4, and 5 are always at node J. All DOFs are always in global directions.

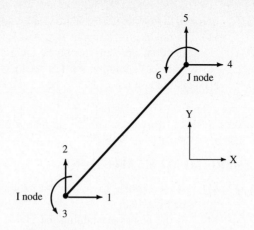

Global DOF definition for beam element.

The force transformation matrix may be used to recover the element forces once the final displacements are found using the MEMFRC command. The element forces are defined by

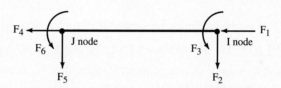

Definition of force results for beam.

The directions for the axial forces (F_1 and F_4) are always positive going from nodes I to J. The shears are positive at 90° counterclockwise from the axial forces. The moments are positive if counterclockwise.

ADDK M1+ M2 M3 N=N1

where   M1   is the global stiffness matrix (previously defined by the ZERO command).
        M2   is the element stiffness matrix to be added.
        M3   is an integer array in which column N1 contains the global degree of freedom numbers to which the element degrees of freedom correspond (matrix containing the **LV** vectors).
        N1   is the column of M3 to be used as the **LV** vector.

This command adds an element stiffness matrix to the total global stiffness matrix. This is a general command that works with a stiffness matrix of any size. If the matrix M2 is N × N, the matrix M3 must have N rows.

ADDL M1+ M2 M3 N=N1

where   M1   is the global load matrix. (The matrix *must* have been defined previously by the ZERO or LOAD command.)

M2   is the element load matrix to be added.

M3   is an integer array in which column N1 contains the global degree of freedom numbers to which the element degrees of freedom correspond (matrix containing the **LV** vectors).

N1   is the column of M3 to be used as the **LV** vector.

This command adds an element load matrix to the total global load matrix. This is a general command that works with a load matrix of any size. If the matrix M2 is N × N, the matrix M3 must have N rows.

MEMFRC M1 M2 M3 M4+ N=N1

where   M1   is the element force transformation array, usually from FRAME, PLATE, or PLANE.

M2   is the matrix containing the global joint displacements.

M3   is the integer array in which column N1 contains the global DOF numbers that correspond to the element DOFs. (This is as used in ADDK and ADDL.)

M4   is the element force array created. The force directions are defined under the FRAME command.

N1   is the column number of M3 to be used as the **LV** vector.

This command forms the array M4, which contains the local element forces.

PLATE M1 M2 M3 N=N1,N2,N3,N4,N5,N6,N7,N8,N9 E=E1 P=P1 H=H1

where   M1   is the resulting stiffness matrix.

M2   is the resulting force transformation matrix.

M3   is a 9 × 2 matrix containing the coordinates for the plate element.

Ni   are the node numbers for the element. These must be from 1 to 9. The numbers N1 through N9 refer to the row where the coordinate is found in the M3 matrix.

E1   is Young's modulus.

P1   is Poisson's ratio.

H1   is thickness.

The command PLATE forms the 27 × 27 stiffness and the 20 × 27 force transformation matrix for an isoparametric plate. The plate is a variable three- to nine-node element (N2, N4, N5, N6, N8 can be zero). The plate is a Mindlin plate.

MEMBRANE M1 M2 M3 N=N1,N2,N3,N4,N5,N6,N7,N8,N9 E=E1 P=P1 H=H1

where   M1   is the resulting stiffness matrix.

M2   is the resulting force transformation matrix.

M3   is a 9 × 2 matrix containing the coordinates for the plate element.

Ni   are the node numbers for the element. These must be from 1 to 9. The numbers N1 through N9 refer to the row where the coordinate is found in the M3 matrix.

E1   is Young's modulus.

P1   is Poisson's ratio.

H1   is thickness.

The command MEMBRANE forms the 18 × 18 stiffness and the 12 × 18 force transformation matrix M2 for an isoparametric membrane element. The membrane is a variable three- to nine-node element (N2, N4, N5, N6, N8 can be zero).

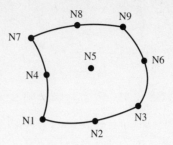

Nodal connectivity sequence for nine-node element.

File Operations

The FILE operations allow matrices to be stored on disk so that memory can be freed. These commands are useful only for very large problems.

FILE M1

This command saves the matrix M1 in a disk file with the name "INPUT.EXT,"

where   INPUT   is the name of the current input file. The default is INPUT.
        M1   is the matrix to be saved. The file extension EXT is made up of the first three characters of the matrix name.

RFILE M1

This command reads into memory the matrix M1, which is stored in a disk file. The database directory must contain an entry for the matrix M1, shown using the LIST command. If no entry exists, one must be created by the DEFINE command.

DEFINE M1+ TP R=R1 C=C1

where   M1   is the matrix for which the directory is to be created.
        TP   is the matrix type (TP=R if the matrix is real; TP=I if the matrix is integer).
        R1   is the number of rows in M1.
        C1   is the number of columns in M1.

This command defines the directory for a matrix stored on a disk file. This command is needed only if the database does not contain the matrix directory.

Plotting Command

This command allows the contents of an array to be plotted. The *Y* axis of the plot comprises the values of the rows of the array. The *X* axis comprises uniform values indicating the column position in the array.

PLOT M1 N=N1 R=R1,R2,...

where   M1   is the matrix to be plotted.

N1   is the number of rows of N1 that are to be plotted.

Ri   is the row number to be plotted from M1. There must be N1 entries for Ri.

This command generates a screen plot of the rows of matrix M1.

Looping Operations

CAL-90 allows looping operations. A set of commands in the INPUT file separated by a LOOP and either NEXT or IF command allows for looping with conditional return to interactive mode. Normally, after a set of commands is issued, a RETURN command is used to shift from a macro to interactive. The NEXT or IF command must be used when a LOOP command is used. The following looping commands allow for a fixed number of loops or conditional returns within a looping structure.

LOOP SEPAR N=N1

where   SEPAR   is the separator, following immediately after the LOOP command in a batch file. This can be a one- to six-character string, including numbers.

N1   is the number of times the operations should be performed. When used in conjunction with the IF command, the value of N1 specifies an upper limit for the possible number of loops to ensure termination of looping if there is no convergence.

This command is used to loop repeatedly over a group of operations. This can be used either in conjunction with the command NEXT to loop a specific number of times or with the command IF to control looping with logical comparisons. LOOP can only be used interactively.

NEXT

This command is a looping control command. A loop begins with the LOOP command and a separator and ends at the NEXT or IF command. Looping continues on to the command after the NEXT command in the input file when completed executing the commands following the looping separator SEPAR, N1 times, where N1 is specified by the LOOP command.

IF M1 L1 M2

where   M1   is a real array.

L1   is a logical comparison statement. L1 can be GT, LT, GE, LE, or EQ. These are the same as the FORTRAN statements.

M2   is a real array.

This command is also a looping control command. A loop begins with LOOP and a separator and can end with NEXT or IF. Looping continues on to the command following

IF (in the input file) when the logical comparison L1 of the (1, 1) terms of the arrays M1 and M2 in a looping operation is satisfied or if N1 loops have been performed. If the specified statement is false, the statement causes repeated execution of the commands following the looping separator.

Dynamics Commands

Special dynamics operations are included for performing structural dynamics operations. The following commands can be used in conjunction with the direct stiffness commands to perform dynamic analysis.

EIGEN M1− M2+ M3− T=T1

where   M1   is the N × N stiffness matrix K.

M2   is the N × N result matrix containing the eigenvectors Φ.

M3   is the row or column matrix containing the diagonal values of the mass matrix M. The values of the mass matrix *must* be positive. The results, the eigenvalues ω^2, will be written back onto M3.

T1   is the approximate number of significant figures for the eigenvalues. The default for T1 is 4.

The command EIGEN solves the eigenproblem:

$$(M - \omega^2 K)\Phi = 0$$

FUNG M1 M2+ T=T1 L=L1,L2

where   M1   is a 2 × K matrix with the time values in the top row and the function values in the second row. The values of M1 do *not* have to be at equal spacing.

M2   is the resulting matrix of equal-function-value results. It can have two forms, depending on the value of L2.

L1   is the number of time steps that the resulting matrix will contain.

L2   is the flag to tell which type of resulting function is required. If L2=1, M2 will be a 1 × L1 row matrix containing the function values at time increment T1. If L2=2, M1 will be a 2 × L1 matrix in which the first row contains the time values and the second row contains the corresponding function values.

T1   is the time step between the resulting function values.

The command FUNG forms the matrix M2, which contains values at equal time intervals of the function specified by the matrix M1. The total time for which M2 will contain values is T1*L1.

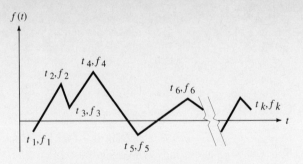

Input time function at uneven intervals.

STEP M1− M2 M3 M4− M5+ M6 M7 T=T1 L=L1,L2 P=P1,P2,P3

where   M1   is the N × N stiffness matrix.

M2   is the N × N mass matrix.

M3   is the N × N damping matrix.

M4   is an N × 3 matrix of initial values, one row for each DOF. The initial values corresponding to columns 1, 2, and 3 are the displacement, velocity, and acceleration.

M5   is the N × L2 result matrix of displacements. Column i represents the displacement at the time i*L1*T1.

M6   is the N × 1 load distribution matrix.

M7   is the 1 × K matrix representing multipliers at equal time increments T1 (from FUNG). The load on each DOF is M6(i)*M7. L1 is the interval for displacement results, output at every L1 time steps.

L2   is the total number of displacement vectors to be output. Therefore, the total time for which results are calculated is L1*L2*T1. The number of columns of M7, K, must be greater than L1*L2.

Pi   are the integration parameters for the Newmark integration method. P1 is δ. P2 is α. P3 is θ.

Integration Type	δ	α	θ
Newmark's average acceleration	$\frac{1}{2}$	$\frac{1}{4}$	1.0
Linear acceleration (DEFAULT)	$\frac{1}{2}$	$\frac{1}{6}$	1.0
Wilson's θ method (low damping)	$\frac{1}{2}$	$\frac{1}{6}$	1.42
Wilson's θ method (high damping)	$\frac{1}{2}$	$\frac{1}{6}$	2.00

This command performs a step by step dynamic analysis for a structure. It solves the equation

$$M\ddot{v} + C\dot{v} + Kv = f(t)$$

DYNAM M1 M2 M3 M4 M5+ N=N1 T=T1

where M1 is the row or column matrix containing the structural frequencies, ω.

        M2 is a row or column matrix containing the modal damping terms, ξ.

        M3 is the N $\times$ 1 load multipliers for each mode. These should be Φ^{T} times a factor of load applied to each DOF.

        M4 is the 2 $\times$ K matrix containing the time-dependent load multipliers. M4 is the same as the input matrix for the FUNG command. The load applied to each mode is M3(i)*M4.

        T1 is the time increment for evaluation, Δt.

        N1 is the number of displacements to be generated. Therefore, the total time for which the results will be calculated is N1*T1.

The command DYNAM evaluates the set of uncoupled dynamic equations of motion in the time domain. The equation solved is

$$\ddot{X} + 2\xi\omega\dot{X} + \omega^2 X = \Phi^{\mathrm{T}}f(t)$$

The input required consists of the results from the EIGEN command. The DYNAM command uses the exact solution for a single-degree-of-freedom system for nonuniform time steps and linear varying load. This is *not* the same as a step-by-step solution.

FREQMD M1 M2 M3 M4 M5+ N=N1

where M1 is the row or column matrix containing the structural frequencies, ω^2.

        M2 is the row or column matrix containing the modal damping ratios, ξ.

        M3 is the N $\times$ 1 vector of load multipliers for each mode.

        M4 is the load vector in the frequency domain. It is a K $\times$ 2 matrix in the form returned by the DFT command.

        M5 is the N $\times$ K matrix of the resulting generalized displacements in the frequency domain. *Note*: This response can be transformed to the time domain using the IDFT command.

        N1 is the number of modes used in calculating the response. The default is the total number available. The loads for each mode are calculated as the modal multiplier (M3) times the frequency loading (M4).

The FREQMD command calculates the dynamic response for a set of uncoupled second-order differential equations associated with the mode superposition method in the frequency domain.

FREQTL M1 M2 M3 M4 M5 M6+

where M1 is the N $\times$ N stiffness matrix.

        M2 is the N $\times$ N mass matrix.

        M3 is the N $\times$ N damping matrix.

        M4 is the N $\times$ 1 load distribution multiplier for each DOF.

        M5 is the load vector in the frequency domain. It is a K $\times$ 2 matrix in the form returned by the DFT command. The load for each DOF is calculated as the load multiplier (M4) times the frequency load (M5).

M6   is the N × K matrix which contains the resulting displacements, in the frequency domain. The result can be transformed to the time domain by the IDFT command.

The FREQTL command calculates the dynamic response of the coupled set of second-order differential equations in the frequency domain.

Simpal Interaction Commands

The following commands allow for interaction with the SIMPAL analysis program. These commands allow for the sharing of global and element stiffness, mass, and force vectors.

SIMDIR

The SIMDIR command will give a list of all the control parameters of the current SIMPAL problem. This includes the list of the active element groups.

GETSIM K=K1+ M=M1+ L=L1+

where   K1   is the matrix where the stiffness will be placed.
         M1   is the matrix where the mass will be placed.
         L1   is the matrix where the loads will be placed.

The command GETSIM will bring the GLOBAL stiffness, mass, and load matrices created in SIMPAL into the CAL database. If any of the fields are left blank, the corresponding matrix will not be read. *Note*: This command works only for a single block stiffness matrix.

PUTSIM K=K1 M=M1 L=L1

where   K1   is the matrix to be saved as the global stiffness.
         M1   is the matrix to be saved as the global mass.
         L1   is the matrix to be saved as the SIMPAL loads.

The command PUTSIM will save the matrices as GLOBAL stiffness mass and load matrices in the SIMPAL database. Any matrix left blank will not be saved in the SIMPAL database. *Note*: This command works only for a single block stiffness matrix.

PUTELM M1 M2 M3 M4 M5

where   M1   is the matrix to be used as the element stiffness.
         M2   is an integer array that gives the global DOF number in SIMPAL to which the element is associated. M2 *must* be created by the LOADI command.
         M3   is the element load matrix. M3 must be the number of element DOFs by the number of load cases.
         M4   is the mass matrix. M4 is a row or column vector that contains the diagonalized mass values.
         M5   is a three-character extension name for the element to be saved under. This extension must be unique; names like K1, K2, K3, and so on, will not conflict with any already in SIMPAL.

The command PUTELM stores the given matrices as an element to be used in the current SIMPAL analysis.

GETELM M1+ M2+ M3+ M4+ M5 N=N1

where   M1   will contain the element stiffness.

            M2   will be an integer array that contains the global DOF number from SIMPAL to which the element DOF is associated.

            M3   will contain the element loads.

            M4   will be a column vector that contains the diagonal mass values.

            M5   is the extension name of the element group from which the element is to be recovered.

            N1   is the element number that is to be recovered.

The command GETELM retrieves the element matrices for a single element in the SIMPAL database. All the recovered arrays will be a standard size. This size is determined by SIMPAL. Elements that have fewer DOFs than this maximum size will be stored in the upper left corner of the returned matrices. The matrix of the exact size can be extracted by use of the DUPSM command.

Fourier Transform Commands

These two commands allow a Fourier and an inverse Fourier transform to be performed so that dynamics can be performed in the frequency domain.

DFT M1− M2+ T=T1

where   M1   is the input time function. It is a row or column vector containing the function values at *equal* time intervals, T1. This can be created by the FUNG command.

            M2   is the generated matrix containing the function values in the frequency domain. It has the same number of rows as the input vector but has two columns. The first column is the function amplitudes. The second column is the frequencies. The first term of the repeated frequency is the cosine amplitude and the second is the sine.

            T1   is the time interval at which the input function values are specified.

The command DFT preforms the discrete Fourier transform from the time domain to the frequency domain.

IDFT M1− M2+

where   M1   is the array of frequency amplitudes in order of increasing frequency. The cosine term should precede the sine term for each frequency component. This is a column or row vector that has the same form as the first column of the result from the DFT command. This matrix *can* be the result from the DFT command.

            M2   is the array created containing the amplitude values of the time function at equal time intervals.

The command IDFT preforms the inverse discrete Fourier transform of a frequency-domain function to the time domain. The results are output at equal time intervals.

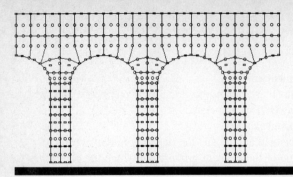

SSTAN
Users' Guide

SSTAN is a small, simple, general-purpose program for static load analysis of three-dimensional structural systems. SSTAN allows for load combinations. This linearly combines the results for different loading according to a given set of load factors. It has minor nonlinearities, including zero compression or tension truss members and $P-\Delta$ effects for frame members. Its major purpose is to provide finite-element analysis capability for educational use and for designers of small structures. Since its major use is for small structures, state-of-the-art numerical techniques have been omitted to make the program simple to use and understand. Also, practical options such as large capacity and sophisticated generation are not included.

Using SSTAN and STANPLOT

The general analysis program included with this book is called SSTAN. It is a three-dimensional finite-element program that has a graphics postprocessor. It is written to run on a microcomputer and designed to be used as an educational tool. SSTAN runs best with a math coprocessor but runs without one. It is a batch program which assumes that the structural information is contained in a file named "INPUT." The results are put into a file named "OUTPUT." SSTAN uses the same free-form input as CAL-90. Therefore, comment lines, arithmetic statements, and header lines are all valid. Since SSTAN is a batch program, all of the input needs to be prepared in advance and then the program runs a complete analysis without additional user interaction. The input is prepared and stored in the INPUT file. Then the program is run by typing

C:\> sstan

SSTAN can be stored in any directory on the PC; however, it is an overlaid program and therefore needs to have a path to the directory in which it is stored or it must be in the current directory. SSTAN will use all free memory within the 640K of base memory on a PC. To gain maximum capacity, remove any TSRs or drivers from lower memory. The INPUT file must always be in the current directory and must be an ASCII file—called nondocument mode in some word processors. A summary of the required input follows. The first line in the file is an analysis title. *It must be the first line.* It is also important that the title does not begin with any of the header names, such as COORDINATE, FRAME, and so on. The second line is the analysis control information line. It is of the form

N1,N2,N3,N4

where N1 is the number of nodes in the structure and N2 is the number of different element types in the structure. Again, this means that if you are using both truss and beams, there are two, regardless of the number of member properties. N3 is the number of load cases and N4 is the number of load combinations.

The rest of the INPUT is on a free-formatted header basis. All data are arranged by groups and signified by a header. For example, nodal coordinates are signified by the header COORDINATES. These data are order independent; that is, they can be placed in the INPUT file in any order. All data groups must end with a blank line.

The following headers and groups are available in SSTAN:

1. *COORDINATES.* This specifies the nodal coordinates of the structure to be analyzed. Note that the program assumes that the structure is three-dimensional.

2. *BOUNDARY.* This specifies the boundary conditions or nodal DOFs. Nodes can be either released or fixed only in the global *X–Y–Z* coordinate system.

3. *TRUSS.* This specifies the truss element data. Trusses can have initial tension, rigid-end offsets, and the ability not to take either compression or tension. Zero-compression members can be used for slender bracing members. Zero-tension members can be used for gap (uplift) problems.

4. *BEAM.* This specifies the bending member data. These elements can be used for beams as well as columns. Beams can have uniform loads applied to the member, rigid-end offsets, and can include $P–\Delta$ effects.

5. *PLATE, MEMBRANE, SHELL, BRICK, and AXISYMMETRIC.* These additional types of members, which constitute what are called finite elements, are linear in behavior.

6. *LOADS.* This specifies the concentrated loads applied to the structure. They can be concentrated loads or moments and applied to any load case.

7. *COMBINATIONS.* This section allows linear combinations of the basic load cases. This option is not valid when using noncompression/tension trusses or $P–\Delta$ effects.

The coordinate system assumed in SSTAN is a right-hand-rule system. All displacements, forces, and moments are given in this right-handed system. The results for element forces are given at the nodes of a member in the local coordinate system.

Running STANPLOT

STANPLOT is the graphics postprocessor that works in conjunction with SSTAN. STANPLOT is also an overlaid program and has the same restrictions about the path as SSTAN. STANPLOT also uses a math coprocessor as well as VGA graphics and a mouse. STANPLOT also requires the PLOT.FON file to be located in the root directory.

Input Conventions

The input data are prepared through the use of an ASCII editor and placed in a file named ''INPUT.'' SSTAN uses a free-format interpreter for all input. Therefore, only blanks or commas are needed to separate data. The input uses a header format for groups of information. For example, the header "COORDINATES" is used to specify the nodal coordinates. Each group of data requires a header to define the group and its associated data following the header. The problem definition data comprise the only group that does *not* have a header. It consists of a problem title and the control information. The problem definition group *must* be the first two lines in the file. A group with a header can be placed at any location within the input file. The following are the conventions used for individual data input:

A "C" in column 1 of any line will cause the line to be echoed as a comment on the console.

Upper- or lowercase letters may be used interchangeably. The system is *not* case sensitive.

A backslash (\) at any location on a line indicates that from that point on, the next line will be interpreted as a continuation of the current line. A total of 4000 characters or 50 full lines may be continued.

A colon (:) indicates the end of information on a line. Information entered to the right of the colon is ignored by the program. Therefore, the colon can be used to provide additional comments within the input file.

Most data are specified by a letter= and then the data. For example, I= 53, 67.5 is one possible input. The letter and = *must not* have any space between them. All other data are separated by spaces and/or commas.

If a blank identifier is specified, no letter=, the data strings are assumed to be the first data on the line. All letter= data can be put in any order on the line after the blank identifier data.

If fewer data exist than are specified, the values returned will be either a zero or a blank, according to the routine used.

Real numbers do not require decimal points.

E formats with + or − exponents are accepted.

Simple arithmetic statements may be used within the input for real numbers. The functions that can be used are +, −, *, /. The order of evaluation is sequential, not hierarchical as in the FORTRAN language. Parentheses are *not* acceptable.

Input Data

The first step in the analysis of a structure by computer is to number all the joints and members in the system. The next step is to identify the displacement degrees of freedom for all the joints. In addition, each joint's coordinates and each member must be specified. The definition of positive joint displacements and loads is as follows:

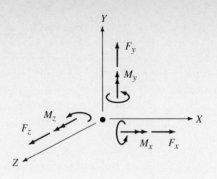

(Follows right-hand rule, global coordinate system for point loads and displacements.)

The following sequence of input data lines provides a numerical definition of the structure. The first two lines of the input file *must* be the problem definition group. It consists of

Heading Line (*First line*)

This line is used as a title for the structure to be analyzed. It can be up to 80 characters long and *must* be the first line of the file.

Master Control Line (*Second line*)

NJT,NTE,NLC,NCM P=PD C=CM T=TL M=MX

where   NJT   is the total number of joints in the structural system.
        NTE   is the number of different types of elements (maximum of 7).
        NLC   is the number of different load cases.
        NCM   is the number of load case combinations.
        PD   is the flag to turn on $P-\Delta$ effects in frames. PD = 0: no $P-\Delta$ effects (DEFAULT); PD = 1: $P-\Delta$ effects included.
        CM   is the flag to include noncompression/nontension truss members. CM = 0: all truss members take compression and tension (DEFAULT); CM = 1: specified truss member cannot take compression/tension.
        TL   is the convergence tolerance for the nonlinear solutions: $P-\Delta$ and noncompression/nontension truss members (DEFAULT = 0.01).
        NX   is the maximum number of cycles allowed for the nonlinear solution (DEFAULT = 100).

This *must* be the second line of the file.

Joint Coordinate Data

This section defines the nodal coordinates in a positive Cartesian coordinate system. The following lines will define the nodal coordinates:

COORDINATES

ND X=X1 Y=Y1 Z=Z1 G=GF,GL,GI S=S1,S2,S3,S4,S5,S6

where   ND   is the node number for the coordinates.

           X1   is the X coordinate for the node ND.

           Y1   is the Y coordinate for node ND.

           Z1   is the Z coordinate for node ND.

           GF   is the first node in the generation sequence. (*Note*: Its coordinate must have already been specified.)

           GL   is the last node in the generation sequence. (*Note*: Its coordinate must have already been specified.)

           GI   is the increment for the node numbers that are generated at equal increments between nodes GF and GL (DEFAULT = 1). GF, GL, and GI can be left blank if no generation is desired.

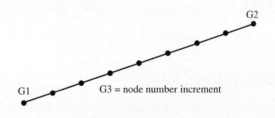

Linear coordinate generation definition.

           Si   are the rectangular surface node generation sequences. S1, S2, S3, and S4 are four corner nodes in the surface set of nodes to generate. S5 is the number of nodes to generate in the S1-to-S2 direction. S6 is the number of nodes to generate in the S1-to-S3 direction.

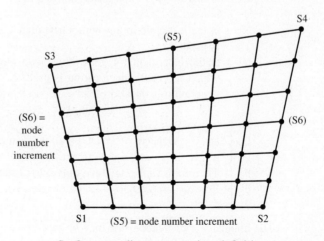

Surface coordinate generation definition.

where   S1   is the node number of corner node 1 for surface coordinate generation.

           S2   is the node number of corner node 2 for surface coordinate generation.

           S3   is the node number of corner node 3 for surface coordinate generation.

           S4   is the node number of corner node 4 for surface coordinate generation.

S5   is the node number increment for generation between corners 1 and 2 and between corners 3 and 4.

S6   is the node number increment for generation between corners 1 and 3 and between corners 2 and 4.

If both linear and surface generation parameters appear on the same line of input, the linear coordinate generation will be performed first, followed by the surface coordinate generation.

The node specification line may be repeated as many times as is necessary to define all the nodes.

The G= or S= (generation) specification can be the only data on the line.

Any node that is not specified will default to the origin (0, 0, 0).

This section *must* end with a blank line.

Boundary Condition Data

This section defines the active degrees of freedom for the structure to be analyzed. All the degrees of freedom must be specified as fixed or released, according to the boundary conditions and the types of elements connected to the node. The following lines of input describe the state of the DOF:

BOUNDARY

NF,NL,NI DOF=D_x,D_y,D_z,θ_x,θ_y,θ_z

where   NF   is the first node in a generation sequence for which the DOF specification is used.

NL   is the last node in a generation sequence for which the DOF specification is used.

NI   is the increment for generating node numbers between NF and NL for which the DOF specification is used. NL and NI can be left blank if no generation is desired.

$\left.\begin{array}{l} D_x \\ D_y \\ D_z \\ \theta_x \\ \theta_y \\ \theta_z \end{array}\right\}$ are the respective six DOFs allowed at a node. They correspond to the global X, Y, and Z directions. Each DOF can be either fixed or released. Therefore, DOF_i can have only one of the two following values: $DOF_i =$ F is for fixed; $DOF_i = $ R is for released.

The boundary specification line may be repeated as many times as is necessary to define the boundary conditions of all the nodes.

Any node that is not defined will default to the fully fixed condition.

The last specification given will override any previous ones.

This section *must* end with a blank line.

Three-Dimensional Truss Members

This section is used to specify three-dimensional truss members. These members can be designated as not able to have compression or tension, pretensioned, and having rigid-end offsets. The required data are as follows:

TRUSS   *Header for Truss Elements*

NM,NP

where   NM   is the number of truss members.
             NP   is the number of material properties.

Material Property Lines (*NP lines*)

MID A=A E=E

where   MID   is the material property set number.
            A   is the axial area, *A*.
            E   is the modulus of elasticity, *E*.

Truss Member Connectivity (*Repeat as many times as needed*)

NN,NI,NJ M=MP C=C1 G=G1,G2,G3,G4 I=IX,IY,IZ J=JX,JY,JZ T=T1,T2,T3,...

where   NN   is the member identification number.
            NI   is the joint number I.
            NJ   is the joint number J.
            MP   is the material property set number.
            C1   is the flag for whether this member can take compression or tension. C1 = 0 means that the member takes compression (DEFAULT); C1 = 1 means that the member cannot take compression; C1 = −1 means that the member cannot take tension. (*Note*: This option must be turned on from the master control line.)
            Ti   is the tension in the member for load case i.
            IX,IY,IZ   are the end eccentricities in the global *X*, *Y*, and *Z* coordinates, for the *i* node.
            JX,JY,JZ   are the end eccentricities, in the global *X*, *Y*, and *Z* coordinates, for the *j* node.
            G1   is the number of additional members to be generated.
            G2   is the member ID increment for members generated.
            G3   is the I joint increment for members generated.
            G4   is the J joint increment for members generated.

This section *must* be terminated with a blank line.

Three-Dimensional Frame Members

The beam element is a full three-dimensional beam with the ability to apply a uniform load along its length (bending about the strong axis). The three-dimensional beam is used to model members that have bending stiffness about the two principal axes, torsional and axial stiffness. Shear deformations are defaulted to be zero. Members must be prismatic and are connected between two points, I and J. A dummy joint number is used to specify the directions of the principal bending axes. The member can have pins at either or both ends. The following lines of input are required to define beam elements:

BEAM   *Header for Bending Elements*

NM,NP

where   NM   is the number of frame members.
        NP   is the number of material types.

Material Property Lines (*NP Lines*)

MID I=I2,I3 J=J A=A,AS3,AS2 E=E G=G

where   MID   is the material property set number.
        A   is the axial area.
        AS3   is the shear area in the local 3 direction.
        AS2   is the shear area in the local 2 direction.
        J   is the torsional moment of inertia about the local 1 axis.
        I2   is the moment of inertia about the 2 axis, I22 (strong axis).
        I3   is the moment of inertia about the 3 axis, I33 (weak axis).
        E   is the modulus of elasticity (Young's modulus), E.
        G   is the shear modulus used to calculate the torsional and shear stiffness (DEFAULT = $E/[2(1 + v)]$ with $v = 0.3$, G = $E/2.6$).

Frame Member Connectivity (*Repeat as many times as needed*)

NN,NI,NJ,NK M=MP G=G1,G2,G3,G4,G5 P=P1 I=IX,IY,IZ J=JX,JY,JZ L=L1,L2,L3,...

where   NN   is the member number.
        NI   is node number I.
        NJ   is node number J.
        NK   is node number K.
        MP   is the material property set for this member (DEFAULT = previous).
        Li   is the magnitude of the uniform distributed load acting on this member for load case i. The load is assumed acting parallel to the local 3 axis in the negative direction.
        P1   is the moment release or hinge option. P1 = 1 is a hinge at node I; P1 = 2 is a hinge at node J; P1 = 3 is a hinge at node I and node J.
        IX,IY,IZ are the end eccentricities for the *i* node.
        JX,JY,JZ are the end eccentricities for the *j* node.
        G1   is the number of additional members that are needed to generate (DEFAULT = 0).
        G2   is the member number increment for members generated.
        G3   is the generation increment for node I (DEFAULT = 0).

G4   is the generation increment for node J (DEFAULT = 0).
G5   is the generation increment for node K (DEFAULT = 0).
G1 through G5 can be left blank if no generation is desired.

This section *must* be terminated with a blank line.

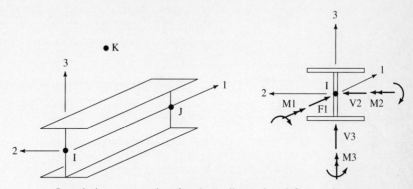

Local sign convention for three-dimensional frame elements.

Axis 1 is the I–J direction; axis 3 is normal to axis 1 and is in the I–J–K plane; axis 2 is normal to the plane 2 formed by the 1 and 3 axes and completes the right-hand rule.

Axisymmetric Element

These elements must be defined in the *Y–Z* plane. In addition, for the axisymmetric option, it is assumed that the axis of rotation is the *Z* axis.

AXISYM   *Header for Axisymmetric Elements*

NM,NP

where   NM   is the number of plane members.
        NP   is the number of material types.

Element Material Lines (NP lines)

MID E=E U=U G=G

where   MID   is the material property set number.
        E1   is the modulus of elasticity.
        U1   is Poisson's ratio.
        G1   is the shear modulus [DEFAULT = E/2*(1 + U)].

Axisymmetric Element Connectivity (Repeat as many times as needed)

NN N=N1,N2,N3,N4,N5,N6,N7,N8,N9 M=MP H=H G=G1,G2 Q=Q1,Q2,Q3,Q4

where   NN   is the element number.
        N1 to N9   are the node numbers for the current element in the order shown below.
        MP   is the material property set number to be used.

Q1,Q2,Q3,Q4   is an alternative method of describing an element. It can only be used to define four node elements. If Q= is used, N= cannot be used on the same line. Q1 through Q4 are the four nodes corresponding to N1, N3, N7, and N9.

G1,G2   G1 and G2 are element generation parameters. G1 is the number of elements to generate in the r direction. G2 is the number of elements to generate in the s direction. The element being specified is counted as one of the elements being generated.

This section *must* be terminated with a blank line.

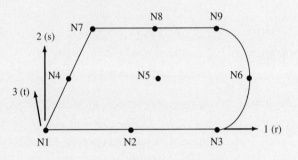

Note: This element is a variable three- to nine-node element. Nodes N2, N4, N5, N6, and N8 may be specified as 0 and thus reduce the element to four nodes. By setting only some of these to 0, the element can be four to nine nodes. If N2, N4, N5, N6, and N8 = 0 and N3 = N9, it becomes a three-node element.

Membrane–Plate–Shell Elements

These elements are all variable three- to nine-node elements and have identical input. The only difference is the header line. The PLANE element is a membrane element and has only two in-plane displacements. The PLATE element has only the two rotations and the normal displacement. The SHELL element is the combination of these two elements. A small normal rotational stiffness is provided for plates and shells to avoid numerical instability. All elements can be oriented in general three-dimensional space.

PLANE

or

PLATE   *Use the Appropriate Header for the Type of Elements Required.*

or

SHELL

NM,NP

where   NM   is the number of membrane/plate/shell elements.
          NP   is the number of material types.

Element Material Lines (*NP lines*)

MID E=E1 U=U1 G=G1 W=W1

where   MID   is the material property set number.
          E1   is the modulus of elasticity.
          U1   is Poisson's ratio.
          G1   is the shear modulus [DEFAULT = E/2*(1 + U)].
          W1   is the weight density of the material. The self weight of the element will
                      be included in load case 1 *only*.

Membrane/Plate/Shell Element Connectivity (*Repeat as many times as needed*)

NN N=N1,N2,N3,N4,N5,N6,N7,N8,N9 M=MP H=H G=G1,G2 L=L1,L2,L3,...

or

NN Q=Q1,Q2,Q3,Q4 M=MP H=H G=G1,G2 L=L1,L2,L3,...

where   NN   is the element number.
          N1 to N9   are the node numbers for the current element in the order shown
                                     below.
          MP   is the material property set number to be used.
          Li   is the uniform load to apply to the PLATE or SHELL for load
                                     case i.
          H   is the element thickness.
          Q1,Q2,Q3,Q4   is an alternative method of describing an element. It can only
                                     be used to define four-node elements. If Q= is used, N= cannot
                                     be used on the same line. Q1 through Q4 are the four nodes
                                     corresponding to N1, N3, N7, and N9.
          G1,G2   G1 and G2 are element generation parameters. G1 is the number
                                     of elements to generate in the r direction. G2 is the number of
                                     elements to generate in the s direction. The element being
                                     specified is counted as one of the elements being generated.

This section *must* be terminated with a blank line.

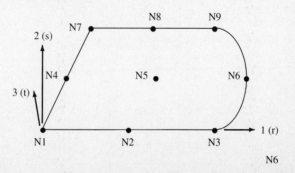

N6

This element is a variable three- to nine-node element. Nodes N2, N4, N5, N6, and N8 may be specified as 0 and thus reduce the element to four nodes. By setting only some of these to 0, the element can be four to nine nodes. If N2, N4, N5, N6, and N8 = 0 and N3 = N9, it becomes a three-node element.

Brick Element

The brick element is a variable 8- to 20-node three-dimensional solid element. It models full three-dimensional stress states. It uses a 14-point integration scheme and produces stresses at the corner nodes.

BRICK   *Header for Brick Element*

NM,NP

where   NM   is the number of brick elements.
        NP   is the number of material types.

Element Material Lines (NP lines)

MID E=E U=U

where   MID   is the material property set number.
        E   is the modulus of elasticity.
        U   is Poisson's ratio.

Brick Element Connectivity (Repeat as many times as needed)

NN N=N1,N2,N3,...,N18,N19,N20 M=MP Q=Q1,Q2,Q3,... F=F1,F2,F3,...

where   NN   is the element number.
        N1 to N20 are the node numbers for this element.
        MP   is the material property set number for this element.
        Fi   is the face on which the uniform load of Qi acts for load case i. The face is defined by the table below. It can have a value between 1 and 6. It is defined in local coordinates.
        Qi   is the uniform surface load to be used for load case i. The load acts on the face defined by the corresponding Fi.

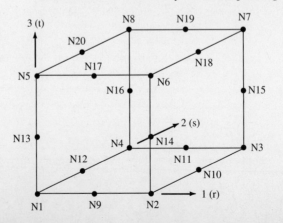

Faces for Distributed Loading

Face Number	Local Plane	Nodes in Loaded Face
1	RS	N5, N17, N6, N18, N7, N19, N8, N20
2	−RS	N1, N9, N2, N10, N3, N11, N4, N12
3	ST	N2, N10, N3, N15, N7, N18, N6, N14
4	−ST	N1, N12, N4, N16, N8, N20, N5, N13
5	TR	N3, N15, N7, N19, N8, N16, N4, N11
6	−TR	N1, N9, N2, N14, N6, N17, N5, N13

Note: This element is a variable 8- to 20-node element. Nodes N9 through N20 may be specified as 0 and thus reduce the element to eight nodes. By setting only some of these to 0, the element can be 8 to 20 nodes.

Concentrated Joint Loads

One line must be supplied for each loaded joint and each load case. The final line must be a blank line. If there are no concentrated loads, this section can be skipped.

LOADS

NF,NL,NI L=LC F=FX,FY,FZ,MX,MY,MZ

where NF   is the first node in the generation sequence on which the load will act.

NL   is the last node in the generation sequence on which the load will act (DEFAULT = NF).

NI   is the increment for the generation of node numbers between nodes NF and NL on which the loads will act (DEFAULT = 1).

LC   is the load case for which the load is to be active (DEFAULT = 1).

FX   is the load corresponding to the X translational DOF.

FY   is the load corresponding to the Y translational DOF.

FZ   is the load corresponding to the Z translational DOF.

MX   is the load corresponding to the X rotational DOF (moment).

MY   is the load corresponding to the Y rotational DOF (moment).

MZ   is the load corresponding to the Z rotational DOF (moment).

The load specification line may be repeated as many times as is necessary to define all the loads acting on the structure.

Any loads not specified will default to zero.

This section *must* end with a blank line.

Load Combinations

This section allows for the linear combination of individual load cases. Most building codes require various combinations of live, dead, and wind loads. If these loadings are specified as separate load cases within SSTAN, the results can be combined in the required factored combinations. The output from these load cases will show as additional load cases in STANPLOT. The following data are required to combine basic load cases:

COMBINE

NCB C=C1,C2,C3,C4,...

where   NCB   is the combination number.
　　　　Ci   is the load factor to apply for load case i.

There should be one line for each load combination requested. A total of NCM lines is given in line 2 of the input.

There should be as many Ci's as there are load cases (NLC from line 2 of the input).

This section should end with a blank line.

Index

Analysis programs,–215
Antisymmetry—structural,–279–284
Assembly,–185–187, 190, 217
Axisymmetric element,–353–366
    examples,–356–362, 362–366
    results,–362
    theory,–353

Bandwidth,–*see Equation solution*
Beam element,–157, 201
Beam functions,–294
Boundary conditions,–8
    hinges,–273–275
    imposed displacements,–75,
        277–279
    inclined supports,–270
    P-Δ effects,–289
    rigid-end offsets,–284–287
    springs,–270

CAL-90,–36–38, *Appendix A*
    direct stiffness,–189–195
    examples
        frame,–166–168, 170,
            190–198
        truss,–175, 199–201
    usage,–36
    users' guide,–407–421

Compatibility,–9
Consistent deformations,–51–54, 57
    examples
        beam,–54–57
        shear,–77–78
        temperature
            beam,–74
            truss,–71
        truss,–58–62
        two redundants,–63, 67
    multiple redundants,–62
    shear,–76–78
    solution procedure,–53
    support displacements,–75
    temperature,–70, 72
    truss,–57
Contragradient law,–153
Constitutive law,–8

Degrees of freedom,–*see DOF*
Direct stiffness,–*see Stiffness*
Displaced shapes,–102–106
Displacements,–5
    constrained,–16
    dependent,–6
    independent,–6
    internal,–45, 152
DOF,–95, 179–181, 202, 216
Dummy unit load—example,–48–
  51

Elastic,–9
Elements
    axisymmetric,–353–366
    beam,–201
    membrane,–325–344
    plate,–371–392
    shell,–371, 383
    solid,–397–403
    truss,–133–135
End eccentricities,–*see Boundary conditions*
Equation solution,–25–31, 217
    backward substitution,–26, 28
    banded storage,–28–31
    factorization,–26
    frontal method,–35, 217
    Gauss elimination,–26–28, 217
    pivot,–27
    profile storage,–34
    renumbering,–31–34
Equilibrium,–8
Errors,–236, 249–251, 317–319, 321–322, 333, 336, 338

Finite elements,–*Chapter 8*
    axisymmetric,–353–366
    beam,–201
    membrane,–325–344
    plate,–371–392
    shape functions,–312, 317–320
    shell,–371, 383
    solid,–397–403
    theory,–311–328
    truss,–133–135
Fixed-end moments,–89
Flexibility,–12, 16
    method,–12
    coefficients,–13
Forces,–5
    constrained,–16
    dependent,–7
    independent,–7
    internal,–44, 152

Forces (continued)
    recovery,–160–161, 188, 218, 254, 362
    results,–221, 233, 251, 254

Gauss quadrature,–320, 323, 337
Gauss elimination,–*see Equation solution*
Geometric stiffness,–*see Stiffness*

Hinges,–273

Indeterminacy
    kinematic,–18
    static,–18
Integration,–320, 323, 337

Linear algebra,–*Chapter 2*
    addition,–21
    identity,–23
    inverse,–24
    matrix,–21
    multiplication,–22
    subtraction,–22
    transpose,–24
Linear elastic,–9
Load cases,–218
Loading,–136, 156, 159, 218
    distributed,–137–140, 159
    finite element,–340, 357, 374, 375
    projected,–267

Mapping errors,–321
Matrix algebra,–21–28
Maxwell's reciprocal law,–17, 85
Membrane element,–325–344
    examples,–330–336
Mesh correctness,–338–340

Model correctness,–250–252, 339
Modeling,–4, 239, *Chapter 7*

Nodes,–5
Nonlinear,–8, 253, 289–301
Numerical integration,–319–321

P-Δ,–*see Stiffness, geometric*
Plate element,–371–392
    examples,–375–382
    results,–375
    theory—Kirchhoff,–371
    theory—Mindlin,–372

Rigid-end offsets,–284–287

Shape function
    beam functions,–294
    errors,–317–319
    one-dimensional,–312, 316, 318
    two-dimensional,–327
Shear
    area,–78
    deformation,–76
    locking,–337
Shell elements,–371, 383–392
Slope deflection,–89–102
    equation,–94
    examples
        continuous beam,–96–98
        frames,–107–110
        modified,–100–102
        slanted member frames,–110–116
    modified equation,–99
    procedure,–94
Small angle,–10
Small displacement,–10, 103
Solid element,–397–403
    examples,–399
    loads,–399
    results,–398
Solution,–*see Equation solution*

Solution errors,–*see Errors*
SSTAN,–*Appendix B*
    capabilities,–219
    examples
        axisymmetric plate,–358
        bridge,–384
        cantilever plate,–375
        continuous beam,–231
        flat slab,–378
        four-node membrane,–330
        hinged member,–275
        nine-node membrane,–333
        notched beam,–341
        P-Δ analysis,–299
        rigid-end offsets,–287
        solid element cantilever,–399
        steel collar,–362
        three-dimensional frame,–245
        two-dimensional truss,–228
        two-story braced frame,–237
        zero compression members,–252
    solution errors,–236
    users' guide,–423–436
STANPLOT,–221–227, 344, 383
Stiffness,–11, 15
    assembly,–185–187, 190
    by definition,–116–119, 166
        frame,–130–132
        procedure,–119
        slanted frame,–119–130
        truss,–133–135
    coefficients,–14
    definition of,–117, 119
    direct,–156
        frame example,–162–167, 168–173
        procedure,–159, 162, 188
        truss example,–173–176
    finite element,–313–314
    geometric,–289–295
        consistent,–292–295
        example,–295–299
        procedure,–299
    global coordinates,–184

Stiffness (continued)
 method,–11
 slope deflection,–89
 spring,–11
 three-dimensional beam,–203
 transformation,–*see Transformation matrices*
 truss,–175, 180
 two-dimensional beam,–157,
 179
Superposition,–10
Support displacements,–75, 277
Supports,–*see Boundary conditions*
Symmetry
 matrix,–17
 rules,–280
 structural,–279–284

Temperature deformations
 beams,–72
 truss,–70

Transformation matrices,–151, 153, 156,
 176, 180, 188, 203
 element force,–160
 full,–153, 156, 168
 location,–182–185
 rotational,–177–179, 180–
 182
 rotational three-dimensional,–
 203–207
 two-part,–176, 188

Virtual displacements,–43
Virtual work,–43–48
 beam,–46–48
 example—beam,–48–51
 external,–44
 internal,–45
 stiffness matrix,–156
 temperature deformation,–70
 truss,–45

DATE DUE

APR 2 6 2002		

DEMCO 38-297

201.24 Warning of copyright for software lending by nonprofit libraries.

(a)   *Definition.* A Warning of Copyright for Software Rental is a notice under paragraph (b)(2)(A) of section 109 of the Copyright Act, title 17 of the United States Code, as amended by the Computer Software Rental Amendments Act of 1990, Public Law 101-650. As required by that paragraph, the Warning of Copyright for Software Rental shall be affixed to the packaging that contains the computer program which is lent by a nonprofit library for nonprofit purposes.

(b)   *Contents.* A Warning of Copyright for Software Rental shall consist of a verbatim reproduction of the following notice, printed in such size and form and affixed in such manner as to comply with paragraph (c) of this section.

Notice: Warning of Copyright Restrictions

The copyright law of the United States (Title 17, United States Code) governs the reproduction, distribution, adaptation, public performance, and public display of copyright material.

Under certain conditions specified in law, nonprofit libraries are authorized to lend, lease, or rent copies of computer programs to patrons on a nonprofit basis and for nonprofit purposes. Any person who makes an unauthorized copy or adaptation of the computer program, or redistributes the loan copy, or publicly performs or displays the computer program, except as permitted by title 17 of the United States Code, may be liable of copyright infringement.

This institution reserves the right to refuse to fulfill a loan request if, in its judgement, fulfillment of the request would lead to violation of the copyright law.

(c)   *Form and manner of use.* A Warning of Copyright for Software Rental shall be affixed to the packaging that contains the copy of the computer program, which is the subject of a library loan to patrons, by means of a label cemented, gummed, or otherwise durably attached to the copies or to a box, reel, cartridge, cassette, or other container used as a permanent receptacle for the copy of the computer program. The notice shall be printed in such manner as to be clearly legible, comprehensible, and readily apparent to a casual user of the computer program.

For further information, contact Dorothy Schrader, General Counsel, Library of Congress, Washington, D.C. 20559/(202) 707-8380.

[PACNET]